Current Topics in Developmental Biology

Volume 62

Developmental Vascular Biology

Series Editor

Gerald P. Schatten
Director, PITTSBURGH DEVELOPMENTAL CENTER
Deputy Director, Magee-Women's Research Institute
Professor and Vice-Chair of Ob-Gyn Reproductive Sci. & Cell Biol.-Physiology
University of Pittsburgh School of Medicine
Pittsburgh, PA 15213

Editorial Board

Peter Grüss
Max-Planck-Institute of Biophysical Chemistry
Göttingen, Germany

Philip Ingham
University of Sheffield, United Kingdom

Mary Lou King
University of Miami, Florida

Story C. Landis
National Institutes of Health
National Institute of Neurological Disorders and Stroke
Bethesda, Maryland

David R. McClay
Duke University, Durham, North Carolina

Yoshitaka Nagahama
National Institute for Basic Biology, Okazaki, Japan

Susan Strome
Indiana University, Bloomington, Indiana

Virginia Walbot
Stanford University, Palo Alto, California

Founding Editors

A. A. Moscona
Alberto Monroy

Current Topics in Developmental Biology
Volume 62
Developmental Vascular Biology

Edited by

Gerald P. Schatten
Director, PITTSBURGH DEVELOPMENTAL CENTER
Deputy Director, Magee-Women's Research Institute
Professor and Vice-Chair of Ob-Gyn-Reproductive
  Sci. & Cell Biol.-Physiology
University of Pittsburgh School of Medicine
Pittsburgh, PA 15213

ELSEVIER
ACADEMIC
PRESS

AMSTERDAM • BOSTON • HEIDELBERG • LONDON
NEW YORK • OXFORD • PARIS • SAN DIEGO
SAN FRANCISCO • SINGAPORE • SYDNEY • TOKYO

Cover Photo Credit: Courtesy of Diether Lambrechts and Peter Carmeliet
Flanders Interuniversity Institute for Biotechnology
Center for Transgene Technology and Gene Transfer
KU Leuven, Leuven, B-3000, Belgium

Elsevier Academic Press
525 B Street, Suite 1900, San Diego, California 92101-4495, USA
84 Theobald's Road, London WC1X 8RR, UK

This book is printed on acid-free paper.

Copyright © 2004, Elsevier Inc. All Rights Reserved.

No part of this publication may be reproduced or transmitted in any form or by any means, electronic or mechanical, including photocopy, recording, or any information storage and retrieval system, without permission in writing from the Publisher.

The appearance of the code at the bottom of the first page of a chapter in this book indicates the Publisher's consent that copies of the chapter may be made for personal or internal use of specific clients. This consent is given on the condition, however, that the copier pay the stated per copy fee through the Copyright Clearance Center, Inc. (www.copyright.com), for copying beyond that permitted by Sections 107 or 108 of the U.S. Copyright Law. This consent does not extend to other kinds of copying, such as copying for general distribution, for advertising or promotional purposes, for creating new collective works, or for resale. Copy fees for pre-2004 chapters are as shown on the title pages. If no fee code appears on the title page, the copy fee is the same as for current chapters.
0070-2153/2004 $35.00

Permissions may be sought directly from Elsevier's Science & Technology Rights Department in Oxford, UK: phone: (+44) 1865 843830, fax: (+44) 1865 853333, E-mail: permissions@elsevier.com.uk. You may also complete your request on-line via the Elsevier homepage (http://elsevier.com), by selecting "Customer Support" and then "Obtaining Permissions."

For all information on all Academic Press publications
visit our Web site at www.academicpress.com

ISBN: 0-12-153162-7

PRINTED IN THE UNITED STATES OF AMERICA
04 05 06 07 08   9 8 7 6 5 4 3 2 1

Contents

Contributors   xi
Preface   xiii

1

Blood Vessel Signals During Development and Beyond
Ondine Cleaver

    I. Introduction   2
    II. Vascular Development   2
    III. Vascular Cell–Cell Interactions   3
    IV. Cell Biology of ECs   4
    V. Heterogeneity of ECs   8
    VI. EC Interactions During Development   9
    VII. EC Interactions in Adult Tissues   21
    VIII. Evidence Against a Requirement for Endothelial Signals   26
    IX. Conclusions and Perspectives   27
        References   29

2

HIFs, Hypoxia, and Vascular Development
Kelly L. Covello and M. Celeste Simon

    I. Introduction   37
    II. Oxygen-Regulated Gene Expression   38
    III. HIF in Developmental O_2 Homeostasis   42
    IV. HIF and Disease   44
    V. Conclusion   47
        References   47

3

Blood Vessel Patterning at the Embryonic Midline
Kelly A. Hogan and Victoria L. Bautch

    I. Introduction   56
    II. Vascular Development and Patterning   56
    III. Signaling Pathways Implicated in Vascular Patterning   61
    IV. Axial Structures Implicated in Vascular Patterning   67
    V. Conclusions and Future Directions   75
        References   77

4

Wiring the Vascular Circuitry: From Growth Factors to Guidance Cues
Lisa D. Urness and Dean Y. Li

    I. Introduction   87
    II. Basic Signaling Pathways Required for Forming Blood Vessels   88
    III. The Formation of Arterial and Venous Networks   90
    IV. Effects of Local Cues on Arterial–Venous Identity   95
    V. Genetic Evidence for Vascular Guidance   96
    VI. Vascular Endothelial Growth Factors Are Attractive Guidance Cues   98
    VII. Guidance as a Balance of Attractive and Repulsive Cues   100
    VIII. Neural Guidance Pathways   100
    IX. Neural and Vascular Guidance Mechanisms Share Common Signaling Pathways   104
    X. Neuro-Vascular Development as Interdependent Processes   108
    XI. Independent Versus Interdependent Guidance of Vessels and Nerves   112
    XII. Future Directions in Vascular Guidance   113
    XIII. Summary   113
        References   114

5

Vascular Endothelial Growth Factor and Its Receptors in Embryonic Zebrafish Blood Vessel Development
Katsutoshi Goishi and Michael Klagsbrun

    I. Introduction   128
    II. Vascular Growth Factors   128

- III. Blood Vessel Development 129
- IV. Advantages of the Zebrafish Model for Vascular Research 130
- V. Methods for Analyzing the Zebrafish Vasculature 132
- VI. Vascular Endothelial Growth Factor 136
- VII. VEGF Receptors 138
- VIII. Neuropilins 139
- IX. Semaphorins 142
- X. Future Perspectives 143
- References 144

6

Vascular Extracellular Matrix and Aortic Development
Cassandra M. Kelleher, Sean E. McLean, and Robert P. Mecham

- I. Introduction 154
- II. Vessel Wall Formation and Structure 155
- III. The Vascular Extracellular Matrix 156
- IV. Collagens 158
- V. The Elastic Fiber 162
- VI. Fibulins 166
- VII. EMILIN/Multimerin Family 168
- VIII. Fibronectin 169
- IX. The Basement Membrane 169
- X. Proteoglycans 172
- XI. Matricellular Proteins 175
- XII. Correlation of Matrix Gene Expression Profile with Cytoskeletal Markers 179
- XIII. Conclusions 180
- References 180

7

Genetics in Zebrafish, Mice, and Humans to Dissect Congenital Heart Disease: Insights in the Role of VEGF
Diether Lambrechts and Peter Carmeliet

- I. Conserved Body Plan Architecture in Vertebrates 190
- II. Vascular Malformations 194
- III. Genetic Causes of CHDs and Vascular Anomalies 195
- IV. The Use of Animal Models 198
- V. Role of VEGF in Cardiovascular Development 201

VI. Expression Patterns of VEGF in Mice Reveal Other Distinct Biological Functions   203
VII. The Role of VEGF During Heart Septation   204
VIII. VEGF: a Connector Between the Developing Vascular and Neuronal System   205
IX. A Role for VEGF in Arterial EC Specification   207
X. Normal Ontogenesis but Abnormal Remodeling of PAAs in Mice Lacking VEGF164   207
XI. Association of VEGF Gene Variations with Cardiovascular Defects in DGS   211
XII. Conclusion   215
References   215

8

Development of Coronary Vessels
Mark W. Majesky

I. Introduction   225
II. Formation of the Proepicardium   226
III. Formation of the Epicardium   228
IV. Vasculogenesis in the Subepicardium   231
V. Role of Growth Factors in Coronary Vasculogenesis   233
VI. Epicardial to Mesenchymal Transformation (EMT)   234
VII. Epicardial EMT Leads to Changes in Gene Expression   237
VIII. Bidirectional Epicardial-Myocardial Signaling   239
IX. Establishment of a Coronary Circulation: Making Connections to the Aorta   241
X. Maturation of Coronary Vessels and Formation of a Tunica Media   243
XI. Development of the Coronary Lymphatic Vessels   246
XII. Coronary Vessel Anomalies   247
XIII. Coronary Development and Disease   249
References   250

9

Identifying Early Vascular Genes Through Gene Trapping in Mouse Embryonic Stem Cells
Frank Kuhnert and Heidi Stuhlmann

I. Introduction   261
II. Gene Discovery Through Mutagenesis Screens: Chemical Mutagenesis Versus Gene Trap Mutagenesis   262

Contents

    III. Entrapment Vectors   263
    IV. Gene Trap Screens   266
    V. ES Cell *In Vitro* Differentiation as a Model System to Identify Trapped Genes Expressed in Vascular Lineages   269
    VI. Designs of Cardiovascular Entrapment Screens   272
    VII. Conclusions and Future Directions   276
         References   277

Index   283
Contents of Previous Volumes   295

Contributors

Numbers in parentheses indicate the pages on which the authors' contributions begin.

Victoria L. Bautch (55), Departments of Biology, Carolina Cardiovascular Biology Center, and Curriculum in Genetics and Molecular Biology, University of North Carolina at Chapel Hill, Chapel Hill, North Carolina 27599

Peter Carmeliet (189), Flanders Interuniversity Institute for Biotechnology, Center for Transgene Technology and Gene Therapy, KU Leuven, Leuven, B-3000, Belgium

Ondine Cleaver (1), Howard Hughes Medical Institute and Department of Molecular and Cellular Biology, Harvard University, Cambridge, Massachusetts 02138

Kelly L. Covello (37), Abramson Family Cancer Research Institute and Department of Cell and Developmental Biology, University of Pennsylvania School of Medicine, Philadelphia, Pennsylvania 19104

Katsutoshi Goishi (127), Vascular Biology Program and Department of Surgery, Children's Hospital and Harvard Medical School, Boston, Massachusetts 02115

Kelly A. Hogan (55), Department of Biology, University of North Carolina at Chapel Hill, Chapel Hill, North Carolina 27599

Cassandra M. Kelleher (153), Washington University School of Medicine, Department of Cell Biology and Physiology, St. Louis, Missouri 63110

Michael Klagsbrun (127), Vascular Biology Program, Department of Surgery, and Department of Pathology, Children's Hospital and Harvard Medical School, Boston, Massachusetts 02115

Frank Kuhnert (261), Department of Cell Biology, Division of Vascular Biology, The Scripps Research Institute, La Jolla, California 92037 and Department of Hematology, Stanford University Medical Center, Stanford, California 94305

Diether Lambrechts (189), Flanders Interuniversity Institute for Biotechnology, Center for Transgene Technology and Gene Therapy, KU Leuven, Leuven, B-3000, Belgium

Dean Y. Li (87), Program in Human Molecular Biology and Genetics, Departments of Medicine and Oncological Science, University of Utah School of Medicine, Salt Lake City, Utah 84112

Mark W. Majesky (225), Departments of Medicine and Genetics, Carolina Cardiovascular Biology Center, University of North Carolina at Chapel Hill, Chapel Hill, North Carolina 27599

Sean E. McLean (153), Washington University School of Medicine, Department of Cell Biology and Physiology, St. Louis, Missouri 63110

Robert P. Mecham (153), Washington University School of Medicine, Department of Cell Biology and Physiology, St. Louis, Missouri 63110

M. Celeste Simon (37), Abramson Family Cancer Research Institute, Department of Cell and Developmental Biology, and Howard Hughes Medical Institute, University of Pennsylvania School of Medicine, Philadelphia, Pennsylvania 19104

Heidi Stuhlmann (259), Department of Cell Biology, Division of Vascular Biology, The Scripps Research Institute, La Jolla, California 92037

Lisa D. Urness (87), Division of Cardiology, University of Utah School of Medicine, Salt Lake City, Utah 84112

Preface

This volume of Current Topics in Developmental Biology, entitled, "Developmental Vascular Biology," showcases important findings across the field of vessel formation. During embryogenesis, the cardiovascular system is the first organ system to form and function, and so this volume appropriately points out that understanding vascular development is important to a multitude of other developmental questions, all with translational importance.

Blood Vessel Signals During Development and Beyond by Ondine Cleaver of Harvard explores endothelial cell signaling and development, examining the evidence for how vascular cells might drive the fate of adjacent cells. While the clinical implications of endothelial cell signaling and cell differentiation are tremendous, ultimately – the author concludes – the answers to these questions will not be achieved without the development of superior models which can isolate endothelial signaling and cell development from confounding issues of tissue metabolism.

In HIFs, Hypoxia, and Vascular Development, Kelly Covello and Celeste Simon of the University of Pennsylvania review the role of the transcription factor hypoxia inducible factor (HIF) in various physiological responses to oxygen depletion at all stages of embryogenesis and postnatal life. HIF plays roles in both normal functioning and disease states, and so manipulating it may provide new clinical approaches for pulmonary and vascular diseases, as well as for the treatment of cancer.

Blood Vessel Patterning at the Embryonic Midline by Kelly Hogan and Victoria Bautch of the University of North Carolina examines how VEGF and other signals direct vascular formation in early development. Again, the clinical implication of elucidating these mechanisms is profound, as the creation of new blood vessels *in vitro* will be one of tissue engineering's most critical achievements.

Wiring the Vascular Circuitry: From Growth Factors to Guidance Cues by Lisa Urness and Dean Li of the University of Utah is a comprehensive look at the signaling pathways involved in formation of vascular networks, as well as an intriguing look at the signals shared by vessel and nerve network development. Employing the tools of neurobiologists, the authors conclude, will bring us closer to understanding how vascular networks are guided.

Vascular Endothelial Growth Factor and Its Receptors in Embryonic Zebrafish Blood Vessel Development by Katsutoshi Goishi and Michael Klagsbrun of Harvard rightly promotes the zebrafish as an emerging model for the study of the regulation of vessel formation, and reviews the work done to date. A robust zebrafish model of vasculature, on the near horizon, will be an important step before initiating experiments in mice or trials in the clinic.

In Vascular Extracellular Matrix and Aortic Development by Cassandra Kelleher, Sean McLean, and Robert Mecham of Wahington University, the authors examine the impact of the extracellular matrix not only upon the unique characteristics of the endothelial cells of which vessels are composed, but also – and perhaps most interestingly – upon the dynamics of cell growth, movement, and differentiation.

Genetics in Zebrafish, Mice, and Humans to Dissect Congenital Heart Disease: Insight in the Role of VEGF by Diether Lambrechts and Peter Carmeliet of the Flanders Interuniversity Institute for Biotechnology illuminates the role of vascular endothelial growth factor in developmental defects. Indeed, in a dose-dependent manner that may account for the wide variety of congenital cardiac malformations, when VEGF gene transcription and translation is considerably impaired, the incidence of cardiac defects is increased.

In Development of Coronary Vessels by Mark Majesky of the University of North Carolina, the author comprehensively reviews vertebrate epicardial origins and development, starting with the earliest point of proepicardium formation and progressing through the establishment of coronary circulation and the maturation of smooth-muscle lined vessels. An important translational implication is that seemingly small defects in vessel formation may signpost the locations of adult coronary disease.

Finally, Identifying Early Vascular Genes Through Gene Trapping in Mouse Embryonic Stem Cells by Frank Kuhnert and Heidi Stuhlmann of The Scripps Research Institute and Stanford University illustrates a powerful technique for introducing insertional mutations into murine embryonic stem cells *in vitro*, thus identifying genes crucial to the differentation of cells into vascular lineages.

This volume has benefited from the ongoing cooperation of a team of participants who are jointly responsible for the content and quality of its material. The authors deserve the full credit for their success in covering their subjects in depth yet with clarity, and for challenging the reader to think about these topics in new ways. The members of the Editorial Board are thanked for their suggestions of topics and authors. I also thank Leah Kauffman for her fabulous editorial insight and Anna Vacca for her exemplary administrative support. Finally, we are grateful to everyone at the Pittsburgh Development Center of Magee-Womens Research Institute here at the University of Pittsburgh School of Medicine for providing intellectual and infrastructural support for Current Topics in Developmental Biology.

Jerry Schatten
Pittsburgh Development Center, Pennsylvania

1
Blood Vessel Signals During Development and Beyond

Ondine Cleaver
Howard Hughes Medical Institute and
Department of Molecular and Cellular Biology
Harvard University
Cambridge, Massachusetts 02138

- I. Introduction
- II. Vascular Development
- III. Vascular Cell–Cell Interactions
- IV. Cell Biology of ECs
  - A. Fenestrae
  - B. Secretion
  - C. Basement Membrane (BM)
- V. Heterogeneity of ECs
- VI. EC Interactions During Development
  - A. Mural Cells
  - B. Adipocytes
  - C. Bone
  - D. Hematopoietic Cells
  - E. Neurons
  - F. Heart
  - G. Kidney
  - H. Prostate
  - I. Liver
  - J. Adrenal Gland
  - K. Pancreas
- VII. EC Interactions in Adult Tissues
  - A. Leukocytes
  - B. Hematopoietic Stem/Progenitor Cells
  - C. Neural Stem/Progenitor Cells
  - D. Heart
  - E. Liver
- VIII. Evidence Against a Requirement for Endothelial Signals
- IX. Conclusions and Perspectives
  - Acknowledgments
  - References

I. Introduction

The most widely recognized function of the vascular system is to supply tissues with nutrients and gas exchange, and carry away wastes via blood circulation. However, blood vessels also perform many rather surprising functions. The endothelium, a single layer of cells that lines the lumen of all blood vessels, regulates a myriad of processes that ensure adult tissue homeostasis, including the control of vascular tone, blood coagulation, fibrinolysis and wound healing, acute inflammation, antigen presentation, and leukocyte recruitment (Gerritsen, 1987). Recent experimental evidence indicates that blood vessels may also provide signals that drive the differentiation of organs and tissues in the developing embryo.

The principal question this review attempts to explore is: To what extent do blood vessels act as sources of signaling molecules that drive cell fate decisions or the differentiation states of adjacent cells? This single question raises many more. How can a thin layer of endothelial cells (ECs) communicate with surrounding cells? What do we know about the secretory capacity of ECs or about EC membrane-tethered signaling molecules? How does the basement membrane (BM) or the extracellular matrix (ECM) modulate endothelial intercellular communication? Does the vascular wall also contribute developmental signals? When, during embryogenesis, do these interactions take place? Are there regional signals along different portions of the vasculature, or are signals uniformly distributed along all vessels?

This review places the issue of blood vessel signaling within the wider context of vascular development and examines the ultrastructure of ECs that reflects their signaling potential. In addition, this review provides examples of blood vessel signals identified during embryogenesis and in adult tissues, and offers perspectives on important contradictory observations. Last, this review proposes future avenues of investigation that would advance our understanding of vascular signaling potential and examines its relevance to a wide range of clinical and experimental applications.

II. Vascular Development

The vertebrate cardiovascular system assembles and becomes functional prior to the emergence of most other organs. Initially, endothelial precursors called *angioblasts* emerge as scattered cells in the mesoderm and aggregate to form cords that presage future blood vessels. These EC aggregates reorganize to form patent tubes, which then carry blood. The primitive vasculature is said to form via *vasculogenesis*, or the *de novo* formation of vessels, and is referred to as the *primary vascular plexus* (Risau and Flamme, 1995). Subsequently, this initial network of vessels is remodeled, extended, and elaborated via *angiogenesis*, or the sprouting, bridging, or growth of new vessels originating from pre-existing vessels (Risau, 1997).

The endothelium of the primary vascular plexus is initially "bare" and lies in direct contact with cells of emerging organs and tissues prior to their differentiation, uniquely positioning it to contribute early developmental signals. The basic architecture of the initial plexus, such as in the early E8 mouse embryo, consists primarily of two large paired dorsal aortae, a heart, and the cardinal veins. The dorsal aortae are prominent vessels that run along the longitudinal axis of the embryo. Anteriorly, they connect to the endocardium of the heart via the aortic arches, and posteriorly, they join the cardinal veins that return blood to the heart via the common cardinal veins. Posterior to the heart, the aortae lie in immediate contact with the underlying endoderm, the overlying paraxial mesoderm, and the lateral plate mesoderm. This system of large vessels is then connected to many smaller tapering vessels, which break up in a hierarchical manner into capillary beds. As development proceeds, large vessels recruit thick layers of vascular smooth muscle cells (VSMCs or mural cells) and ECM, while smaller vessels become associated with more sparse support cells (pericytes).

III. Vascular Cell–Cell Interactions

The formation of the vasculature requires a large number of cell–cell interactions that must occur in a precise spatio-temporal sequence (Darland and D'Amore, 2001). Investigations into the nature of these interactions reveal the critical importance of cues within the endothelial microenvironment. ECs are acutely sensitive to signals from surrounding non–endothelial cell lineages during all stages of endothelial assembly, elaboration, differentiation, and quiescence.

In particular, there has been keen interest in understanding how local angiogenic molecules drive the migration and assembly of the ECs into blood vessels at precise locations in embryonic tissues. Of fundamental importance is vascular endothelial growth factor (VEGF), a powerful angiogenic molecule that has been shown to be chemotactic and mitogenic to vascular angioblasts, to alter endothelial behavior, morphology, and survival, and to drive both angiogenesis and vasculogenesis (Ferrara *et al.*, 2003). As of the writing of this review, over 10,000 papers investigating various aspects of VEGF were available on PubMed, a publication search engine (www.ncbi.nlm.nih.gov). Therefore, experimental research is forging an increasingly refined understanding of the molecular mechanisms by which tissues communicate with and drive differentiation of vascular endothelium.

Research into the potential reciprocal interaction—how blood vessels affect surrounding tissues—is only beginning to draw attention (Fig. 1).

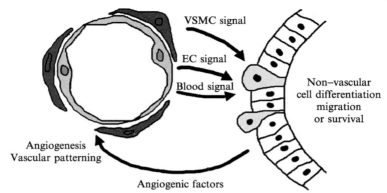

Figure 1 Reciprocal signals exchanged between blood vessels and surrounding tissues. Nonvascular cells produce angiogenic factors, such as vascular endothelial growth factor (VEGF), that regulate the proliferation, patterning, and survival of blood vessels. Blood vessels, in turn, can contribute endothelial-, mesenchymal-, or blood-derived factors to surrounding cells. These signals may control survival, proliferation, migration, or differentiation of nonvascular cell lineages. Reciprocal signals between blood vessels and tissues or organs are likely to ensure their interdependent development during embryogenesis.

Scattered observations have long noted that endothelium is not a passive tissue, and more recent studies have identified specific vascular signaling events important during organ development or tissue maintenance in adults. However, the molecular basis of many of these cell–cell interactions remains to be elucidated. At this point, many blood vessel–derived secreted (paracrine signal) or cell-tethered molecules (juxtacrine signal) have been identified, but their developmental roles have not been characterized. It will be increasingly important to understand the cellular and molecular basis of endothelial impact on its microenvironment, especially as angiogenesis is a fundamental issue in both cancer research and tissue engineering fields.

IV. Cell Biology of ECs

When considering blood vessels as a potential source of signals, we cannot avoid taking a close look at the cellular ultrastructure of ECs. In general, ECs in adult tissues are highly polarized cells with an apical (luminal) membrane, which faces the circulating blood and withstands shear stress, and a basal (abluminal) membrane, which contacts the BM, vascular wall, and tissue components, or the interstitial space (Fig. 2). It is interesting that ECs are coupled to each other and to mural cells via gap junctions, and the

1. Blood Vessel Signals

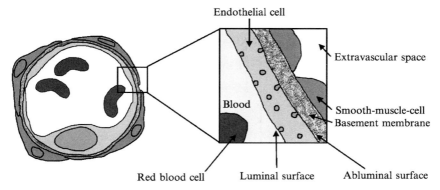

Figure 2 Mature ECs are highly polarized cells. The luminal side faces blood circulation, while the abluminal side contacts the basement membrane (BM) and other tissue components. Many soluble agents freely cross the vascular wall, including the BM and smooth muscle layers.

vascular wall sheath is a relatively permeable structure (Emerson and Segal, 2000). In addition, certain structural attributes of the EC suggest a capacity for active, organ-specific communication with cells of surrounding tissues.

A. Fenestrae

The fenestrae represent one endothelial structure that allows tissue-specific communication between tissues and blood (Fig. 3). Fenestrae are small channels or pores within the endothelium, which vary in shape, size, composition, and function from organ to organ. For instance, secretion of endocrine hormones into the blood from pancreatic islets occurs through fenestrae in islet capillaries that are small and contain a permeable diaphragm (Fig. 3A; Bearer and Orci, 1985). Filtration of blood plasma, in contrast, occurs via fenestrae in kidney glomerular capillaries. These fenestrae contain fibrous sieve plugs, which may be the key site of glomerular filtration (Fig. 3B; Rostgaard and Qvortrup, 2002). In the liver, endothelium of the sinusoids has large fenestrae that lack diaphragms (Fig. 3C). It is striking that fenestrae are absent from brain EC, as a consequence of the blood–brain barrier, formed by a continuous endothelial layer that blocks passive entry of molecules into nervous tissue (Fig. 3D; Engelhardt, 2003). Similarly, lung endothelium is continuous; however, it is characterized by active trnascytosis and the presence of abundant vesicles called caveolae (Fig. 3E). Thus, organ-specific fenestral configuration of ECs allows for organ-specific permeability and communication between blood and tissue cells (Nikolova and Lammert, 2003).

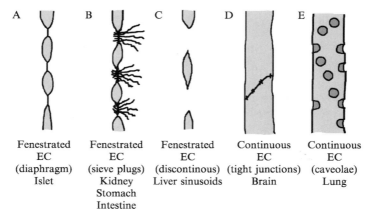

Figure 3 Ultrastructure of ECs in different organs. (A) Endocrine cells of pancreatic islets secrete hormones into the blood via permeable fenestrae in the islet capillary endothelium. (B) Kidney glomerular endothelium has fenestrae that contain sieve plugs. Podocytes extend elaborate foot processes that cover the capillary surface and act as a filtration barrier between the blood and urinary collecting tubules. (C) Endothelium of the liver sinusoids has large fenestrae, or cytoplasmic holes, of variable diameter. These fenestrae can contract and dilate, and are likely to function in lipoprotein metabolism. (D) Brain capillary endothelium is continuous and does not contain fenestrations. (E) Lung capillary endothelium is continuous and characterized by active transcytosis, via abundant caveolae.

B. Secretion

Less understood, perhaps, is the secretory apparatus of endothelial cells that might play a role in the production of signaling molecules. Endothelial cells do not contain obvious secretory granules like endocrine cells or other active secretory cells (Inagami et al., 1995). They are, however, known to produce an impressive number of factors and do exhibit many structures suggestive of secretion (Fig. 4; Vapaatalo and Mervaala, 2001).

One structure abundant in ECs, and long known to mediate endothelial secretion, is the Weibel–Palade body (WPB) (Hannah et al., 2002). WPBs are rod-shaped, lysosome-related secretory organelles that release von Willebrand factor (vWF), a glycoprotein important in haemostasis and blood clotting (Fig. 4A). In addition, WPBs also traffic other factors to the cell surface, including P-selectin, a cell-surface protein involved in leukocyte recruitment during inflammation (Hattori et al., 1989). Another structure that may be involved in EC secretion is the caveolae (Fig. 4B; Minshall et al., 2003; Razani et al., 2002a; Stan, 2002). Caveolae (also called plasmalemmal bodies) are spherical or flask-shaped vesicles, abundant in ECs, which are involved in many cellular functions, such as endocytosis, transcytosis, signal transduction, trafficking, and potentially, secretion. They have been

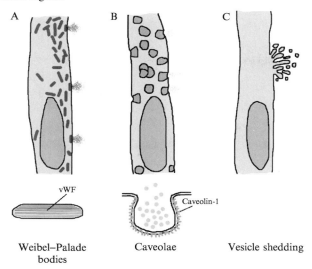

Figure 4 Secretory structures in ECs. (A) Weibel–Palade bodies are rod-shaped organelles, approximately 100–300 nm thick and up to 4 μm long. They primarily contain von Willebrand factor (vWF), organized as proteinaceous tubules, which is a critical clotting factor. (B) Caveolae are spherical or flask-shaped organelles, approximately 50–100 nm in size, which can associate in grapelike clusters. They mediate endocytosis and exocytosis at both the luminal and abluminal endothelial surfaces, transporting a large number of different components, including glycoshpingolipids, nitric oxide sytnase (eNOS), elements of the Ca+ translocation pathway, and many other factors. (C) Vesicle shedding from the cell surface. ECs secrete soluble or membrane associated factors via vesicles that bud from the EC cell surface and are released into the extracellular space.

postulated to compartmentalize signaling molecules and act as "signalosomes" (Razani *et al.*, 2002b). Membrane bound IL-8, for instance, is shuttled via transcytosis of caveolae from the abluminal to the luminal side of ECs, and its presence is critical for leukocyte-EC recruitment during inflammatory responses (Middleton *et al.*, 1997). Alternatively, it is possible that ECs secrete molecules via the shedding of vesicles, which is a selective process widespread in both normal and tumor cells (Fig. 4C; Taylor and Black, 1986). One *in vitro* study demonstrates that matrix metalloproteinases are released as membrane associated proteins within exocytosed vesicles (Taraboletti *et al.*, 2002). Strikingly, this release is rapid, localized to distinct foci along the EC, and can be stimulated by angiogenic and serum factors. Thus, ultrastructural observations support the hypothesis that ECs may either secrete paracrine factors or actively shuttle molecules to the cell surface, although further analysis of the cellular aspects of embryonic endothelium is essential to understand its potential role as a signaling tissue during development.

C. Basement Membrane (BM)

An additional structure relevant when discussing endothelial signaling is the underlying BM (Fig. 2). The BM is a dense, sheetlike structure of approximately 50–100 nm in thickness, which is usually found basolateral to many tissues, including the endothelium (Kalluri, 2003). In general, all cells are thought to produce BM components, which include type IV collagen, laminin, heparan-sulfate proteoglycans (HSPGs), nidogen/entactin, and a variety of other ECM-like materials. Functionally, BMs provide tissues with structural support; however, they are also known to regulate cell behavior. BM dynamics are striking during physiological and pathological angiogenesis, when ECs degrade underlying BM to allow the growth of new blood vessel sprouts. Some BM components have been identified as either pro- or anti-angiogenic. It is important that the BM is permeable to many molecules and is found at active sites of hormonal secretion (islet endocrine cells), filtration (kidney glomerulus), and metabolic transport (alveolar epithelium). However, some growth factors bind BM components, altering its adhesiveness or mechanical integrity (Ingber and Folkman, 1989). In addition, ECs are known to make direct contacts with adjacent cells, despite the intervening presence of BM. An example is the myoendothelial gap junctions between ECs and VSMCs, which allow electrical and chemical coupling of the cells and control vasomotor tone (Emerson and Segal, 2000). Thus, when evaluating EC signaling at the interface of endothelial and non-endothelial tissues, it will be critical to understand BM composition in embryonic and adult tissues and its potential modulation of EC-derived paracrine or juxtacrine signals.

V. Heterogeneity of ECs

Vascular heterogeneity is a final issue that must be mentioned when discussing the interface of blood vessels and surrounding tissues. Although ECs share many common attributes that define them as the lining of the vasculature, they also exhibit astonishing morphological, functional, and molecular differences across different organs and blood vessels of varying sizes (Chi et al., 2003; Cleaver and Melton, 2003; Garlanda and Dejana, 1997). There are even sex-specific molecular differences between analogous vessels, such as in the gonads (Brennan et al., 2002). In fact, it has been said that there are as many different types of ECs as there are different organs and tissues (Suter and Majno, 1965). Endothelial cell diversity is further mirrored by variability in the composition and thickness of the associated BM (Kalluri, 2003; Nikolova and Lammert, 2003). This heterogeneity is likely to originate in the early embryo since precursors to different blood vessels, such as veins and arteries, exhibit molecular differences at the very onset of vascular morphogenesis (Lawson and Weinstein, 2002). Thus, the striking variability of cellular and extracellular properties makes blood vessels uniquely suited to communicate regionally with cells of underlying tissues and organs.

Recent discoveries demonstrate that organ-specific angiogenic molecules, are likely to act in conjunction with the more widespread VEGF to coordinate establishment and maintenance of different vascular beds. Of particular interest is the novel growth factor endocrine gland-vascular endothelial growth factor, or EG-VEGF, which can act as an angiogenic mitogen and induce fenestration in specific classes of ECs (LeCouter *et al.*, 2001). This molecule is restricted to steroidogenic glands, suggesting that it may promote local specialization of the endocrine endothelium. Other tissue-specific angiogenic molecules recently identified include Bv8 in the testes and Myodullin in skeletal muscle (LeCouter *et al.*, 2003a; Pisani *et al.*, 2004).

Little is known about the generation of endothelial and angiogenic molecule diversity during embryogenesis. Therefore, it is conceivable that additional organ-specific angiogenic molecules are yet to be found. In addition, it is rather certain that regionally specific vascular cross-talk exists across organs and tissues in both embryonic and adult tissues.

VI. EC Interactions During Development

Do blood vessels signal to developing tissues? The notion that tissues exchange signals when morphogenetic movements bring them in transient contact is a long acknowledged phenomenon in developmental biology. It is therefore not completely unreasonable to predict that ECs, or associated vascular cells, have the potential to participate in inductive interactions during development. However, the study of blood vessel signals to surrounding tissues presents inherent methodological difficulties *in vivo*, given that all cells require metabolic sustenance and gas exchange from blood vessels, in addition to potential direct signals from vascular cellular components. *In vitro* studies can bypass this problem and uncouple fundamental metabolic requirements from true paracrine or juxtacrine signals. However, *in vitro* systems present their own challenges, in that cell interactions are assayed in a fundamentally non-analogous environment. This review will list a number of studies, both *in vitro* and *in vivo*, that nonetheless lend support to the prediction that blood vessels provide signals to differentiating cells during development, as well as some studies that argue against it.

A. Mural Cells

A classic example of endothelial signaling is the heterotypic cell–cell interaction that exists between ECs and cells of the vascular wall. It was long hypothesized that EC signals were likely to drive mural cell recruitment, simply because of the observed developmental sequence of events. In the embryo, the nascent vessels initially form by vasculogenesis, as angioblasts coalesce into naked endothelial tubes. These primitive tubes are subsequently stabilized by the deposition of a BM and the recruitment of mural cells, such as VSMCs. This process occurs as vessels mature and become relatively quiescent. The end result is the formation of a supportive, multi-layered sheath surrounding the endothelial tubes, which varies in composition and size between different blood vessels.

At least four molecular pathways have been identified as critical regulators of VSMC recruitment to endothelium: platelet derived growth factor-B (PDGF-B) and its receptor PDGFR-β; sphingosine-1-phosphate-1 (S1P1) and its receptor endothelial differentiation sphingolipid G-protein-coupled receptor-1 (EDG1); Ang1 and its receptor Tie2; and transforming growth factor (TGF)-β (Jain and Booth, 2003). Disruption of the PDGF-B, S1P1–EDG1, and Ang1–Tie2 pathways all lead to failures of mural cell migration to blood vessel walls, resulting in vessel dilation and hemorrhaging.

PDGF is likely to mediate a direct endothelial paracrine signaling event. PDGF is preferentially secreted by the abluminal surface of the endothelial cells, and its receptors are expressed in VSMC precursors (Holmgren *et al.*, 1991; Shinbrot *et al.*, 1994; Zerwes and Risau, 1987). *In vitro* coculture experiments demonstrate that ECs can induce robust migration of mural cell progenitors and that this migration is blocked on addition of PDGF-B neutralizing antibodies (Hirschi *et al.*, 1998). It is interesting that different EC signals are responsible for driving VSMC precursor differentiation. Antibodies to TGF-β blocked EC induction of the smooth muscle markers αSM-actin and SM-myosin in mural progenitor cells. PDGF and TGF-β are, therefore, two compelling candidates that may mediate direct paracrine signaling molecules secreted by ECs, which regulate the behavior and differentiation states of surrounding cells.

B. Adipocytes

During adipose tissue development, vascular networks form conspicuously prior to adipocyte differentiation (Crandall *et al.*, 1997). In addition, a recent study demonstrates that gross adipose tissue mass is positively regulated via its vasculature (Rupnick *et al.*, 2002). These two studies, however, do not address whether blood vessels signal directly to adipose tissue or whether the vasculature is required by adipocytes for metabolic support. *In vitro* experiments, however, demonstrate that microvascular ECs secrete extracellular matrix components that promote preadipocyte proliferation and differentiation (Hutley *et al.*, 2001; Varzaneh *et al.*, 1994). It is interesting that not all ECs have the same paracrine ability. Conditioned media from ECs of subcutaneous and omental origins stimulate higher rates of adipocyte proliferation than conditioned media from ECs of dermal origin (Hutley *et al.*, 2001).

Further experiments show that the EC-adipocyte paracrine signal is likely to occur downstream of the VEGF-A pathway, with VEGF-A binding its receptors on ECs and with adipocytes being secondarily stimulated to proliferate, presumably via EC signaling. Blocking antibodies to vascular endothelial growth factor receptor 2 (VEGFR2 or flk-1), which is expressed uniquely on ECs, lead to a coordinate reduction of angiogenesis and preadipocyte differentiation (Fukumura *et al.*, 2003). In addition, EC-conditioned media supports preadipocyte differentiation *in vitro*, while supernatant from ECs blocked with the anti-VEGFR2 antibody cannot. These *in vitro* experiments argue strongly for a non–cell autonomous effect of the blocking antibodies, indicating that ECs relay a paracrine, secreted signal to preadipocytes and that this event occurs downstream of the VEGF pathway.

C. Bone

EC signaling has long been suspected to be critical to the development of bone (Trueta and Amato, 1960). Both embryonic bone development and postnatal longitudinal bone growth occur via a sequence of events involving the precise coupling of cartilage production and regression (chondrogenesis), vascular invasion, and bone formation (osteogenesis) (Gerber and Ferrara, 2000). It is interesting that resting cartilage is distinctly avascular and produces strongly antiangiogenic factors. However, as chondrocytes begin to proliferate and differentiate, they also begin to secrete angiogenic molecules that stimulate the invasion of subchondral blood vessels into the growth plate. Blood vessel invasion then coincides with changes in the ECM, chondrocyte apoptosis, and invasion of bone precursor cells, such as osteoclasts and osteoblasts.

In vitro studies show that ECs can directly stimulate both chondrocytes and osteoblasts to proliferate and differentiate (Bittner *et al.*, 1998; Villars *et al.*, 2002). Another study directly correlates the vascular invasion of cartilage with chondrocyte apoptosis (Gerber *et al.*, 1999b). In addition, suppression of VEGF signaling, within bone, also results in significant bone development abnormalities. Systemic administration of soluble VEGF receptor proteins leads to a failure of blood vessel invasion into bone and a coordinate impairment of bone formation. Specifically, hypertrophic chondrocytes fail to regress as a result of decreased chondroclast recruitment. This suggests that hypertrophic chondrocyte VEGF-A expression is essential for angiogenesis in the cartilage and that blood vessels are required for the clearing of chondrocytes and the recruitment of osteoclasts. However, it will be critical to eliminate the formal possibility that VEGF-A may interact directly with chondrocytes, since expression of VEGF receptors in this cell type has been reported, albeit at low levels (Pufe *et al.*, 2002).

D. Hematopoietic Cells

The existence of signals between endothelial signals and hematopoietic cells might also be surmised given their continued intimate association throughout development and adulthood. *In vitro* experiments demonstrate that EC cultures can support hematopoietic cells, including erythroid and myeloid lineages (Fennie *et al.*, 1995; Ohneda *et al.*, 1998). Moreover, cell transplantation studies in zebrafish *cloche* mutants indicate that hematopoietic cells are likely to require signals from ECs (Parker and Stainier, 1999). In these experiments, wild-type cells are transplanted into both wild-type and *cloche* mutant tissues. *Cloche* mutants lack most ECs during early development (Stainier *et al.*, 1995). Subsequent to transplantation, the contribution of wild-type cells to blood lineages is assessed. Results indicate that transplanted cells are less likely to express blood markers, such as *gata-1*, when transplanted into *cloche* mutant tissues than in wild-type tissues. This suggests that *cloche* has a non-autonomous role during red-blood-cell differentiation. In other words, ECs are required for blood cell differentiation.

E. Neurons

There are many analogies between the nervous and vascular systems, from their similar highly branched morphology, to their reciprocal influence on one another and coordinated patterning (Carmeliet, 2003). The astrocyte-endothelial interface exemplifies well this neurovascular interrelationship. Astrocytes have long been known to influence ECs (Abbott, 2002). However, a number of *in vitro* studies demonstrate that reciprocal signals from ECs can have a profound influence on astrocyte precursor morphology, behavior, proliferation, or physiological function (Estrada *et al.*, 1990; Mi *et al.*, 2001; Yoder, 2002). For instance, Leukemia inhibitory factor (LIF) released by ECs of the optic nerve has been shown to induce astrocyte differentiation (Mi *et al.*, 2001).

In the embryo, ECs have also been shown to influence cell fate decisions by neural precursors. Neural crest cell differentiation into sympathetic neurons

1. Blood Vessel Signals

occurs in close proximity to the dorsal aorta endothelium (Groves et al., 1995; Stern et al., 1991). This suggests that aortic ECs might signal the neural crest during their ventral migration across the endothelial surface. *In vitro* studies support this hypothesis, showing that explanted aortic endothelium can indeed drive neural crest differentiation (Reissmann et al., 1996). In addition, the bone morphogenetic proteins, BMP-2, -4, and -7, which are normally expressed by the aorta, can induce neural crest differentiation both *in vitro* and *in vivo* (Reissmann et al., 1996; Schneider et al., 1999; Shah et al., 1996). In contrast, *in vitro* experiments show that TGF-β1 drives neural crest to take on a smooth muscle fate. These experiments therefore demonstrate that endothelial signals, such as BMPs, can act in a paracrine manner on adjacent cells influencing their developmental fate decisions.

Another vascular signal that affects neurons is the glial-derived neurotrophic factor Artemin (ARTN) (Honma et al., 2002). ARTN is expressed by VSMCs in exquisitely restricted portions of the vasculature and in a dymanic fashion along migration routes for sympathetic neuroblasts and in areas of sympathetic neural projections. ARTN guides proper innervation of a subset of neurons by gradually shifting its expression distally as axons extend toward their targets (Fig. 5). ARTN-deficient mice have defects in the sympathetic nervous system, including abnormal axonal projections and aberrant superior cervical ganglion (SCG) positioning. When ARTN-soaked beads are implanted in the embryo near the sympathetic chain, robust axonal outgrowth is induced in a directional manner. Thus, ARTN

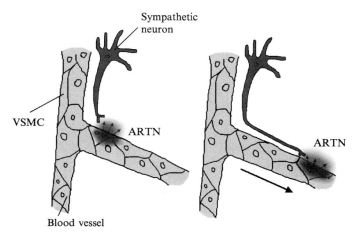

Figure 5 Artemin (ARTN) is a chemotactic factor that patterns sympathetic neurons. ARTN is secreted from blood vessel smooth-muscle cells (VSMCs) in a dynamic fashion and guides the extension of sympathetic neuron axons to the superior cervical ganglion (SCG) (adapted from Carmeliet, 2003).

is a blood-vessel-derived factor, expressed by VSMCs, required for proper neural patterning.

F. Heart

Signaling by ECs is essential during both cardiac development and adult cardiac function. Mouse or zebrafish mutants that lack the endocardium (endothelial lining of the heart) show a failure of myocardial maturation (Shalaby et al., 1995; Stainier et al., 1995). In particular, trabeculation of the myocardium depends on the endocardial expression of neuregulin, a diffusible factor that binds receptors ErbB2 (HER-2) and ErbB4 (HER-4) on adjacent myocardium (Kramer et al., 1996; Meyer and Birchmeier, 1995). Neuregulin-deficient mice exhibit embryonic lethality around E11.5, most likely as a result of failure of myocardial maturation and depressed contractility. Thus, a distinct paracrine signal from the endothelium is required for proper morphogenesis of heart muscle.

It is interesting that reciprocal signals from non-endothelial tissues are required for proper development of the endothelium in the heart. This is well established during cardiac cushion development. In particular, transforming growth factor-β (TGF-β), bone morphogenetic proteins (BMPs), and VEGF expression by the myocardium are essential for the endothelial-to-mesenchymal transition that results in endocardial cushion formation (Brown et al., 1999; Dor et al., 2001; Ramsdell and Markwald, 1997; Yamagishi et al., 1999).

G. Kidney

EC signaling is also likely to occur in the kidney glomerulus. Histological observations place glomerular podocytes in immediate contact with incoming ECs at the developing vascular cleft of the S-shape renal primordium (Saxen and Sariola, 1987). Both *in vitro* and *in vivo* experiments suggest that reciprocal signals exist between glomerular podocytes and capillary ECs. Cocultures of endothelium and either collecting duct cells or tubule epithelial cells show a direct inductive effect of endothelial-derived signals (Linas and Repine, 1999; Stoos et al., 1992). *In vivo*, zebrafish mutant embryos show that when the vasculature is absent or disrupted, differentiation of glomerular nonvascular cell types is affected. In *cloche* mutants, which lack ECs in the developing kidney, podocyte foot processes develop normally but are effaced, the BM is discontinuous, and renal mesangial cells are missing (Majumdar and Drummond, 1999). It is interesting that in zebrafish embryos that lack proper blood flow, ECs fail to invade podocyte clusters and the glomerular primordia fail to migrate and fuse normally (Serluca et al., 2002).

In addition, there appears to be an exquisite sensitivity to VEGF-A dosage and consequent EC composition in the developing kidney. Globally increasing VEGF-A levels in kidney organ cultures leads to EC proliferation and to a coordinate proliferation of tubule epithelium (Tufro et al., 1999). In converse experiments, global inhibition of VEGF-A function with either soluble VEGF receptor proteins or neutralizing antibodies, severely disrupts mesangial cell development (Gerber et al., 1999a; Kitamoto et al., 1997).

A recent study used the Cre-lox system to alter VEGF-A levels in the glomerulus in vivo (Eremina et al., 2003). Using VEGF-loxP+/+, Nephrin-Cre+/− mice, VEGF-A expression is eliminated specifically in glomerular podocytes. Predictably, endothelial recruitment is significantly impaired, remaining ECs lack fenestrations, and on later glomerular maturation, ECs are completely lost. This EC depletion results in podocyte foot process effacement and failure of normal mesangial cell differentiation. Mice with decreased VEGF levels in the podocytes exhibit severe glomerular defects. Postnally, ECs first undergo endotheliosis (or swelling), followed by loss of fenestrations and necrosis. Podocytes are coordinately affected and are completely lost by 9 weeks of age. These data suggest that glomerular ECs are required by non-vascular kidney cells. However, it is unclear whether EC signals directly regulate glomerular differentiation or whether blood vessels promote survival via metabolic support.

H. Prostate

A number of studies suggest that vascular endothelium is a primary androgen target in the prostate and plays a role in regulating prostate tissue mass. Initial experiments demonstrated that castration of rats induces involution of the ventral prostate, which is conspicuously preceded by reduction of blood flow and EC apoptosis (Shabsigh et al., 1998). However, testosterone stimulation of castrated rats results in a burst of endothelial proliferation and blood vessel growth, which is later followed by a subsequent regrowth of prostate glandular epithelium (Franck-Lissbrant et al., 1998). This angiogenic response was shown to be a result of testosterone induction of VEGF in the rat ventral prostate epithelium (Haggstrom et al., 1999). A recent study demonstrated that pigment epithelium-derived factor (PEDF) is a key inhibitor of stromal vasculature and epithelial tissue growth, and is likely to function as a counterbalance to VEGF signals in normal prostate tissues (Doll et al., 2003). PEDF-deficient mice exhibit excessive blood vessels and hyperplasia of the prostate and pancreas. It remains to be seen whether prostate epithelial inhibition is caused by a direct or an indirect effect of

PEDF via ECs. In this system, although the coordinate growth of vascular and prostate epithelial growth suggests that these tissues may be actively communicating, it is unclear whether the observed growth results from more than simply metabolic sustenance. *In vitro* experiments would help determine if the cellular component of blood vessels supports prostate epithelium via direct signaling events.

I. Liver

The liver primordium is one of the first visceral organs to emerge, and its development is initiated as the primary vascular plexus becomes functional. Endodermal cells fated to become the liver begin to form a thickened, multilayered epithelium at about E8.5d.p.c. Cells of this endodermal epithelium then delaminate and migrate into the septum transversum mesenchyme at E9.5d.p.c, intermingling with primitive sinusoidal ECs, which form capillary-like structures between the migrating hepatic cords (Fig. 6; Sherer, 1975). Later, the liver is largely composed of sheets of hepatocytes lined with fenestrated sinusoid ECs and interrupted at regular intervals by portal tracts containing bile ducts and larger blood vessels. Hepatocytes are thus in continuous, immediate contact with ECs.

Both *in vivo* and *in vitro* experiments demonstrate that ECs contribute signals to liver development (Matsumoto *et al.*, 2001). In one experiment,

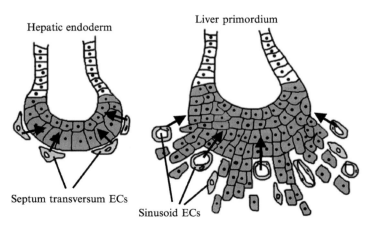

Figure 6 Signals from ECs of the septum transversum drive mouse liver development. Ventral endoderm initiates hepatic development independent of vascular cues. Subsequently, ECs signal to liver progenitors, driving them to delaminate from the endoderm and invade the mesoderm of the septum transversum. Both growth and differentiation of hepatocytes depend on EC signals (adapted from Nikolova and Lammert, 2003).

1. Blood Vessel Signals

liver morphology is assessed in *VEGFR2 (flk-1)* null embryos, which lack most endothelial precursors and consequently fail to develop a vasculature. Although liver development is initiated normally in these mutants, hepatic endodermal cells fail to delaminate and migrate into the septum transversum. To exclude possible secondary effects, such as impaired outgrowth or lack of blood flow in these mutants, E9.5 liver buds are cultured *in vitro*. Explants from wild-type mice show that hepatic differentiation proceeds normally when endothelium is present. However, explants from mice that lack ECs fail to express hepatic differentiation markers, such as albumin. In additional experiments, angiogenesis inhibitors are used to suppress endothelial growth in liver explants from wild-type embryos and to assess whether ECs were needed later during liver expansion. In these hepatic explants, few ECs develop and albumin expression is significantly reduced, suggesting that the presence of ECs is required both for hepatic outgrowth and for continued hepatic morphogenesis.

There is evidence that blood vessels may also influence the development of intrahepatic biliary ducts (IHBDs). Histological observations of human diseases collectively termed ductal plate malformations reveal that abnormal biliary ducts are often correlated with defects in adjacent portal blood vessels. In addition, the development of bile ducts always occurs in association with the portal vein and is spatially and temporally coordinated with the aggregation of portal myofibroblasts (Fig. 7; Lemaigre, 2003). Conspicuously, IHBDs only form immediately adjacent to portal veins, but not hepatic veins or arteries, hinting at differences in signaling capabilities of the different hepatic blood vessels.

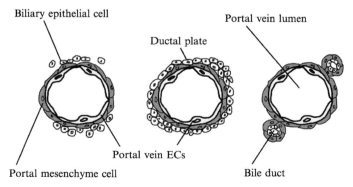

Figure 7 Formation of intrahepatic bile ducts (IHDBs) in close association with portal veins. It is likely that signals from either endothelial cells or portal mesenchyme drives the recruitment of biliary epithelial cells to the ductal plate and the subsequent formation of IHBDs (adapted from LeMaigre, 2003).

Compelling evidence of crosstalk between ducts and blood vessels is found in humans and mice defective in Notch signaling. This is evident in individuals affected with Alagille syndrome. This disease is characterized by a number of organ abnormalities, including abnormal portal tracts that often lack bile ducts, but have supernumerary arteries (Emerick and Whitington, 2002). It is interesting that this syndrome has been associated with haploinsufficiency of Jagged-1, a Notch receptor ligand (Li *et al.*, 1997; Oda *et al.*, 1997). Mice deficient in the Notch2 gene alone, or doubly mutant in the Notch-2 and Jagged-1 genes, have many abnormalities reminiscent of Alagille syndrome, including absence of bile ducts (McCright *et al.*, 2002). These data suggest a paracrine signaling event from the endothelium, or its associated mesenchyme, to the bile duct epithelium because neither gene is expressed in the bile duct epithelium itself (Jagged-1 is expressed in the ECs and non-EC cells of the arteries and portal vein, and Notch-2 is expressed in mesenchymal cells that surround the portal tracts).

A reciprocal interaction has been noted during portal tract development, showing that bile duct epithelium signals back to portal blood vessels. Vascular defects are evident in mice mutant for HNF-1β or HNF-6, both expressed strongly in the epithelium of the bile duct, but not the adjacent blood vessels. HNF-1β null mice have no arteries in the mature portal tracts (Coffinier *et al.*, 2002), and HNF-6 null mice have excess hepatic arteries in newborn livers (Clotman *et al.*, 2003). These provide yet another example of the reciprocal nature of signaling events that is observed between organ-specific cell lineages and local blood vessels.

J. Adrenal Gland

Blood vessel endothelium may also drive cell differentiation in the highly vascularized adrenal gland. Dense capillary beds can be found in both the inner medulla, which consists of catecholamine-secreting chromaffin cells, and the outer cortex, which consists of steroidogenic zona glomerulosa cells (Dobbie and Symington, 1966). Culture of an adrenal medulla-derived cell line (PC12) on bovine adrenal medullary ECs results in chromaffin-like differentiation of the PC12 cells (Mizrachi *et al.*, 1990). Culture of zona glomerulosa cells with bovine adrenal ECs results in aldosterone release from the steroidogenic cells (Rosolowsky *et al.*, 1999). It is interesting that PC12 cells require contact with the ECs for their differentiation, while EC-conditioned media is sufficient to stimulate aldosterone release from the zona glomerulosa cells. Thus, it is clear that, at least *in vitro*, ECs can communicate with a variety of adrenal cell types via different pathways, mediated by either secreted or cell-surface associated factors.

K. Pancreas

Endothelial signals have also been identified during the development of the pancreas. Histological observations place major milestones of pancreatic specification and differentiation at sites in the endoderm that are in immediate contact with major embryonic blood vessels (Lammert et al., 2001; Pictet et al., 1972; Slack, 1995). Initially, the pancreatic anlage appears as dorsal and ventral domains of pancreatic duodenal homeobox-1 (*Pdx-1*) expression. The dorsal domain of *Pdx-1* expression is initiated in the endoderm adjacent to the fusing dorsal aortate (Fig. 8), and the ventral domains first appear as two ventrolateral regions of the endoderm that lie in proximity to the two ventral vitelline veins. These pancreatic domains then evaginate as buds precisely at locations of endoderm contact with endothelium. Later, during pancreatic development, smaller vessels perfuse the pancreatic mesenchyme and endocrine differentiation of insulin- and glucagon-expressing cells occurs in close contact with ECs.

Three experiments demonstrate that endothelial signals are critical to pancreatic development in mouse (Lammert et al., 2001). In the first experiment, *in vivo* embryonic dissection of frog embryos is carried out to remove angioblasts from the intermediate mesoderm and to transiently prevent

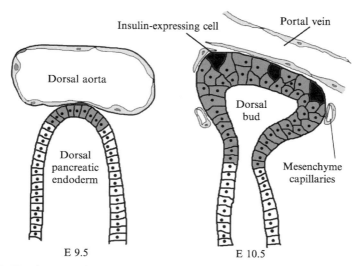

Figure 8 Development of the dorsal pancreatic bud in mice occurs in close association with vascular structures such as the dorsal aorta and the portal vein. Experimental evidence demonstrates that dorsal aorta endothelium is required for outgrowth of the pancreatic bud, expression of the gene *Ptf1a*, and endocrine gene expression, such as insulin and glucagon (adapted from Nikolova and Lammert, 2003).

dorsal aorta formation. These aortaless embryos fail to undergo pancreatic differentiation, as assayed by the expression of the endocrine progenitor marker *Pax6* and the differentiation marker insulin. However, when the aorta is given time to regenerate, presumably due to anterior-posterior EC migration, insulin expression is restored. This experiment indicates that EC signals are required for endocrine pancreatic differentiation.

In the second experiment, the converse approach is taken to increase contact of endothelium and pancreatic endoderm. VEGF-A is overexpressed in transgenic mice under the control of the *Pdx-1* promoter, resulting in hypervascularization of the developing pancreas. These transgenic mice develop both hyperplasia of the pancreatic islets within the pancreas and ectopic insulin-expressing cells in regions of the stomach and the duodenum that express the VEGF-A transgene. Excess endocrine cells can be observed in both transgenic embryos and adults. Both these experiments demonstrate that proper dosage of vascular endothelium within an organ must be tightly regulated, since an imbalance transduces significant effects on surrounding endocrine cell types.

In the third experiment, *in vitro* recombination experiments demonstrate that endothelium provides direct inductive signals on prepancreatic endoderm, independent of blood-born nutrients or circulatory gas exchange. In these experiments, dorsal prepancreatic endoderm is recombined *in vitro* with a number of isolated embryonic tissues, including the aortic endothelium, notochord, neural tube, and somitic mesoderm. Only endothelium can induce tissue growth and insulin expression in the endoderm, demonstrating that endothelial signals are sufficient for the pancreatic program, at least in a relatively prepatterned endoderm.

More recent experiments in mice further demonstrate that endothelial signals are required during pancreas development (Yoshitomi and Zaret, 2004). In this study, the pancreas is analyzed in *VEGFR2 (flk-1)* null mice, which lack all vascular endothelium. The authors find that the dorsal pancreatic bud in these embryos fails to evaginate normally and grow into the mesenchyme. In addition, the dorsal endoderm fails to express the gene *Ptf1a* early, and both insulin and glucagon later. *Pdx-1* expression is initiated normally in the dorsal bud; however, it is not maintained in either the mutant mice or explant recombinants. It is interesting that the ventral bud does not appear to require signals from the vitelline vein endothelium for its initial outgrowth. *In vitro* experiments further demonstrate that wild-type aortic endothelium can rescue the expression of *Ptf1a* in *VEGFR2*-deficient endoderm. This study underscores the critical requirement of aortal endothelial signals to the endoderm and specifically identifies EC induction of the *Ptf1a* gene, as well as pancreatic morphogenesis and differentiation.

Together, these experiments strongly suggest that endothelial signals are required during pancreas development. It is interesting to note, however, that the initiation of *Pdx-1* expression is normal in the dorsal pancreas of the *VEGFR2* null mutant mice (Yoshitomi and Zaret, 2004). This observation clearly demonstrates that initial pancreatic specification is likely to be initiated by early signals generated from other tissues, such as the notochord, but that further signals from the aortic endothelium are required for later pancreatic outgrowth and endocrine differentiation. Similarly, *Pdx-1* expression is detected by PCR in aortaless frog embryos (Cleaver, unpublished observations). This supports work carried out using chick embryonic manipulation and *in vitro* recombinations, which show that notochord signals are required during pancreas specification (Kim *et al.*, 1997). As seen in many other developmental interactions, it appears likely that endothelium provides inductive signals that are part of a step-wise series of sequential inductive events required during organogenesis.

An additional point of interest is that there appears to be particular windows of time during development in which endothelial signals are required for the pancreatic endocrine cell program. For instance, this is evident when VEGF expression is ablated specifically in islets using a conditional VEGF mutant (Lammert *et al.*, 2003). Mice lacking VEGF in differentiated endocrine cells form islets that appear relatively normal, despite the marked absence of a capillary network. However, the average islet size does appear to be reduced in these mutants. It is therefore likely that, early during embryogenesis, endocrine precursors in the developing endoderm require signals from the large vessels of the vascular plexus and are less dependent on the later islet microvasculature. Conversely, when differentiated islets are hypervascularized late during development, using a transgene expressing VEGF under the control of the insulin promoter, islets develop normally and no hyperplasia of endocrine cells is observed (Lammert *et al.*, 2001). Endothelial signals are thus required early during endocrine cell specification or differentiation, and absence or excess of blood vessels within the islets later has only modest repercussions on further endocrine cell development.

VII. EC Interactions in Adult Tissues

A. Leukocytes

Endothelial-leukocyte intercellular communication is an excellent example of a juxtacrine interaction involving ECs, which occurs in adult tissues. Generally, the role of endothelium during an infection response is an exceedingly complicated one, involving many cell surface molecules and chemokines that initiate and regulate endothelial permeability, coagulation, fibrinolysis, and leukocyte recruitment and diapedesis (or transendothelial migration). It is well beyond the scope of this review to describe the large number of factors produced by the endothelium that control all these processes (Hordijk, 2003; Keller et al., 2003; Muller, 2003), however, a few key molecules can be mentioned to illustrate juxtacrine endothelial cell–cell interactions.

Initially, during infection, leukocytes are recruited to sites of inflammation. Since leukocytes cannot actively swim, they initially bind to locally activated endothelium near a site of infection. Therefore, ECs respond to local inflammatory stimuli by expressing membrane adhesion molecules and chemokines that essentially capture circulating leukocytes. Leukocytes first adhere to the endothelium via microvillous processes and then engage in a sequence of transient tethering events that manifest as a rolling movement, driven by the hemodynamic shear forces of flowing blood. Adhesion is then stabilized, bringing the leukocyte to a stop. Leukocytes then approach the borders of ECs and exit the vascular lumen by passing between them (diapedesis or extravasation), making their way to the site of infection.

The molecular basis for leukocyte behavior during infection has been the focus of much investigation (Fig. 9). Initial emigration of leukocytes is mediated by localized EC transcytosis and luminal membrane presentation of immobilized IL-8 (Middleton et al., 1997). In addition, selectins (such as

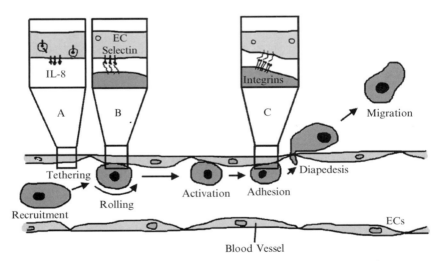

Figure 9 EC juxtacrine interactions: ECs recruit leukocytes during an inflammation response. (A) IL-8 is presented on the luminal surface of ECs and mediates recruitment of passing leukocytes. (B) Attachment of leukocytes is initiated via selectins on the luminal surface of ECs and their cognate ligands on the surface of the leukocytes. Leukocytes make transient attachment and "roll" along the EC surface, pushed along by blood flow. (C) Firm adhesion is then mediated via integrins on the leukocyte surface, binding to their cognate ligands on the surface of ECs. Leukocytes then extravasate, or squeeze, between ECs at their borders, exit the blood circulation, and enter the extravascular space to participate in immune responses.

P- and E-selectins) on ECs, and their sugar counterligands on leukocytes, mediate the initial loose tethering and rolling (Rossiter et al., 1997). The firm adhesion that follows rolling is mediated by $\beta 2$ integrins on leukocytes binding to members of the immunoglobulin superfamily on ECs, such as intercellular adhesion molecule 1 and 2 (ICAM-1 and -2), the vascular cell adhesion molecule (VCAM-1), and mucosal addressin cellular adhesion molecule 1 (MAdCAM-1) (Hynes, 1992). Additional immobilized "arrest chemokines" include secondary lymphoid tissue chemokine (SLC), keratinocyte-derived chemokine (KC), monocyte chemoattractant protein-1 (MCP-1), and, regulated on activation, normal T cell exposed and secreted (RANTES) (Ley, 2003).

It is clear that endothelium plays an active role in intercellular communication during inflammation and the immune response to infection. EC-secreted and cell surface molecules coordinate to cause major behavioral changes in leukocytes. It also demonstrates that the endothelium can communicate via both its basal and its luminal cell surfaces. It is this characteristic of versatility at the cell surface that places endothelial cells among the most active cell types in the body.

B. Hematopoietic Stem/Progenitor Cells

ECs also play an active role in the adult bone marrow stromal microenvironment and provide growth factors that support the maintenance and proliferation of hematopoietic stem and progenitor cells (Bautz et al., 2000; Davis et al., 1997a; Rafii et al., 1995). Early studies demonstrate that ECs can be stimulated, with either tumor necrosis factor alpha (TNFα) or interleukin-1, to produce the hematopoietic growth factor granulocyte/macrophage colony-stimulating factor (GM-CSF) (Broudy et al., 1986, 1987; Sieff et al., 1987). In vitro co-culture studies demonstrate that ECs, derived from both bone marrow and brain, can support long-term hematopoiesis, including myelopoiesis and megakaryocytopoiesis (Davis et al., 1997b; Rafii et al., 1995).

The functional relationship between bone marrow ECs (BMECs) and hematopoietic precursors is revealed in the bone marrow "vascular niche" (Avecilla et al., 2004). In this study, FGF-4 and stromal derived factor-1 (SDF-1) stimulate megakaryocyte adherence to sinusoidal BMECs and rescue thrombopoiesis in thrombopoietin deficient mice. It is important to note that these chemokines act via the BMECs, since they produce no major effects on megakaryocytes alone. When antibodies to VE-cadherin are used to specifically disrupt sinusoidal blood vessels, megakaryocyte adhesion to BMECs is disrupted and both thrombopoiesis and megakaryocyte maturation are inhibited. Therefore, BMEC signals support hematopoietic precursor survival and maturation within the bone marrow. It is interesting that there is a reciprocal communication from hematopoietic precursors to BMECs. Myelosuppression leads to a clear regression of BMECs (Heissig *et al.*, 2002). This is most likely due to loss of VEGF-A, which is expressed by hematopoietic precursors and which supports endothelial maintenance in the sinusoids (Bautz *et al.*, 2000). VEGF-A stimulation of BMECs, in turn, causes expression of GM-CSF, which sustains hematopoietic precursors. Thus, it is likely that once again, endothelial and nonendothelial cells exchange signals and support one another in a tissue-specific manner.

C. Neural Stem/Progenitor Cells

The influence of endothelial signals in the adult is also seen in the central nervous system. Histological observations once again place progenitor cells (in this case neural stem cells, NSCs) and endothelium in close spatiotemporal juxtaposition (Zerlin and Goldman, 1997). In the adult hippocampus, neural and endothelial progenitors proliferate coordinately in tight clusters in the subganule zone, or the subependymal zone of the lateral ventricle (SVZ), a region known to generate new neurons throughout adulthood (Palmer *et al.*, 2000). These aggregates have been termed vascular niches or neuroangiogenic foci, because they are often located at branch points or termini of fine capillaries.

Further evidence for endothelial support of neurogenesis comes from *in vitro* studies of SVZ explants (Leventhal *et al.*, 1999). Rat forebrain SVZ explants co-cultured with ECs exhibit significant neuron maturation, neurite outgrowth, and migration. The endothelial neurotrophic factor was identified as brain-derived neurotrophic factor (BDNF). The same group extended these results by examining the hormonally driven seasonal neurogenesis in the adult songbird brain (Louissaint *et al.*, 2002). They found that proliferation of neurons is closely associated with angiogenesis within the higher vocal center (HVC) of songbirds. They found that testosterone upregulates both VEGF and VEGFR2 in the HVC and induces endothelium to proliferate and to secrete increased levels of BDNF (Fig. 10). BDNF secretion, in turn, stimulates the migration of NSCs and drives the maturation and survival of neurons in the ventricular zone. This study provides another elegant example of paracrine cross-talk between endothelium and surrounding cells, important during cell differentiation in the adult.

D. Heart

The importance of EC derived signals can also be observed in the adult heart. Both endocardial endothelial (EE) cells and microvascular endothelial (MVE) cells produce autacoids and transduce signals that modulate cardiac

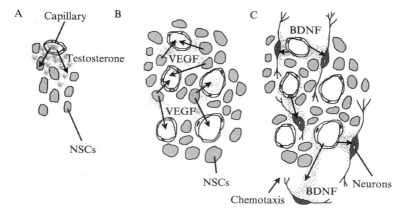

Figure 10 Reciprocal paracrine signaling between blood vessels and neural stem cells (NSCs) in the higher vocal center (HVC) of songbird brains. Seasonal increases in testosterone lead to expression of VEGF by neural precursors in the HVC. Increase in VEGF expression leads to an increase in endothelial proliferation and angiogenic growth in the HVC. An increased number of blood vessels in the HVC leads to an increase in the endothelial levels of brain-derived neurotrophic factor (BDNF). Endothelial BDNF, in turn, stimulates the migration and recruitment of neural precursors into the HVC and possibly their maturation and survival.

performance. One classical study demonstrated that selective denudation of EE using isolated cardiac myocardium resulted in measurable changes in myocardial contractility (Brutsaert et al., 1988). Since then, both EE and MVE have been shown to release diffusible factors that have direct effects on cardiomyocyte function and contractility, such as nitric oxide (NO), prostanoids, endothelin, natriuretic peptides, angiotensin II, kinins, and others (Brutsaert, 2003; Kuruvilla and Kartha, 2003; Shah, 1996). Dysfunction of endothelium in the heart results in a number of valvular heart diseases (Leask et al., 2003), which demonstrates a dependence of proper cardiac function on endothelial signals.

E. Liver

A recent study has linked EC signals with liver proliferation and protection in the adult. VEGF-A from liver cells is shown to induce hepatic ECs to produce growth and survival factors that signal back to liver cells (LeCouter et al., 2003b). Administration of circulating VEGF-A to mice results in enlarged livers, due to increased proliferation of both hepatocyes and non-parenchymal liver cells. Surprisingly, the increased hepatocyte cell division is not dependent on the proliferation of sinusoidal endothelial cells. In addition, the authors show that VEGF-A does not act directly on hepatocytes.

Instead, liver ECs are stimulated to secrete growth factors, such as hepatocyte growth factor (HGF) and interleukin-6 (IL-6), which in turn stimulate hepatocyte proliferation. Both *in vitro* and *in vivo* assays show that VEGF-A acts through VEGFR1 in this case. A VEGFR1 (Flt-1)–specific ligand stimulates hepatocyte proliferation in the absence of EC proliferation, whereas a VEGFR2 (Flk-1)–specific ligand stimulates EC proliferation in the absence of hepatocyte proliferation. Thus, VEGF-A is able to activate two distinct pathways: one that acts through VEGFR1 and induces growth factor release, and one that acts through VEGFR2 and induces EC cell division.

In addition, LeCouter *et al.*, investigated the role of VEGF-A in adult liver regeneration. They used the liver toxin carbon tetrachloride (CCl_4) to generate an experimental model of liver regeneration. Injection of mice with VEGF-A neutralizing antibodies significantly increased hepatocyte necrosis, demonstrating that VEGF-A protects against liver damage in an angiogenesis-independent, but EC-dependant, manner. In the converse experiment, adenoviral overexpression of ligands specific for either VEGFR1 or VEGFR2 lessened the severity of hepatocyte necrosis.

VIII. Evidence Against a Requirement for Endothelial Signals

Important observations that argue against the importance of vascular signaling during visceral organ development have been made in zebrafish (Field *et al.*, 2003a,b). *Cloche* mutant embryos lack detectable ECs in the vicinity of the liver and pancreas primordia, and differentiation of these organs appears to proceed normally. This is in contrast to liver and pancreas development in mice, which initiate normally but require signals from ECs for subsequent differentiation (Lammert *et al.*, 2001; Matsumoto *et al.*, 2001; Yoshitomi and Zaret, 2004).

This striking discrepancy awaits satisfactory explanation. It is possible that species-specific differences exist and zebrafish have not evolved a requirement for additional signals from endothelium during organ development. Unlike mouse embryos, zebrafish can survive up to a week without a functional vasculature. Alternatively, it is possible that the zebrafish pancreas, like the mouse ventral pancreas, does not depend on signals from the endothelium. The dorsal and ventral pancreatic buds in mice have been shown to require very different molecular pathways despite the relative similarity of their final tissue composition. Yet another possibility is that some form of endothelial precursor exists in *cloche* embryos in the vicinity of developing organs, which cannot be identified using current markers. *Cloche* embryos do contain a population of proliferative ECs in the posterior trunk that express a number of vascular markers and form functional vessels

(Stainier *et al.*, 1995; Thompson *et al.*, 1998). However, current observations seem to suggest that these vessels develop more posteriorly than the liver and pancreas primordial. Finally, it is possible that the source of signals driving fish liver and pancreas development are simply either different or redundant in fish. Although less likely, it is also formally possible that the *cloche* genetic lesion somehow rescues endodermal derivatives from their requirement for EC signals. Identification of the *cloche* gene product will help the characterization of these mutant fish at the molecular level and may help shed light on the differential requirement for EC signals between different organisms. Understanding these species-specific differences may yield interesting insights into vascular development and into the evolutionary relationship between ECs and the non-vascular parenchyma.

IX. Conclusions and Perspectives

At this point, our understanding of endothelial signals during development and adult tissue maintenance is largely restricted to observations that correlate a requirement for EC juxtaposition to responses in non-vascular cells, such as morphological events, cell migration or differentiation, or expression of tissue-specific genes. Generally, signaling between EC and surrounding tissues consists of a complex set of sequential reciprocal interactions. In one direction, tissues express angiogenic molecules, such as VEGF, driving the proliferation, migration, and assembly of ECs during both vasculogenesis and angiogenesis. In turn, vascular cellular components, including ECs and VSMCs, send reciprocal signals back to surrounding tissues, driving their specification, differentiation, or patterning. It is thus likely that the exchange of signals between vascular and non-vascular tissues assures that they grow and elaborate together, establishing a foundation for their continued physical and physiological relationship.

When considering how best to approach the issue of vascular signaling, it quickly becomes clear that there are inherent difficulties in studying endothelial signaling *in vivo*, such as in organisms with absent or disrupted vasculature. For instance, mouse embryos that lack vascular endothelium die very early due to a lack of nutrition–gas exchange in their tissues (Shalaby *et al.*, 1995). From cellular responses resulting from the absence of specific vascular signals, it is therefore difficult to distinguish cell death from cellular responses resulting from the absence of specific vascular signals. Converse experiments, which include tissues with experimentally expanded vasculature, also present unique challenges. For instance, hypervascularized tissues resulting from VEGF ligand overexpression have the complication that non-vascular tissues may also respond to VEGF. Expression of VEGF receptors (such as VEGFR2, VEGFR1, and neuropilin) has now been reported in other cell types, such as some neurons and osteoblasts (Haigh *et al.*, 2003; Harper *et al.*, 2001; Robinson *et al.*, 2001; Zelzer *et al.*, 2002). Perhaps the most clearly interpretable experiments that show direct signaling between tissues are those done *in vitro* with isolated endothelial and non-vascular tissues. *In vitro* experiments, however, have their own range of caveats, including questions of tissue purity and culture conditions. Yet a final approach consists of targeted ablation of candidate genes specifically within endothelial or smooth muscle compartments of blood vessels. Alterations in surrounding tissues are thus clearly secondary and result from impaired vascular signaling. However, it is reasonable to predict that ablation of genes within the vasculature may simultaneously compromise blood vessel integrity, presenting the same interpretation difficulties as "vasculo-compromised" animal models. The true challenge for future studies is, therefore, to develop better experimental methods to study the issue of endothelial signaling and to uncouple tissue metabolic requirements for vascular blood circulation from actual developmental signals from blood vessels themselves.

In addition, it will be critical to define the molecular nature of organ- and tissue-specific endothelial signals in both embryonic and adult tissues. Presently, we are only starting to characterize the transcriptional profile of embryonic ECs and to define the progression of cellular events during vascular wall formation. From a purely developmental biology standpoint, it will be fascinating to map the origins of endothelial diversity and to understand how developing blood vessels may influence surrounding ectodermal, mesodermal, and endodermal lineages as they make cell fate decisions. Understanding the functional interface between the vasculature and surrounding cells, in both the embryo and the adult, will undoubtedly have a significant impact on many health-related fields of study and clinical applications. For instance, it may be possible to exploit EC heterogeneity for the development and delivery of tissue-targeted therapies. It may also advance experimental methods for building new vascular networks in both engineered and ischemic tissues. Understanding blood vessel communication with surrounding cells will also certainly impact cancer research. We know that tumors cannot grow without a blood supply and that they promote their own vascularization via angiogenesis from nearby quiescent vessels. Thus, tumor vessels are currently a prime target for clinical suppression of tumor growth. Understanding their interaction with underlying tumor cells will likely contribute to the development of targeted therapies. In addition, further characterization of signaling events within "vascular niches" of stem cell differentiation, both in neural and hematopoietic tissues, may prove globally relevant to adult stem cell populations in other tissues, or to *in vitro* directed differentiation of stem cells in the laboratory.

Acknowledgments

I am very grateful to Yuval Dor, Eckhard Lammert, Tom Carroll, and Ben Stanger for insightful discussions and critical reading of the manuscript. I am also indebted to Doug Melton for long-standing support and encouragement. I apologize to the numerous primary authors not cited due to space constraints. The author was supported by the Howard Hughes Medical Institute.

References

Abbott, N. J. (2002). Astrocyte-endothelial interactions and blood-brain barrier permeability. *J. Anat.* **200,** 629–638.
Avecilla, S. T., Hattori, K., Heissig, B., Tejada, R., Liao, F., Shido, K., Jin, D. K., Dias, S., Zhang, F., Hartman, T. E., Hackett, N. R., Crystal, R. G., Witte, L., Hicklin, D. J., Bohlen, P., Eaton, D., Lyden, D., De Sauvage, F., and Rafii, S. (2004). Chemokine-mediated interaction of hematopoietic progenitors with the bone marrow vascular niche is required for thrombopoiesis. *Nat. Med.* **10,** 64–71.
Bautz, F., Rafii, S., Kanz, L., and Mohle, R. (2000). Expression and secretion of vascular endothelial growth factor-A by cytokine-stimulated hematopoietic progenitor cells. Possible role in the hematopoietic microenvironment. *Exp. Hematol.* **28,** 700–706.
Bearer, E. L., and Orci, L. (1985). Endothelial fenestral diaphragms: A quick-freeze, deep-etch study. *J. Cell Biol.* **100,** 418–428.
Bittner, K., Vischer, P., Bartholmes, P., and Bruckner, P. (1998). Role of the subchondral vascular system in endochondral ossification: Endothelial cells specifically derepress late differentiation in resting chondrocytes *in vitro*. *Exp. Cell Res.* **238,** 491–497.
Brennan, J., Karl, J., and Capel, B. (2002). Divergent vascular mechanisms downstream of Sry establish the arterial system in the XY gonad. *Dev. Biol.* **244,** 418–428.
Broudy, V. C., Kaushansky, K., Harlan, J. M., and Adamson, J. W. (1987). Interleukin 1 stimulates human endothelial cells to produce granulocyte-macrophage colony-stimulating factor and granulocyte colony-stimulating factor. *J. Immunol.* **139,** 464–468.
Broudy, V. C., Kaushansky, K., Segal, G. M., Harlan, J. M., and Adamson, J. W. (1986). Tumor necrosis factor type alpha stimulates human endothelial cells to produce granulocyte/macrophage colony-stimulating factor. *Proc. Natl. Acad. Sci. USA* **83,** 7467–7471.
Brown, C. B., Boyer, A. S., Runyan, R. B., and Barnett, J. V. (1999). Requirement of type III TGF-beta receptor for endocardial cell transformation in the heart. *Science* **283,** 2080–2082.
Brutsaert, D. L. (2003). Cardiac endothelial-myocardial signaling: Its role in cardiac growth, contractile performance, and rhythmicity. *Physiol. Rev.* **83,** 59–115.
Brutsaert, D. L., Meulemans, A. L., Sipido, K. R., and Sys, S. U. (1988). Endocardial control of myocardial performance. *Adv. Exp. Med. Biol.* **226,** 609–615.
Carmeliet, P. (2003). Blood vessels and nerves: Common signals, pathways and diseases. *Nat. Rev. Genet.* **4,** 710–720.
Chi, J. T., Chang, H. Y., Haraldsen, G., Jahnsen, F. L., Troyanskaya, O. G., Chang, D. S., Wang, Z., Rockson, S. G., van de Rijn, M., Botstein, D., and Brown, P. O. (2003). Endothelial cell diversity revealed by global expression profiling. *Proc. Natl. Acad. Sci. USA* **100,** 10623–10628.
Cleaver, O., and Melton, D. A. (2003). Endothelial signaling during development. *Nat. Med.* **9,** 661–668.

Clotman, F., Libbrecht, L., Gresh, L., Yaniv, M., Roskams, T., Rousseau, G. G., and Lemaigre, F. P. (2003). Hepatic artery malformations associated with a primary defect in intrahepatic bile duct development. *J. Hepatol.* **39**, 686–692.

Coffinier, C., Gresh, L., Fiette, L., Tronche, F., Schutz, G., Babinet, C., Pontoglio, M., Yaniv, M., and Barra, J. (2002). Bile system morphogenesis defects and liver dysfunction upon targeted deletion of HNF1beta. *Development* **129**, 1829–1838.

Crandall, D. L., Hausman, G. J., and Kral, J. G. (1997). A review of the microcirculation of adipose tissue: Anatomic, metabolic, and angiogenic perspectives. *Microcirculation* **4**, 211–232.

Darland, D. C., and D'Amore, P. A. (2001). Cell–cell interactions in vascular development. *Curr. Top. Dev. Biol.* **52**, 107–149.

Davis, T. A., Black, A. T., Kidwell, W. R., and Lee, K. P. (1997a). Conditioned medium from primary porcine endothelial cells alone promotes the growth of primitive human haematopoietic progenitor cells with a high replating potential: Evidence for a novel early haematopoietic activity. *Cytokine* **9**, 263–275.

Davis, T. A., Black, A. T., and Lee, K. P. (1997b). Soluble factor(s) alone produced by primary porcine microvascular endothelial cells support the proliferation and differentiation of human CD34+ hematopoietic progenitor cells with a high replating potential. *Transplant Proc.* **29**, 2003–2004.

Dobbie, J. W., and Symington, T. (1966). The human adrenal gland with special reference to the vasculature. *J. Endocrinol.* **34**, 479–489.

Doll, J. A., Stellmach, V. M., Bouck, N. P., Bergh, A. R., Lee, C., Abramson, L. P., Cornwell, M. L., Pins, M. R., Borensztajn, J., and Crawford, S. E. (2003). Pigment epithelium-derived factor regulates the vasculature and mass of the prostate and pancreas. *Nat. Med.* **9**, 774–780.

Dor, Y., Camenisch, T. D., Itin, A., Fishman, G. I., McDonald, J. A., Carmeliet, P., and Keshet, E. (2001). A novel role for VEGF in endocardial cushion formation and its potential contribution to congenital heart defects. *Development* **128**, 1531–1538.

Emerick, K. M., and Whitington, P. F. (2002). Partial external biliary diversion for intractable pruritus and xanthomas in Alagille syndrome. *Hepatology* **35**, 1501–1506.

Emerson, G. G., and Segal, S. S. (2000). Endothelial cell pathway for conduction of hyperpolarization and vasodilation along hamster feed artery. *Circ. Res.* **86**, 94–100.

Engelhardt, B. (2003). Development of the blood–brain barrier. *Cell Tissue Res.* **314**, 119–129.

Eremina, V., Sood, M., Haigh, J., Nagy, A., Lajoie, G., Ferrara, N., Gerber, H. P., Kikkawa, Y., Miner, J. H., and Quaggin, S. E. (2003). Glomerular-specific alterations of VEGF-A expression lead to distinct congenital and acquired renal diseases. *J. Clin. Invest.* **111**, 707–716.

Estrada, C., Bready, J. V., Berliner, J. A., Pardridge, W. M., and Cancilla, P. A. (1990). Astrocyte growth stimulation by a soluble factor produced by cerebral endothelial cells in vitro. *J. Neuropathol. Exp. Neurol.* **49**, 539–549.

Fennie, C., Cheng, J., Dowbenko, D., Young, P., and Lasky, L. A. (1995). CD34+ endothelial cell lines derived from murine yolk sac induce the proliferation and differentiation of yolk sac CD34+ hematopoietic progenitors. *Blood* **86**, 4454–4467.

Ferrara, N., Gerber, H. P., and LeCouter, J. (2003). The biology of VEGF and its receptors. *Nat. Med.* **9**, 669–676.

Field, H. A., Dong, P. D., Beis, D., and Stainier, D. Y. (2003a). Formation of the digestive system in zebrafish. II. Pancreas morphogenesis. *Dev. Biol.* **261**, 197–208.

Field, H. A., Ober, E. A., Roeser, T., and Stainier, D. Y. (2003b). Formation of the digestive system in zebrafish. I. Liver morphogenesis. *Dev. Biol.* **253**, 279–290.

Franck-Lissbrant, I., Haggstrom, S., Damber, J. E., and Bergh, A. (1998). Testosterone stimulates angiogenesis and vascular regrowth in the ventral prostate in castrated adult rats. *Endocrinology* **139**, 451–456.

Fukumura, D., Ushiyama, A., Duda, D. G., Xu, L., Tam, J., Krishna, V., Chatterjee, K., Garkavtsev, I., and Jain, R. K. (2003). Paracrine regulation of angiogenesis and adipocyte differentiation during *in vivo* adipogenesis. *Circ. Res.* **93**, 88–97.

Garlanda, C., and Dejana, E. (1997). Heterogeneity of endothelial cells. Specific markers. *Arterioscler Thromb. Vasc. Biol.* **17**, 1193–1202.

Gerber, H. P., and Ferrara, N. (2000). Angiogenesis and bone growth. *Trends Cardiovasc. Med.* **10**, 223–228.

Gerber, H. P., Hillan, K. J., Ryan, A. M., Kowalski, J., Keller, G. A., Rangell, L., Wright, B. D., Radtke, F., Aguet, M., and Ferrara, N. (1999a). VEGF is required for growth and survival in neonatal mice. *Development* **126**, 1149–1159.

Gerber, H. P., Vu, T. H., Ryan, A. M., Kowalski, J., Werb, Z., and Ferrara, N. (1999b). VEGF couples hypertrophic cartilage remodeling, ossification and angiogenesis during endochondral bone formation. *Nat. Med.* **5**, 623–628.

Gerritsen, M. E. (1987). Functional heterogeneity of vascular endothelial cells. *Biochem. Pharmacol.* **36**, 2701–2711.

Groves, A. K., George, K. M., Tissier-Seta, J. P., Engel, J. D., Brunet, J. F., and Anderson, D. J. (1995). Differential regulation of transcription factor gene expression and phenotypic markers in developing sympathetic neurons. *Development* **121**, 887–901.

Haggstrom, S., Lissbrant, I. F., Bergh, A., and Damber, J. E. (1999). Testosterone induces vascular endothelial growth factor synthesis in the ventral prostate in castrated rats. *J. Urol.* **161**, 1620–1625.

Haigh, J. J., Morelli, P. I., Gerhardt, H., Haigh, K., Tsien, J., Damert, A., Miquerol, L., Muhlner, U., Klein, R., Ferrara, N., Wagner, E. F., Betsholtz, C., and Nagy, A. (2003). Cortical and retinal defects caused by dosage-dependent reductions in VEGF-A paracrine signaling. *Dev. Biol.* **262**, 225–241.

Hannah, M. J., Williams, R., Kaur, J., Hewlett, L. J., and Cutler, D. F. (2002). Biogenesis of Weibel–Palade bodies. *Semin. Cell Dev. Biol.* **13**, 313–324.

Harper, J., Gerstenfeld, L. C., and Klagsbrun, M. (2001). Neuropilin-1 expression in osteogenic cells: Down-regulation during differentiation of osteoblasts into osteocytes. *J. Cell Biochem.* **81**, 82–92.

Hattori, R., Hamilton, K. K., Fugate, R. D., McEver, R. P., and Sims, P. J. (1989). Stimulated secretion of endothelial von Willebrand factor is accompanied by rapid redistribution to the cell surface of the intracellular granule membrane protein GMP-140. *J. Biol. Chem.* **264**, 7768–7771.

Heissig, B., Hattori, K., Dias, S., Friedrich, M., Ferris, B., Hackett, N. R., Crystal, R. G., Besmer, P., Lyden, D., Moore, M. A., Werb, Z., and Rafii, S. (2002). Recruitment of stem and progenitor cells from the bone marrow niche requires MMP-9 mediated release of kit-ligand. *Cell* **109**, 625–637.

Hirschi, K. K., Rohovsky, S. A., and D'Amore, P. A. (1998). PDGF, TGF-beta, and heterotypic cell-cell interactions mediate endothelial cell-induced recruitment of 10T1/2 cells and their differentiation to a smooth muscle fate. *J. Cell Biol.* **141**, 805–814.

Holmgren, L., Glaser, A., Pfeifer-Ohlsson, S., and Ohlsson, R. (1991). Angiogenesis during human extraembryonic development involves the spatiotemporal control of PDGF ligand and receptor gene expression. *Development* **113**, 749–754.

Honma, Y., Araki, T., Gianino, S., Bruce, A., Heuckeroth, R., Johnson, E., and Milbrandt, J. (2002). Artemin is a vascular-derived neurotropic factor for developing sympathetic neurons. *Neuron* **35**, 267–282.

Hordijk, P. (2003). Endothelial signaling in leukocyte transmigration. *Cell Biochem. Biophys.* **38**, 305–322.
Hutley, L. J., Herington, A. C., Shurety, W., Cheung, C., Vesey, D. A., Cameron, D. P., and Prins, J. B. (2001). Human adipose tissue endothelial cells promote preadipocyte proliferation. *Am. J. Physiol. Endocrinol. Metab.* **281**, E1037–E1044.
Hynes, R. O. (1992). Integrins: Versatility, modulation, and signaling in cell adhesion. *Cell* **69**, 11–25.
Inagami, T., Naruse, M., and Hoover, R. (1995). Endothelium as an endocrine organ. *Annu. Rev. Physiol.* **57**, 171–189.
Ingber, D. E., and Folkman, J. (1989). Mechanochemical switching between growth and differentiation during fibroblast growth factor-stimulated angiogenesis *in vitro*: Role of extracellular matrix. *J. Cell Biol.* **109**, 317–330.
Jain, R. K., and Booth, M. F. (2003). What brings pericytes to tumor vessels? *J. Clin. Invest.* **112**, 1134–1136.
Kalluri, R. (2003). Basement membranes: Structure, assembly and role in tumour angiogenesis. *Nat. Rev. Cancer* **3**, 422–433.
Keller, T. T., Mairuhu, A. T., de Kruif, M. D., Klein, S. K., Gerdes, V. E., ten Cate, H., Brandjes, D. P., Levi, M., and van Gorp, E. C. (2003). Infections and endothelial cells. *Cardiovasc Res.* **60**, 40–48.
Kim, S. K., Hebrok, M., and Melton, D. A. (1997). Notochord to endoderm signaling is required for pancreas development. *Development* **124**, 4243–4252.
Kitamoto, Y., Tokunaga, H., and Tomita, K. (1997). Vascular endothelial growth factor is an essential molecule for mouse kidney development: Glomerulogenesis and nephrogenesis. *J. Clin. Invest.* **99**, 2351–2357.
Kramer, R., Bucay, N., Kane, D. J., Martin, L. E., Tarpley, J. E., and Theill, L. E. (1996). Neuregulins with an Ig-like domain are essential for mouse myocardial and neuronal development. *Proc. Natl. Acad. Sci. USA* **93**, 4833–4838.
Kuruvilla, L., and Kartha, C. C. (2003). Molecular mechanisms in endothelial regulation of cardiac function. *Mol. Cell Biochem.* **253**, 113–123.
Lammert, E., Cleaver, O., and Melton, D. (2001). Induction of pancreatic differentiation by signals from blood vessels. *Science* **294**, 564–567.
Lammert, E., Gu, G., McLaughlin, M., Brown, D., Brekken, R., Murtaugh, L. C., Gerber, H. P., Ferrara, N., and Melton, D. A. (2003). Role of VEGF-A in vascularization of pancreatic islets. *Curr. Biol.* **13**, 1070–1074.
Lawson, N. D., and Weinstein, B. M. (2002). Arteries and veins: Making a difference with zebrafish. *Nat. Rev. Genet.* **3**, 674–682.
Leask, R. L., Jain, N., and Butany, J. (2003). Endothelium and valvular diseases of the heart. *Microsc. Res. Tech.* **60**, 129–137.
LeCouter, J., Kowalski, J., Foster, J., Hass, P., Zhang, Z., Dillard-Telm, L., Frantz, G., Rangell, L., DeGuzman, L., Keller, G. A., Peale, F., Gurney, A., Hillan, K. J., and Ferrara, N. (2001). Identification of an angiogenic mitogen selective for endocrine gland endothelium. *Nature* **412**, 877–884.
LeCouter, J., Lin, R., Tejada, M., Frantz, G., Peale, F., Hillan, K. J., and Ferrara, N. (2003a). The endocrine-gland-derived VEGF homologue Bv8 promotes angiogenesis in the testis: Localization of Bv8 receptors to endothelial cells. *Proc. Natl. Acad. Sci. USA* **100**, 2685–2690.
LeCouter, J., Moritz, D. R., Li, B., Phillips, G. L., Liang, X. H., Gerber, H. P., Hillan, K. J., and Ferrara, N. (2003b). Angiogenesis-independent endothelial protection of liver: Role of VEGFR-1. *Science* **299**, 890–893.
Lemaigre, F. P. (2003). Development of the biliary tract. *Mech. Dev.* **120**, 81–87.

Leventhal, C., Rafii, S., Rafii, D., Shahar, A., and Goldman, S. A. (1999). Endothelial trophic support of neuronal production and recruitment from the adult mammalian subependyma. *Mol. Cell Neurosci.* **13**, 450–464.
Ley, K. (2003). Arrest chemokines. *Microcirculation* **10**, 289–295.
Li, L., Krantz, I. D., Deng, Y., Genin, A., Banta, A. B., Collins, C. C., Qi, M., Trask, B. J., Kuo, W. L., Cochran, J., Costa, T., Pierpont, M. E., Rand, E. B., Piccoli, D. A., Hood, L., and Spinner, N. B. (1997). Alagille syndrome is caused by mutations in human jagged1, which encodes a ligand for Notch1. *Nat. Genet.* **16**, 243–251.
Linas, S. L., and Repine, J. E. (1999). Endothelial cells regulate proximal tubule epithelial cell sodium transport. *Kidney Int.* **55**, 1251–1258.
Louissaint, A., Jr., Rao, S., Leventhal, C., and Goldman, S. A. (2002). Coordinated interaction of neurogenesis and angiogenesis in the adult songbird brain. *Neuron* **34**, 945–960.
Majumdar, A., and Drummond, I. A. (1999). Podocyte differentiation in the absence of endothelial cells as revealed in the zebrafish avascular mutant, cloche. *Dev. Genet.* **24**, 220–229.
Matsumoto, K., Yoshitomi, H., Rossant, J., and Zaret, K. S. (2001). Liver organogenesis promoted by endothelial cells prior to vascular function. *Science* **294**, 559–563.
McCright, B., Lozier, J., and Gridley, T. (2002). A mouse model of Alagille syndrome: Notch2 as a genetic modifier of Jag1 haploinsufficiency. *Development* **129**, 1075–1082.
Meyer, D., and Birchmeier, C. (1995). Multiple essential functions of neuregulin in development. *Nature* **378**, 386–390.
Mi, H., Haeberle, H., and Barres, B. A. (2001). Induction of astrocyte differentiation by endothelial cells. *J. Neurosci.* **21**, 1538–1547.
Middleton, J., Neil, S., Wintle, J., Clark-Lewis, I., Moore, H., Lam, C., Auer, M., Hub, E., and Rot, A. (1997). Transcytosis and surface presentation of IL-8 by venular endothelial cells. *Cell* **91**, 385–395.
Minshall, R. D., Sessa, W. C., Stan, R. V., Anderson, R. G., and Malik, A. B. (2003). Caveolin regulation of endothelial function. *Am. J. Physiol. Lung Cell Mol. Physiol.* **285**, L1179–L1183.
Mizrachi, Y., Naranjo, J. R., Levi, B. Z., Pollard, H. B., and Lelkes, P. I. (1990). PC12 cells differentiate into chromaffin cell-like phenotype in coculture with adrenal medullary endothelial cells. *Proc. Natl. Acad. Sci. USA* **87**, 6161–6165.
Muller, W. A. (2003). Leukocyte–endothelial-cell interactions in leukocyte transmigration and the inflammatory response. *Trends Immunol.* **24**, 327–334.
Nikolova, G., and Lammert, E. (2003). Interdependent development of blood vessels and organs. *Cell Tissue Res.* **314**, 33–42.
Oda, T., Elkahloun, A. G., Pike, B. L., Okajima, K., Krantz, I. D., Genin, A., Piccoli, D. A., Meltzer, P. S., Spinner, N. B., Collins, F. S., and Chandrasekharappa, S. C. (1997). Mutations in the human Jagged1 gene are responsible for Alagille syndrome. *Nat. Genet.* **16**, 235–242.
Ohneda, O., Fennie, C., Zheng, Z., Donahue, C., La, H., Villacorta, R., Cairns, B., and Lasky, L. A. (1998). Hematopoietic stem cell maintenance and differentiation are supported by embryonic aorta-gonad-mesonephros region-derived endothelium. *Blood* **92**, 908–919.
Palmer, T. D., Willhoite, A. R., and Gage, F. H. (2000). Vascular niche for adult hippocampal neurogenesis. *J. Comp. Neurol.* **425**, 479–494.
Parker, L., and Stainier, D. Y. (1999). Cell-autonomous and non-autonomous requirements for the zebrafish gene cloche in hematopoiesis. *Development* **126**, 2643–2651.
Pictet, R. L., Clark, W. R., Williams, R. H., and Rutter, W. J. (1972). An ultrastructural analysis of the developing embryonic pancreas. *Dev. Biol.* **29**, 436–467.
Pisani, D. F., Pierson, P. M., Massoudi, A., Leclerc, L., Chopard, A., Marini, J. F., and Dechesne, C. A. (2004). Myodulin is a novel potential angiogenic factor in skeletal muscle. *Exp. Cell Res.* **292**, 40–50.

Pufe, T., Wildemann, B., Petersen, W., Mentlein, R., Raschke, M., and Schmidmaier, G. (2002). Quantitative measurement of the splice variants 120 and 164 of the angiogenic peptide vascular endothelial growth factor in the time flow of fracture healing: A study in the rat. *Cell Tissue Res.* **309**, 387–392.

Rafii, S., Shapiro, F., Pettengell, R., Ferris, B., Nachman, R. L., Moore, M. A., and Asch, A. S. (1995). Human bone marrow microvascular endothelial cells support long-term proliferation and differentiation of myeloid and megakaryocytic progenitors. *Blood* **86**, 3353–3363.

Ramsdell, A. F., and Markwald, R. R. (1997). Induction of endocardial cushion tissue in the avian heart is regulated, in part, by TGFbeta-3-mediated autocrine signaling. *Dev. Biol.* **188**, 64–74.

Razani, B., Combs, T. P., Wang, X. B., Frank, P. G., Park, D. S., Russell, R. G., Li, M., Tang, B., Jelicks, L. A., Scherer, P. E., and Lisanti, M. P. (2002a). Caveolin-1-deficient mice are lean, resistant to diet-induced obesity, and show hypertriglyceridemia with adipocyte abnormalities. *J. Biol. Chem.* **277**, 8635–8647.

Razani, B., Woodman, S. E., and Lisanti, M. P. (2002b). Caveolae: From cell biology to animal physiology. *Pharmacol. Rev.* **54**, 431–467.

Reissmann, E., Ernsberger, U., Francis-West, P. H., Rueger, D., Brickell, P. M., and Rohrer, H. (1996). Involvement of bone morphogenetic protein-4 and bone morphogenetic protein-7 in the differentiation of the adrenergic phenotype in developing sympathetic neurons. *Development* **122**, 2079–2088.

Risau, W. (1997). Mechanisms of angiogenesis. *Nature* **386**, 671–674.

Risau, W., and Flamme, I. (1995). Vasculogenesis. *Annu. Rev. Cell Dev. Biol.* **11**, 73–91.

Robinson, G. S., Ju, M., Shih, S. C., Xu, X., McMahon, G., Caldwell, R. B., and Smith, L. E. (2001). Nonvascular role for VEGF: VEGFR-1, 2 activity is critical for neural retinal development. *Faseb. J.* **15**, 1215–1217.

Rosolowsky, L. J., Hanke, C. J., and Campbell, W. B. (1999). Adrenal capillary endothelial cells stimulate aldosterone release through a protein that is distinct from endothelin. *Endocrinology* **140**, 4411–4418.

Rossiter, H., Alon, R., and Kupper, T. S. (1997). Selectins, T-cell rolling and inflammation. *Mol. Med. Today* **3**, 214–222.

Rostgaard, J., and Qvortrup, K. (2002). Sieve plugs in fenestrae of glomerular capillaries—site of the filtration barrier? *Cells Tissues Organs* **170**, 132–138.

Rupnick, M. A., Panigrahy, D., Zhang, C. Y., Dallabrida, S. M., Lowell, B. B., Langer, R., and Folkman, M. J. (2002). Adipose tissue mass can be regulated through the vasculature. *Proc. Natl. Acad. Sci. USA* **99**, 10730–10735.

Saxen, L., and Sariola, H. (1987). Early organogenesis of the kidney. *Pediatr. Nephrol.* **1**, 385–392.

Schneider, C., Wicht, H., Enderich, J., Wegner, M., and Rohrer, H. (1999). Bone morphogenetic proteins are required *in vivo* for the generation of sympathetic neurons. *Neuron* **24**, 861–870.

Serluca, F. C., Drummond, I. A., and Fishman, M. C. (2002). Endothelial signaling in kidney morphogenesis: A role for hemodynamic forces. *Curr. Biol.* **12**, 492–497.

Shabsigh, A., Chang, D. T., Heitjan, D. F., Kiss, A., Olsson, C. A., Puchner, P. J., and Buttyan, R. (1998). Rapid reduction in blood flow to the rat ventral prostate gland after castration: Preliminary evidence that androgens influence prostate size by regulating blood flow to the prostate gland and prostatic endothelial cell survival. *Prostate* **36**, 201–206.

Shah, A. M. (1996). Paracrine modulation of heart cell function by endothelial cells. *Cardiovasc Res.* **31**, 847–867.

Shah, N. M., Groves, A. K., and Anderson, D. J. (1996). Alternative neural crest cell fates are instructively promoted by TGFbeta superfamily members. *Cell* **85**, 331–343.

Shalaby, F., Rossant, J., Yamaguchi, T. P., Gertsenstein, M., Wu, X. F., Breitman, M. L., and Schuh, A. C. (1995). Failure of blood-island formation and vasculogenesis in Flk-1-deficient mice. *Nature* **376**, 62–66.

Sherer, G. K. (1975). Tissue interaction in chick liver development: A reevaluation. I. Epithelial morphogenesis: the role of vascularity in mesenchymal specificity. *Dev. Biol.* **46**, 281–295.

Shinbrot, E., Peters, K. G., and Williams, L. T. (1994). Expression of the platelet-derived growth factor beta receptor during organogenesis and tissue differentiation in the mouse embryo. *Dev. Dyn.* **199**, 169–175.

Sieff, C. A., Tsai, S., and Faller, D. V. (1987). Interleukin 1 induces cultured human endothelial cell production of granulocyte-macrophage colony-stimulating factor. *J. Clin. Invest.* **79**, 48–51.

Slack, J. M. (1995). Developmental biology of the pancreas. *Development* **121**, 1569–1580.

Stainier, D. Y., Weinstein, B. M., Detrich, H. W., 3rd, Zon, L. I., and Fishman, M. C. (1995). Cloche, an early acting zebrafish gene, is required by both the endothelial and hematopoietic lineages. *Development* **121**, 3141–3150.

Stan, R. V. (2002). Structure and function of endothelial caveolae. *Microsc. Res. Tech.* **57**, 350–364.

Stern, C. D., Artinger, K. B., and Bronner-Fraser, M. (1991). Tissue interactions affecting the migration and differentiation of neural crest cells in the chick embryo. *Development* **113**, 207–216.

Stoos, B. A., Carretero, O. A., Farhy, R. D., Scicli, G., and Garvin, J. L. (1992). Endothelium-derived relaxing factor inhibits transport and increases cGMP content in cultured mouse cortical collecting duct cells. *J. Clin. Invest.* **89**, 761–765.

Suter, E. R., and Majno, G. (1965). Passage of lipid across vascular endothelium in newborn rats. An electron microscopic study. *J. Cell Biol.* **27**, 163–177.

Taraboletti, G., D'Ascenzo, S., Borsotti, P., Giavazzi, R., Pavan, A., and Dolo, V. (2002). Shedding of the matrix metalloproteinases MMP-2, MMP-9, and MT1-MMP as membrane vesicle-associated components by endothelial cells. *Am. J. Pathol.* **160**, 673–680.

Taylor, D. D., and Black, P. H. (1986). Shedding of plasma membrane fragments. *In* "Developmental Biology" (M. Steinberg, Ed.), pp. 33–57. Plenum Press, New York.

Thompson, M. A., Ransom, D. G., Pratt, S. J., MacLennan, H., Kieran, M. W., Detrich, H. W., 3rd, Vail, B., Huber, T. L., Paw, B., Brownlie, A. J., Oates, A. C., Fritz, A., Gates, M. A., Amores, A., Bahary, N., Talbot, W. S., Her, H., Beier, D. R., Postlethwait, J. H., and Zon, L. I. (1998). The cloche and spadetail genes differentially affect hematopoiesis and vasculogenesis. *Dev. Biol.* **197**, 248–269.

Trueta, J., and Amato, V. P. (1960). The vascular contribution of osteogenesis. III. Changes in the growth cartilage caused by experimentally induced ischaemia. *J. Bone Joint Surg.* **42B**, 571–587.

Tufro, A., Norwood, V. F., Carey, R. M., and Gomez, R. A. (1999). Vascular endothelial growth factor induces nephrogenesis and vasculogenesis. *J. Am. Soc. Nephrol.* **10**, 2125–2134.

Vapaatalo, H., and Mervaala, E. (2001). Clinically important factors influencing endothelial function. *Med. Sci. Monit.* **7**, 1075–1085.

Varzaneh, F. E., Shillabeer, G., Wong, K. L., and Lau, D. C. (1994). Extracellular matrix components secreted by microvascular endothelial cells stimulate preadipocyte differentiation *in vitro*. *Metabolism* **43**, 906–912.

Villars, F., Guillotin, B., Amedee, T., Dutoya, S., Bordenave, L., Bareille, R., and Amedee, J. (2002). Effect of HUVEC on human osteoprogenitor cell differentiation needs heterotypic gap junction communication. *Am. J. Physiol. Cell Physiol.* **282**, C775–C785.

Yamagishi, T., Nakajima, Y., Miyazono, K., and Nakamura, H. (1999). Bone morphogenetic protein-2 acts synergistically with transforming growth factor–beta3 during endothelial-mesenchymal transformation in the developing chick heart. *J. Cell. Physiol.* **180**, 35–45.

Yoder, E. J. (2002). Modifications in astrocyte morphology and calcium signaling induced by a brain capillary endothelial cell line. *Glia* **38**, 137–145.

Yoshitomi, H., and Zaret, K. S. (2004). Endothelial cell interactions initiate dorsal pancreas development by selectively inducing the transcription factor Ptf1a. *Development* **131**, 807–817.

Zelzer, E., McLean, W., Ng, Y. S., Fukai, N., Reginato, A. M., Lovejoy, S., D'Amore, P. A., and Olsen, B. R. (2002). Skeletal defects in VEGF (120/120) mice reveal multiple roles for VEGF in skeletogenesis. *Development* **129**, 1893–1904.

Zerlin, M., and Goldman, J. E. (1997). Interactions between glial progenitors and blood vessels during early postnatal corticogenesis: Blood vessel contact represents an early stage of astrocyte differentiation. *J. Comp. Neurol.* **387**, 537–546.

Zerwes, H. G., and Risau, W. (1987). Polarized secretion of a platelet-derived growth factor-like chemotactic factor by endothelial cells *in vitro*. *J. Cell. Biol.* **105**, 2037–2041.

2

HIFs, Hypoxia, and Vascular Development

Kelly L. Covello*,† and M. Celeste Simon*,†,‡
*Abramson Family Cancer Research Institute
†Department of Cell and Developmental Biology
‡Howard Hughes Medical Institute
University of Pennsylvania School of Medicine
Philadelphia, Pennsylvania 19104

I. Introduction
II. Oxygen-Regulated Gene Expression
    A. The Hypoxia Inducible Factor (HIF) Family
    B. HIF Target Genes
    C. HIF Regulation
III. HIF in Development O_2 Homeostasis
    A. Early Vascular Development
    B. Role of Hypoxia and HIF in Vascular Development
IV. HIF and Disease
    A. HIF and Postdevelopment Function
    B. HIF and Cancer
V. Conclusion
    References

Cellular oxygen (O_2) concentrations are tightly regulated to maintain ATP levels required for metabolic reactions in the human body. Responses to changes in O_2 concentrations are primarily regulated by the transcription factor hypoxia inducible factor (HIF). HIF activates transcription of genes that increase systemic O_2 delivery or provide cellular metabolic adaptation under conditions of hypoxia. HIF activity is essential for embryogenesis and various processes in postnatal life, and therefore, HIF levels need to be precisely controlled. Abnormal HIF expression is related to numerous diseases of the vascular system, including heart disease, cancer, and chronic obstructive pulmonary disease. © 2004, Elsevier Inc.

I. Introduction

The ability to maintain oxygen (O_2) homeostasis is crucial for survival of all multicellular organisms. Abnormally high levels of O_2 (hyperoxia) may result in the generation of reactive oxygen species (ROS) that damages cellular or organelle membranes and DNA. Conversely, low levels of

O_2 (hypoxia) can result in insufficient ATP production and the inability to sustain essential cellular functions. As a result, organisms have evolved different mechanisms to maintain cellular O_2 concentration within a narrow physiological range. In some invertebrates, where few cells are present, simple diffusion is sufficient to provide cells with the required O_2 levels. In other invertebrates, such as *Drosophila melanogaster*, O_2 is delivered to tissues via a branching tubular system of the trachea. In contrast, vertebrates have evolved multiple, complex physiological systems for oxygenation that require the hematopoietic, cardiovascular, and respiratory systems.

There are short- and long-term responses to low oxygen levels that occur at the cellular, tissue, and systemic levels. Rapid hypoxic responses include increased glucose uptake (through glucose transporter 1 [Glut-1]); a switch from oxidative phosphorylation to glycolysis (through glycolytic enzymes such as phosphoglycerate kinase [PGK]); increased red-blood-cell production (through hematopoietic cytokines such as erythropoietin [EPO]); and vasodilation (through inducible nitric oxide synthase). However, these responses are only short-term survival strategies. To allow O_2 delivery to new tissues and restore O_2 delivery to starved tissues, new blood-vessel growth and/or remodeling of the existing vasculature must occur. Hypoxia occurs in a number of physiological settings, and the mechanisms mediating hypoxic responses are similar during embryogenesis, postnatal life, and pathological states. The exact structure and regulation of these important systems and hypoxic responses provide the necessary foundation for O_2 homeostasis.

II. Oxygen-Regulated Gene Expression

A. The Hypoxia Inducible Factor (HIF) Family

The primary transcriptional regulator of both cellular and systemic hypoxic responses is the heterodimeric transcription factor hypoxia inducible factor (HIF). HIF heterodimers consist of an alpha (α) subunit (HIFα) and a beta (β) subunit (HIFβ, also known as the aryl hydrocarbon nuclear translocator [ARNT] (Wang *et al.*, 1995; Wenger and Gassmann, 1997). Both subunits contain basic helix-loop-helix (bHLH) and PER-ARNT-SIM homology (PAS) domains that mediate dimerization. HIF dimerization is essential for HIF transcriptional activity.

The HIF protein family is complex in that three genes encode HIFα subunits (HIF-1α, -2α, and -3α) and three genes encode HIFβ subunits (ARNT/HIF-1β, ARNT2, and ARNT3). However, not all family members are involved in activating transcription in the response to hypoxia, and dimerization between different subunits can be limited by expression. Whereas

HIF-1α and HIF-2α can directly activate transcription of target genes, HIF-3α lacks a transcriptional activation domain and appears not to activate transcription (Gu *et al.*, 1998; Jiang *et al.*, 1996; Makino *et al.*, 2001, 2002; Tian *et al.*, 1997, 1998). Of the different β subunits, only ARNT and ARNT2 can form functional HIF complexes to regulate hypoxia-responsive genes. ARNT3/MOP3 does not form HIF complexes; nor can it restore HIF target gene expression in response to low oxygen when expressed in $Arnt^{-/-}$ ES cells (Cowden and Simon, 2002). HIF-1α and ARNT (HIF-1β) appear to be expressed ubiquitously in human and mouse tissues, whereas HIF-2α, HIF-3α, ARNT2, and ARNT3 are more tissue restricted. For example, HIF-2α (also known as endothelial PAS domain protein 1 [EPAS 1], HIF-1-like factor [HLF], HIF-1-related factor [HRF]) mRNA is expressed in endothelial cells, lung, and neural crest derivatives during development (Ema *et al.*, 1997; Flamme *et al.*, 1997; Tian *et al.*, 1997). Post-natally, under conditions of hypoxia, HIF-2α is expressed in other tissues and cell types, such as bone marrow macrophages, kidney epithelial cells, liver parenchyma, cardiac myoctyes, uterine decidual cells, and pancreatic parenchymal cells (Wiesener *et al.*, 2003).

B. HIF Target Genes

During hypoxia, HIF induces the expression of many genes involved in metabolism, cell survival, erythropoiesis, and vascular remodeling (Semenza, 2000). HIF activates gene expression by binding to a 50 base pair cis-acting hypoxia response element (HRE), located in the enhancer and promoter regions of many genes (Pugh *et al.*, 1991; Semenza *et al.*, 1991). The list of target genes activated by HIF continues to increase (Table I) and includes angiogenic factors such as vascular endothelial growth factor (VEGF) and platelet-derived growth factor-β (PDGF-β) (Gerber *et al.*, 1997; Iyer *et al.*, 1998; Ryan *et al.*, 1998; Wood *et al.*, 1996, 1998).

C. HIF Regulation

Since HIF is involved in mediating hypoxic responses in a timely manner, HIF expression needs to be tightly regulated. Both HIF subunits (HIFα and HIFβ) are constitutively transcribed and translated. The ability of HIF to activate transcription is regulated by the stability of the HIFα subunit. As summarized in Fig. 1, during normal oxygen conditions, or "normoxia" (21%), HIFα subunits are ubiquitinated and degraded via the 26S proteasome pathway through interaction with an E3 ubiquitin ligase complex

Table I HIF Target Genes

Oxygen regulated gene	Function	Reference
Glucose transporter 1 (Glut1)	Glucose uptake	(Ebert et al., 1995; Iyer et al., 1998; Okino et al., 1998)
Aldolase A	Glycolysis	(Semenza et al., 1994, 1996)
Phospohglycerate kinase 1	Glycolysis	(Semenza et al., 1994)
Enolase	Glycolysis	(Semenza et al., 1994, 1996)
Lactate dehydrogenase A	Glycolysis	(Firth et al., 1995)
VEGF	Angiogenesis	(Forsythe et al., 1996; Levy et al., 1995; Liu et al., 1995)
Flt-1	Angiogenesis	(Gerber et al., 1997)
iNOS	Vasodilation	(Melillo et al., 1995; Palmer et al., 1998)
Erythropoietin	Erythropoiesis	(Firth et al., 1994; Wang and Semenza, 1993)
Transferrin	Iron transport	(Rolfs et al., 1997)
Transferrin receptor	Iron uptake	(Lok and Ponka, 1999; Tacchini et al., 1999)

containing the von Hippel–Lindau tumor suppressor protein (pVHL) (Cockman et al., 2000; Kallio et al., 1999; Maxwell et al., 1999; Ohh et al., 2000).

The mechanisms and factors involved in the HIF regulatory pathway have been intensely investigated. HIFα subunits are hydroxylated on key proline residues by an oxygen-dependent family of prolyl hydroxylases (PHDs) (Bruick and McKnight, 2001; Epstein et al., 2001; Ivan et al., 2001; Jaakkola et al., 2001; Masson et al., 2001; Yu et al., 2001b). The HIF-PHDs are evolutionarily conserved from nematodes to mammals (Bruick and McKnight, 2001; Epstein et al., 2001). To date, three HIF-PHD homologs with different activities, specificities, and expression patterns have been identified. The hydroxylation reaction is O_2-, 2-oxoglutarate-, Fe^{2+}-, and ascorbate-dependent. The high efficiency of the hydroxylation and subsequent degradation reactions results in poor HIFα stability that exhibits a half life of minutes under normoxic conditions (Salceda and Caro, 1997). However, under hypoxic conditions, the HIF-PHDs are not active and HIFα subunits are stabilized, translocate into the nucleus, and heterodimerize with a HIFβ subunit to form a HIF complex (Kallio et al., 1999; Tanimoto et al., 2000). HIF transactivation is then achieved by the recruitment of coactivators, such as p300/CBP (Arany et al., 1996; Carrero et al., 2000; Ebert and Bunn, 1998; Jiang et al., 1997; Pugh et al., 1997). Under normoxia, HIF and p300 interaction is blocked by an additional hydroxylation reaction that is mediated by a HIF asparaginyl hydroxylase

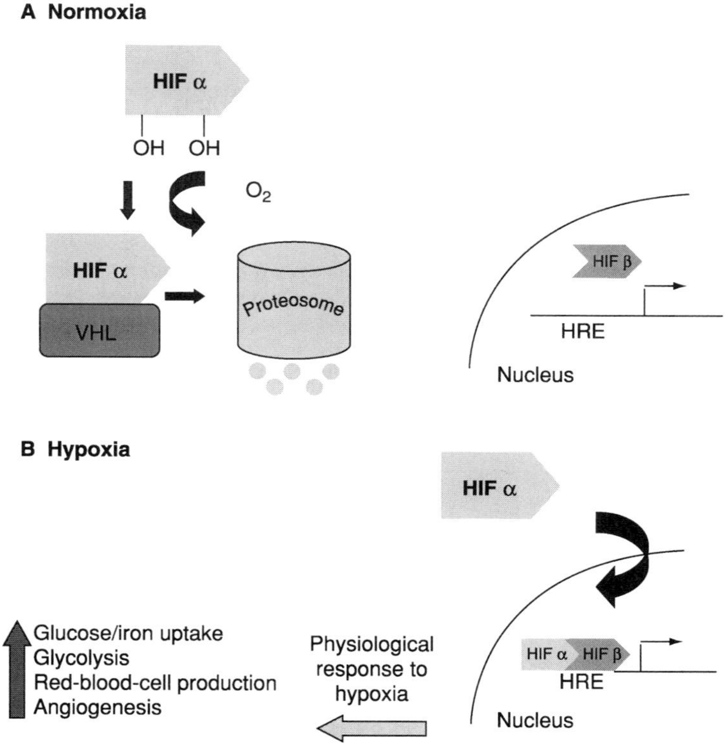

2. HIFs, Hypoxia, and Vascular Development

A Normoxia

B Hypoxia

Figure 1 Simplified model of HIF regulation. Under normoxic conditions, the ODD domain of HIF-α subunit is hydroxylated, which targets it for pVHL-mediated proteosomal degradation. Under hypoxia, HIF-α is stabilized and translocates into the nucleus. Here, HIF-α dimerizes with HIF-1β/ARNT to recruit cofactors and activate gene transcription. (See Color Insert.)

known as FIH (factor inhibiting HIF) (Lando et al., 2002a,b; Mahon et al., 2001). In summary, this regulation ensures that decreases in O_2 lead to a finely tuned and proportional increase in the accumulation of HIFα subunits.

The mechanisms by which cells sense decreased O_2 levels and transduce this signal to induce HIF activity are not clearly understood and remain controversial. HIF-PHDs and mitochondria are two of many proposed oxygen sensors, but the exact location of oxygen sensing has not been shown (Chandel et al., 1998, 2000). In all models, however, the hydroxylation modification of HIFα enables it to be recognized by and interact with the von Hippel–Lindau gene product (pVHL), thus enabling HIFα degradation (Ivan et al., 2001; Jaakkola et al., 2001; Kallio et al., 1999).

III. HIF in Developmental O_2 Homeostasis

A. Early Vascular Development

During embryogenesis, hypoxia induces transcriptional programs essential for normal embryonic development. Initially, diffusion is sufficient to provide cells with the O_2 and nutrients needed for cellular metabolism (Maltepe and Simon, 1998). During gastrulation, however, as cells continue to divide (resulting in embryonic growth), O_2 and nutrients are no longer provided to all cells by passive diffusion. Consequently, development of the hematopoietic and vascular systems begins shortly thereafter.

The vasculature in developing embryos is a dynamic structure that gradually evolves in multiple steps (Risau, 1997). First, vascular channels arise from the formation of angioblasts (or endothelial precursor cells) within the embryo proper. In addition, this process includes the formation of blood islands in the visceral yolk sac. Collectively, this process is known as vasculogenesis. These cells expand and interconnect to form the primary vascular plexus. Next, these primary vessels remodel in a process known as angiogenesis. The vessels undergo branching, sprouting, migration, and proliferation to form a mature vascular network. Mature vessels also require the recruitment of supporting cells, including pericytes and smooth muscle cells. All of these processes are mediated by interactions between endothelial cells, extracellular matrix, and support cells that involve differentiation, proliferation, migration, and cell–cell interactions. Many components important for vascular development have been discovered. For example, it is well established that VEGF and its receptors, Flt-1 and Flk-1, have essential roles in vasculogenesis and angiogenesis (Ferrara and Davis-Smyth, 1997; Ferrara *et al.*, 1996; Fong *et al.*, 1995; Shalaby *et al.*, 1995, 1997). Other factors involved include angiopoietins, Tie-1, Tie-2, and ephrin-B2 (Carmeliet and Jain, 2000).

B. Role of Hypoxia and HIF in Vascular Development

Mutations in genes encoding various growth factors or their receptors important for vascular development, such as VEGF, display severe phenotypes in mouse knockout models. It is interesting that many of these growth factors and receptors are activated by HIF. For example, VEGF is an important cytokine that promotes new vessel growth from preexisting vasculature by causing quiescent vascular endothelial cells to enter the cell cycle, migrate, and induce new vessel growth (Ferrara and Davis-Smyth, 1997). $Vegf^{+/-}$ animals are embryonic lethal and die within 11 or 12 days of embryonic development (E11–12) (Carmeliet *et al.*, 1996; Ferrara *et al.*,

2. HIFs, Hypoxia, and Vascular Development

1996). These embryos are characterized by a rudimentary dorsal aorta, vascular defects, and reduced formation of yolk sac vessels and blood cells. Homozygous null embryos, generated by aggregation of embryonic stem cells and tetraploid embryos, die by E8.5–9.5 exhibiting more severe defects, including a complete absence of the dorsal aorta (Carmeliet et al., 1996).

Flk-1 and Flt-1 are VEGF receptors that are expressed at different times in vascular development. Flk-1 is expressed in earlier hematopoietic and angiogenic precursors, while Flt-1 is expressed later in more differentiated endothelial precursors. $Flk\text{-}1^{-/-}$ mice exhibit embryonic lethality (E8.5–9.5) and defects in hematopoietic and angiogenic lineages (Shalaby et al., 1995, 1997). The embryos lack yolk sac blood islands and vessels in the embryo proper. $Flt\text{-}1^{-/-}$ mice exhibit increased numbers of hemangioblast progenitors that result in vascular disorganization (Fong et al., 1995, 1996, 1999). In summary, direct or indirect HIF target genes, such as VEGF and its cellular receptors Flt-1 and Flk-1, are essential for the formation of the embryonic vasculature.

Additional HIF target genes are involved in other aspects of cardiovascular development. EPO regulates red-blood-cell production in mammals and plays a crucial role in their proliferation, survival, and differentiation (Kieran et al., 1996; Lin et al., 1996; Wu et al., 1995). Embryos lacking Epo and its receptor (EpoR) are embryonic lethal (E13.5) and exhibit decreased numbers of circulating erythrocytes and severe fetal liver hypocellularity. In addition, $Epo^{-/-}$ mice exhibit reduced numbers of proliferating cardiac myocytes, which results in ventricular hypoplasia during cardiac morphogenesis (Wu et al., 1999). In conclusion, mutations in specific HIF target genes affect multiple components of early cardiovascular development.

As stated, hypoxia is a critical signal required for vascular development in the embryo, and, therefore, it is not surprising that HIF and HIF target genes are also involved in vascular development. Indeed, HIF (HIF-1α, HIF-2α, and ARNT/HIF-1β) is essential for embryonic development as well. Mice deficient for ARNT/HIF-1β die within 10.5 days of embryonic development (E10.5) and are characterized by defective vascularization of the yolk sac, brachial arches, and placenta (Adelman et al., 2000; Kozak et al., 1997; Maltepe et al., 1997). Although initial development of the vascular beds remains normal, vessel remodeling throughout the $Arnt^{-/-}$ embryo is abnormal. In addition, $Arnt^{-/-}$ embryos produce decreased numbers of hematopoietic progenitors within the yolk sac, and this phenotype can be rescued by VEGF addition in vitro (Adelman et al., 1999). The placenta is a vascular organ that supplies the embryo with blood and nutrients essential for growth. Placental cell fates are also regulated by HIF-mediated hypoxic responses as $Arnt^{-/-}$ placentas demonstrate an aberrant architecture due to an expansion of giant cell layers at the expense of the labyrinthine and spongiotrophoblast layers (Adelman et al., 2000).

Collectively, these results indicate that HIF-1β is important in development of cardiovascular organs, such as the vasculature, blood, and placenta.

Embryos lacking the gene-encoding *Hif-1*α are embryonic lethal as well and die during midgestation by day 11 of embryonic development (E11) with vascular defects and malformations in cardiovascular morphogenesis (Iyer *et al.*, 1998; Ryan *et al.*, 1998). Initially, vasculogenesis of *Hif-1*$\alpha^{-/-}$ embryos occurs normally but is followed by a marked regression of vascular endothelium by E9.5. *Hif-1*$\alpha^{-/-}$ embryos also exhibit defects in neural tube closure and extensive cell death, especially in the cephalic and brachial regions.

Three independent groups have created *Hif-2*$\alpha^{-/-}$ knockout models and obtained strikingly different results (Compernolle *et al.*, 2002; Peng *et al.*, 2000; Tian *et al.*, 1998). One group demonstrated mid-gestational lethality due to bradycardia, which was attributed to defects in catecholamine biosynthesis (Tian *et al.*, 1998). Another group demonstrated that loss of HIF-2α resulted in some viable mice that died within 24 h due to respiratory distress syndrome. Loss of HIF-2α correlated with defects in VEGF expression, leading to lack of lung surfactant production in pneumocytes (Compernolle *et al.*, 2002). Finally, a third group also observed bradycardia with lethality between E9.5 and E13.5. However, in this group, approximately 50% of homozygous embryos exhibit extensive vascular remodeling defects in the yolk sac and embryo proper (Peng *et al.*, 2000). In these embryos, vasculogenesis occurred normally but exhibited remodeling defects resulting from improper connection of blood vessels, which resulted in enlarged lumens and hemorrhage. Taken together, the mouse models of HIF-1α, HIF-2α, and ARNT indicate that HIF has an important role during postvasculogenesis stages and is required for the remodeling of the primary plexus into a vascular network.

IV. HIF and Disease

A. HIF and Postdevelopment Function

HIF is also involved in responding to changes in O_2 levels during fetal and postnatal life. Analysis of adult *Hif-1*$\alpha^{+/-}$ mice demonstrated that partial HIF-1α deficiency results in impaired responses to chronic hypoxia such as right ventricular hypertrophy, pulmonary artery hypertension, and pulmonary vascular remodeling (Yu *et al.*, 1999). Conditional gene targeting of *Hif-1*α using a floxed allele suggests a role for HIF-1α in mediating signals in naturally hypoxic tissues, such as chondrocytes and macrophages (Cramer, *et al.*, 2003; Schipani *et al.*, 2001; Seagroves *et al.*, 2001). For example, conditional knockout mice lacking HIF-1α in chondrocytes exhibit gross skeletal malformations and die perinatally (Schipani et al., 2001). Recently, HIF-2α postnatal function was investigated by crossing isogenic heterozygous 129S6/SvEvTac *Hif-2α* knockout mice with heterozygous C57BL/6J *Hif-2α* knockout mice to obtain *Hif-2α$^{-/-}$* F1 hybrid mice (Scortegagna et al., 2003a,b). A fraction of the resultant F1 hybrid *Hif2$^{-/-}$* mice survive to adulthood. The surviving *Hif-2α* null mice exhibit pancytopenia, multiorgan pathology, including cardiac hypertrophy and hepatomegaly, and other pathologies related to mitochondrial dysfunction (Scortegagna et al., 2003a,b). In addition, HIF is implicated in pathological angiogenesis, such as neovascularization in ischemic myocardium, hypoxia-induced pulmonary vascular remodeling, and tumor vascularization.

B. HIF and Cancer

Solid tumors undergoing rapid expansion and growth beyond a volume of 1 mm^3 are dependent on sufficient oxygen supply, and many tumors frequently develop a severely hypoxic microenvironment. Therefore, neoangiogenesis is crucial for tumor progression and growth as the remodeling of blood vessels must supply the tumor with oxygen and nutrients (Carmeliet and Jain, 2000). Neoangiogenesis is mediated cell and non-cell autonomously through a variety of angiogenic factors, including VEGF, basic fibroblast growth factor (bFGF), transforming growth factor-β (TGF-β), interleukin 8, and TNF-α (Folkman, 1995). As stated previously, VEGF is an important cytokine that induces new vessel growth from pre-existing vasculature (Ferrara and Davis-Smyth, 1997). Indeed, VEGF mRNA levels correlate with tumor angiogenesis (Cheng et al., 1996; Kondo et al., 2000a,b; Mori et al., 1999; Zhang et al., 1995). VEGF and other factors implicated in tumor angiogenesis are directly or indirectly activated by HIF. Accordingly, it has been hypothesized that upregulation of HIF resulting from an expanding, hypoxic tumor environment can function to promote tumor growth.

In addition to stabilization of HIF-α subunits under low levels of oxygen, HIF dysregulation occurs by other mechanisms. pVHL is highly conserved in multicellular organisms from nematodes to humans. (Gnarra et al., 1997; Iliopoulos et al., 1995; Kamada et al., 2001; Kim et al., 1998; Ohh et al., 1998; Pause et al., 1998). pVHL is a classic tumor suppressor, conforming to the Knudson two-hit model, wherein mutations in this gene lead to the development of a variety of highly vascularized tumors, such as renal clear cell carcinoma, pheochromocytoma, and hemangioblastoma (Kondo and Kaelin, 2001). This tumor suppressor plays a role in numerous cellular processes, including development, tumor growth, cytoskeletal dynamics, cell cycle regulation, and extracellular matrix deposition. Because of its wide variety of cellular roles, pVHL's tumor suppressor function is unlikely to be solely dependent on its ability to degrade HIFα. However, in some cases, mutations within pVHL can disrupt the interaction and regulation of HIFα subunits and lead to constitutive HIFα subunit expression and activity. The resultant tumors are characterized by increased expression of HIF and HIF target genes such as EPO, VEGF, carbonic anhydrase, and glucose transporters. Overproduction of angiogenic factors could explain the hypervascular nature of hemangioblastomas and renal cell carcinomas.

Regardless of the mechanism of HIF expression, HIF has been hypothesized and tested to contribute to tumor progression through various mechanisms. Proposed growth advantageous activities of HIF include aiding the adaptation of cancer cells to the tumor microenvironment via its effects on glucose metabolism and biosynthetic pathways, increasing cell survival through specific targets, and promoting growth through the recruitment of new blood vessels (Akakura, 2001; Akakura et al., 2001; Bruchovsky et al., 1996; Carmeliet et al., 1998; Griffiths et al., 2002; Kung et al., 2000; Ryan et al., 2000; Williams et al., 2002). However, mouse models using xenograft assays to test these ideas have yielded contradictory results. Using multiple independent HIF null cell lines, such as $Hif\text{-}1\ \alpha^{-/-}$ embryonic stem cells, $Hif\text{-}1\ \alpha^{-/-}$ transformed fibroblasts, and hepatoma cells lacking $Hif\text{-}1\ \beta$, some reports demonstrate that HIF-1 α enhances tumor cell survival and proliferation (Maxwell et al., 1997; Ryan et al., 1998, 2000). Conversely, other reports conclude that tumors derived from ES cells lacking $Hif\text{-}1\ \alpha$ can grow faster as a result of decreased rates of apoptosis (Carmeliet et al., 1998; Yu et al., 2001a). The reason for these differences remains unclear, although genetic variation between parental cell lines could explain the discrepancies.

Subsequently, a plethora of evidence has accumulated that suggests that HIF promotes tumor growth. Overexpression of HIF-1 α has been reported in a variety of human tumors, including brain, prostate, breast, and colon cancers (Talks et al., 2000; Turner et al., 2002; Zagzag et al., 2000; Zhong et al., 1998, 1999). HIF overexpression is correlated with high-grade, highly vascularized brain tumors, ovarian carcinomas, and ductal carcinomas in situ (the early preinvasive stage of breast cancer) (Birner et al., 2000; Bos et al., 2001; Zagzag et al., 2000). Likewise, HIF-2α has been reported to be overexpressed in a number of tumor types, including renal clear cell carcinomas, pheochromoctyomas, astrocytomas, and non-small-cell lung cancers (Favier et al., 2001; Giatromanolaki et al., 2001; Jaakkola et al., 2001; Khatua et al., 2003). In some of these highly vascularized tumors, HIF-2α expression positively correlates with VEGF expression, suggesting a role for HIF-2α in tumor angiogenesis (Favier et al., 2001; Talks et al., 2000).

Two independent studies addressed the individual contribution of HIFα subunits to tumorigenesis, and the results support the growing evidence that HIF functions to promote tumor growth in a variety of tumor types. Using a renal carcinoma xenograft model, one group showed that cells reconstituted with wild-type pVHL produced smaller tumors compared to the VHL null RCC line (Maranchie et al., 2002). Intriguingly, competitive inhibition of the HIFα binding site of pVHL resulted in restored tumor growth. Furthermore, stable expression of HIF-1α was insufficient to induce tumors, suggesting that endogenous HIF-2α is responsible (Maranchie et al., 2002). In addition, stabilized HIF-2α, but not HIF-1α, increases RCC tumor growth (Kondo et al., 2002). In summary, HIF overexpression in human cancers and mouse xenograft models suggest an importance of HIF in tumor vascularization and other aspects of tumor progression.

V. Conclusion

HIF is a master regulator of O_2 homeostasis that controls the establishment of essential systems during embryogenesis, as well as the regulation during fetal and postnatal life. One of these processes, postvasculogenesis, is a complex process that involves a number of factors and intricate mechanisms of induction. Modulating HIF levels and activity may provide therapeutic approaches for pathological conditions such as cancer, heart disease, and pulmonary disease.

References

Adelman, D. M., Gertsenstein, M., Nagy, A., Simon, M. C., and Maltepe, E. (2000). Placental cell fates are regulated *in vivo* by HIF-mediated hypoxia responses. *Genes Dev.* **14**, 3191–3203.

Adelman, D. M., Maltepe, E., and Simon, M. C. (1999). Multilineage embryonic hematopoiesis requires hypoxic ARNT activity. *Genes Dev.* **13**, 2478–2483.

Akakura, N. (2001). Significance of constitutive expression of hypoxia-inducible factor-1 alpha (HIF-1 alpha) protein in pancreatic cancer. *Hokkaido Igaku Zasshi* **76**, 375–384.

Akakura, N., Kobayashi, M., Horiuchi, I., Suzuki, A., Wang, J., Chen, J., Niizeki, H., Kawamura, K., Hosokawa, M., and Asaka, M. (2001). Constitutive expression of hypoxia-inducible factor-1alpha renders pancreatic cancer cells resistant to apoptosis induced by hypoxia and nutrient deprivation. *Cancer Res.* **61**, 6548–6554.

Arany, Z., Huang, L. E., Eckner, R., Bhattacharya, S., Jiang, C., Goldberg, M. A., Bunn, H. F., and Livingston, D. M. (1996). An essential role for p300/CBP in the cellular response to hypoxia. *Proc. Natl. Acad. Sci. USA* **93**, 12969–12973.

Birner, P., Schindl, M., Obermair, A., Plank, C., Breitenecker, G., and Oberhuber, G. (2000). Overexpression of hypoxia-inducible factor 1alpha is a marker for an unfavorable prognosis in early-stage invasive cervical cancer. *Cancer Res.* **60**, 4693–4696.

Bos, R., Zhong, H., Hanrahan, C. F., Mommers, E. C., Semenza, G. L., Pinedo, H. M., Abeloff, M. D., Simons, J. W., van Diest, P. J., and van der Wall, E. (2001). Levels of hypoxia-inducible factor-1 alpha during breast carcinogenesis. *J. Natl. Cancer Inst.* **93**, 309–314.

Bruchovsky, N., Snoek, R., Rennie, P. S., Akakura, K., Goldenberg, L. S., and Gleave, M. (1996). Control of tumor progression by maintenance of apoptosis. *Prostate Suppl.* **6**, 13–21.

Bruick, R. K., and McKnight, S. L. (2001). A conserved family of prolyl-4-hydroxylases that modify HIF. *Science* **294**, 1337–1340.

Carmeliet, P., Dor, Y., Herbert, J. M., Fukumura, D., Brusselmans, K., Dewerchin, M., Neeman, M., Bono, F., Abramovitch, R., Maxwell, P., Koch, C. J., Ratcliffe, D., Moons, L., Jain, R. K., Collen, D., and Keshert, E. (1998). Role of HIF-1alpha in hypoxia-mediated apoptosis, cell proliferation and tumour angiogenesis. *Nature* **394**, 485–490.

Carmeliet, P., Ferreira, V., Breir, G., Pollefeyt, S., Kieckens, L., Gertenstein, M., Fahrig, M., Vandenhoeck, A., Harpal, K., Eberhard, C., Pawling, J., Moons, L., Collen, D., Riscu, W., and Nagy, A. (1996). Abnormal blood vessel development and lethality in embryos lacking a single VEGF allele. *Nature* **380**, 435–439.

Carmeliet, P., and Jain, R. K. (2000). Angiogenesis in cancer and other diseases. *Nature* **407**, 249–257.

Carrero, P., Okamoto, K., Coumailleau, P., O'Brien, S., Tanaka, H., and Poellinger, L. (2000). Redox-regulated recruitment of the transcriptional coactivators CREB-binding protein and SRC-1 to hypoxia-inducible factor 1alpha. *Mol. Cell. Biol.* **20**, 402–415.

Chandel, N. S., Maltepe, E., Goldwasser, E., Mathieu, C. E., Simon, M. C., and Schumacker, P. T. (1998). Mitochondrial reactive oxygen species trigger hypoxia-induced transcription. *Proc. Natl. Acad. Sci. USA* **95**, 11715–11720.

Chandel, N. S., McClintock, D. S., Feliciano, C. E., Wood, T. M., Melendez, J. A., Rodriguez, A. M., and Schumacker, P. T. (2000). Reactive oxygen species generated at mitochondrial complex III stabilize hypoxia-inducible factor-1alpha during hypoxia: A mechanism of O_2 sensing. *J. Biol. Chem.* **275**, 25130–25138.

Cheng, S. Y., Huang, H. J., Nagane, M., Ji, X. D., Wang, D., Shih, C. C., Arap, W., Huang, C. M., and Cavenee. W. K. (1996). Suppression of glioblastoma angiogenicity and tumorigenicity by inhibition of endogenous expression of vascular endothelial growth factor. *Proc. Natl. Acad. Sci. USA* **93**, 8502–8507.

Cockman, M. E., Masson, N., Mole, D. R., Jaakkola, P., Chang, G. W., Clifford, S. C., Maher, E. R., Pugh, C. W., Ratcliffe, P. J., and Maxwell, P. H. (2000). Hypoxia inducible factor-alpha binding and ubiquitylation by the von Hippel–Lindau tumor suppressor protein. *J. Biol. Chem.* **275**, 25733–25741.

Compernolle, V., Brusselmans, K., Acker, T., Hoet, P., Tjwa, M., Beck, H., Plaisance, S., Dor, Y., Keshet, E., Lupu, F., Nemery, B., Dewerchin, M., Van Velderhoven, P., Plate, K., Moons, L., Collen, D., and Carmeliet, P. (2002). Loss of HIF-2alpha and inhibition of VEGF impair fetal lung maturation, whereas treatment with VEGF prevents fatal respiratory distress in premature mice. *Nat. Med.* **8**, 702–710.

Cowden, K. D., and Simon, M. C. (2002). The bHLH/PAS factor MOP3 does not participate in hypoxia responses. *Biochem. Biophys. Res. Commun.* **290**, 1228–1236.

Cramer, T., Yamanishi, Y., Clausen, B. E., Forster, I., Pawlinski, R., Mackman, N., Haase, V. H., Jaenisch, R., Corr, M., Nizet, V., Fireslein, G. S., Gerber, H. P., Ferrar, N., and Johnson, R. S. (2003). HIF-1alpha is essential for myeloid cell-mediated inflammation. *Cell* **112**, 645–657.

Ebert, B. L., and Bunn, H. F. (1998). Regulation of transcription by hypoxia requires a multiprotein complex that includes hypoxia-inducible factor 1, an adjacent transcription factor, and p300/CREB binding protein. *Mol. Cell. Biol.* **18**, 4089–4096.

Ebert, B. L., Firth, J. D., and Ratcliffe, P. J. (1995). Hypoxia and mitochondrial inhibitors regulate expression of glucose transporter-1 via distinct Cis-acting sequences. *J. Biol. Chem.* **270**, 29083–29089.

Ema, M., Taya, S., Yokotani, N., Sogawa, K., Matsuda, Y., and Fujii-Kuriyama, Y. (1997). A novel bHLH-PAS factor with close sequence similarity to hypoxia-inducible factor 1alpha regulates the VEGF expression and is potentially involved in lung and vascular development. *Proc. Natl. Acad. Sci. USA* **94**, 4273–4278.

2. HIFs, Hypoxia, and Vascular Development

Epstein, A. C., Gleadle, J. M., McNeill, L. A., Hewitson, K. S., O'Rourke, J., Mole, D. R., Mukherji, M., Metzen, E., Wilson, M. I., Dhanda, A., Tian, Y. M., Masson, N., Hamilton, D. L., Jaakkola, P., Barstead, R., Hodgkin, J., Maxwell, P. H., Pugh, C. W., Schofield, C. J., and Ratcliffe, P. J. (2001). *C. elegans* EGL-9 and mammalian homologs define a family of dioxygenases that regulate HIF by prolyl hydroxylation. *Cell* **107**, 43–54.

Favier, J., Kempf, H., Corvol, P., and Gasc, J. M. (2001). Coexpression of endothelial PAS protein 1 with essential angiogenic factors suggests its involvement in human vascular development. *Dev. Dyn.* **222**, 377–388.

Ferrara, N., Carver-Moore, K., Chen, H., Dowd, M., Lu, L., O'Shea, K. S., Powell-Braxton, L., Hillan, K. J., and Moore, M. W. (1996). Heterozygous embryonic lethality induced by targeted inactivation of the VEGF gene. *Nature* **380**, 439–442.

Ferrara, N., and Davis-Smyth, T. (1997). The biology of vascular endothelial growth factor. *Endocr. Rev.* **18**, 4–25.

Firth, J. D., Ebert, B. L., Pugh, C. W., and Ratcliffe, P. J. (1994). Oxygen-regulated control elements in the phosphoglycerate kinase 1 and lactate dehydrogenase A genes: Similarities with the erythropoietin 3′ enhancer. *Proc. Natl. Acad. Sci. USA* **91**, 6496–6500.

Firth, J. D., Ebert, B. L., and Ratcliffe, P. J. (1995). Hypoxic regulation of lactate dehydrogenase A. *J. Biol. Chem.* **270**, 21021–21027.

Flamme, I., Frohlich, T., von Reutern, M., Kappel, A., Damert, A., and Risau, W. (1997). HRF, a putative basic helix-loop-helix-PAS-domain transcription factor is closely related to hypoxia-inducible factor-1 alpha and developmentally expressed in blood vessels. *Mech. Dev.* **63**, 51–60.

Folkman, J. (1995). Angiogenesis in cancer, vascular, rheumatoid and other disease. *Nat. Med.* **1**, 27–31.

Fong, G. H., Klingensmith, J., Wood, C. R., Rossant, J., and Breitman, M. L. (1996). Regulation of flt-1 expression during mouse embryogenesis suggests a role in the establishment of vascular endothelium. *Dev. Dyn.* **207**, 1–10.

Fong, G. H., Rossant, J., Gertsenstein, M., and Breitman, M. L. (1995). Role of the Flt-1 receptor tyrosine kinase in regulating the assembly of vascular endothelium. *Nature* **376**, 66–70.

Fong, G. H., Zhang, L., Bryce, D. M., and Peng, J. (1999). Increased hemangioblast commitment, not vascular disorganization, is the primary defect in flt-1 knock-out mice. *Development* **126**, 3015–3025.

Forsythe, J. A., Jiang, B. H., Iyer, N. V., Agani, F., Leung, S. W., Koos, R. D., and Semenza, G. L. (1996). Activation of vascular endothelial growth factor gene transcription by hypoxia-inducible factor 1. *Mol. Cell. Biol.* **16**, 4604–4613.

Gerber, H. P., Condorelli, F., Park, J., and Ferrara, N. (1997). Differential transcriptional regulation of the two vascular endothelial growth factor receptor genes. Flt-1, but not Flk-1/KDR, is up-regulated by hypoxia. *J. Biol. Chem.* **272**, 23659–23667.

Giatromanolaki, A., Koukourakis, M. I., Sivridis, E., Turley, H., Talks, K., Pezzella, F., Gatter, K. C., and Harris, A. L. (2001). Relation of hypoxia inducible factor 1 alpha and 2 alpha in operable non-small cell lung cancer to angiogenic/molecular profile of tumours and survival. *Br. J. Cancer* **85**, 881–890.

Gnarra, J. R., Ward, J. M., Porter, F. D., Wagner, J. R., Devor, D. E., Grinberg, A., Emmert-Buck, M. R., Westphal, H., Klausner, R. D., and Linehan, W. M. (1997). Defective placental vasculogenesis causes embryonic lethality in VHL-deficient mice. *Proc. Natl. Acad. Sci. USA* **94**, 9102–9107.

Griffiths, J. R., McSheehy, P. M., Robinson, S. P., Troy, H., Chung, Y. L., Leek, R. D., Williams, K. J., Stratford, I. J., Harris, A. L., and Stubbs, M. (2002). Metabolic changes detected by *in vivo* magnetic resonance studies of HEPA-1 wild-type tumors and tumors deficient in hypoxia-inducible factor-1beta (HIF-1beta): Evidence of an anabolic role for the HIF-1 pathway. *Cancer Res.* **62**, 688–695.

Gu, Y. Z., Moran, S. M., Hogenesch, J. B., Wartman, L., and Bradfield, C. A. (1998). Molecular characterization and chromosomal localization of a third alpha-class hypoxia inducible factor subunit, HIF3alpha. *Gene Expr.* **7**, 205–213.

Iliopoulos, O., Kibel, A., Gray, S., and Kaelin, W. G., Jr. (1995). Tumour suppression by the human von Hippel–Lindau gene product. *Nat. Med.* **1**, 822–826.

Ivan, M., Kondo, K., Yang, H., Kim, W., Valiando, J., Ohh, M., Salic, A., Asara, J. M., Lane, W. S., and Kaelin, W. G., Jr. (2001). HIFalpha targeted for VHL-mediated destruction by proline hydroxylation: Implications for O2 sensing. *Science* **292**, 464–468.

Iyer, N. V., Kotch, L. E., Agani, F., Leung, S. W., Laughner, E., Wenger, R. H., Gassmann, M., Gearhart, J. D., Lawler, A. M., Yu, A. Y., and Semenza, G. L. (1998). Cellular and developmental control of O_2 homeostasis by hypoxia-inducible factor 1 alpha. *Genes Dev.* **12**, 149–162.

Jaakkola, P., Mole, D. R., Tian, Y. M., Wilson, M. I., Gielbert, J., Gaskell, S. J., Kriegsheim, A., Hebestreit, H. F., Mukherji, M., Schofield, C. J., Maxwell, P. H., Pugh, C. W., and Ratcliffe, P. J. (2001). Targeting of HIF-alpha to the von Hippel–Lindau ubiquitylation complex by O_2-regulated prolyl hydroxylation. *Science* **292**, 468–472.

Jiang, B.-H., Rue, E., Wang, G. L., Roe, R., and Semenza, G. L. (1996). Dimerization, DNA binding, and transactivation properties of hypoxia-inducible factor 1. *J. Biol. Chem.* **271**, 17771–17778.

Jiang, B. H., Zheng, J. Z., Leung, S. W., Roe, R., and Semenza, G. L. (1997). Transactivation and inhibitory domains of hypoxia-inducible factor 1alpha. Modulation of transcriptional activity by oxygen tension. *J. Biol. Chem.* **272**, 19253–19260.

Kallio, P. J., Wilson, W. J., O'Brien, S., Makino, Y., and Poellinger, L. (1999). Regulation of the hypoxia-inducible transcription factor 1alpha by the ubiquitin-proteasome pathway. *J. Biol. Chem.* **274**, 6519–6525.

Kamada, M., Suzuki, K., Kato, Y., Okuda, H., and Shuin, T. (2001). Von Hippel–Lindau protein promotes the assembly of actin and vinculin and inhibits cell motility. *Cancer Res.* **61**, 4184–4189.

Khatua, S., Peterson, K. M., Brown, K. M., Lawlor, C., Santi, M. R., LaFleur, B., Dressman, D., Stephan, D. A., and MacDonald, T. J. (2003). Overexpression of the EGFR/FKBP12/HIF-2alpha pathway identified in childhood astrocytomas by angiogenesis gene profiling. *Cancer Res.* **63**, 1865–1870.

Kieran, M. W., Perkins, A. C., Orkin, S. H., and Zon, L. I. (1996). Thrombopoietin rescues *in vitro* erythroid colony formation from mouse embryos lacking the erythropoietin receptor. *Proc. Natl. Acad. Sci. USA* **93**, 9126–9131.

Kim, M., Katayose, Y., Li, Q., Rakkar, A. N., Li, Z., Hwang, S. G., Katayose, D., Trepel, J., Cowan, K. H., and Seth, P. (1998). Recombinant adenovirus expressing Von Hippel–Lindau-mediated cell cycle arrest is associated with the induction of cyclin-dependent kinase inhibitor p27Kip1. *Biochem. Biophys. Res. Commun.* **253**, 672–677.

Kondo, K., and Kaelin, W. G., Jr. (2001). The von Hippel–Lindau tumor suppressor gene. *Exp. Cell Res.* **264**, 117–125.

Kondo, K., Klco, J., Nakamura, E., Lechpammer, M., and Kaelin, W. G., Jr. (2002). Inhibition of HIF is necessary for tumor suppression by the von Hippel–Lindau protein. *Cancer Cell* **1**, 237–246.

Kondo, Y., Arii, S., Furutani, M., Isigami, S., Mori, A., Onodera, H., Chiba, T., and Imamura, M. (2000a). Implication of vascular endothelial growth factor and p53 status for angiogenesis in noninvasive colorectal carcinoma. *Cancer* **88**, 1820–1827.

Kondo, Y., Arii, S., Mori, A., Furutani, M., Chiba, T., and Imamura, M. (2000b). Enhancement of angiogenesis, tumor growth, and metastasis by transfection of vascular endothelial growth factor into LoVo human colon cancer cell line. *Clin. Cancer Res.* **6**, 622–630.

Kozak, K. R., Abbott, B., and Hankinson, O. (1997). ARNT-deficient mice and placental differentiation. *Dev. Biol.* **191**, 297–305.

Kung, A. L., Wang, S., Klco, J. M., Kaelin, W. G., and Livingston, D. M. (2000). Suppression of tumor growth through disruption of hypoxia-inducible transcription. *Nat. Med.* **6**, 1335–1340.

Lando, D., Peet, D. J., Gorman, J. J., Whelan, D. A., Whitelaw, M. L., and Bruick, R. K. (2002a). FIH-1 is an asparaginyl hydroxylase enzyme that regulates the transcriptional activity of hypoxia-inducible factor. *Genes Dev.* **16**, 1466–1471.

Lando, D., Peet, D. J., Whelan, D. A., Gorman, J. J., and Whitelaw, M. L. (2002b). Asparagine hydroxylation of the HIF transactivation domain a hypoxic switch. *Science* **295**, 858–861.

Levy, A. P., Levy, N. S., Wegner, S., and Goldberg, M. A. (1995). Transcriptional regulation of the rat vascular endothelial growth factor gene by hypoxia. *J. Biol. Chem.* **270**, 13333–13340.

Lin, C. S., Lim, S. K., D'Agati, V., and Costantini, F. (1996). Differential effects of an erythropoietin receptor gene disruption on primitive and definitive erythropoiesis. *Genes. Dev.* **10**, 154–164.

Liu, Y., Cox, S. R., Morita, T., and Kourembanas, S. (1995). Hypoxia regulates vascular endothelial growth factor gene expression in endothelial cells. Identification of a 5′ enhancer. *Circ. Res.* **77**, 638–643.

Lok, C. N., and Ponka, P. (1999). Identification of a hypoxia response element in the transferrin receptor gene. *J. Biol. Chem.* **274**, 24147–24152.

Mahon, P. C., Hirota, K., and Semenza, G. L. (2001). FIH-1: A novel protein that interacts with HIF-1alpha and VHL to mediate repression of HIF-1 transcriptional activity. *Genes Dev.* **15**, 2675–2686.

Makino, Y., Cao, R., Svensson, K., Bertilsson, G., Asman, M., Tanaka, H., Cao, Y., Berkenstam, A., and Poellinger, L. (2001). Inhibitory PAS domain protein is a negative regulator of hypoxia-inducible gene expression. *Nature* **414**, 550–554.

Makino, Y., Kanopka, A., Wilson, W. J., Tanaka, H., and Poellinger, L. (2002). Inhibitory PAS domain protein (IPAS) is a hypoxia-inducible splicing variant of the hypoxia-inducible factor-3alpha locus. *J. Biol. Chem.* **277**, 32405–32408.

Maltepe, E., Schmidt, J. V., Baunoch, D., Bradfield, C. A., and Simon, M. C. (1997). Abnormal angiogenesis and responses to glucose and oxygen deprivation in mice lacking the protein ARNT. *Nature* **386**, 403–407.

Maltepe, E., and Simon, M. C. (1998). Oxygen, genes, and development: An analysis of the role of hypoxic gene regulation during murine vascular development. *J. Mol. Med.* **76**, 391–401.

Maranchie, J. K., Vasselli, J. R., Riss, J., Bonifacino, J. S., Linehan, W. M., and Klausner, R. D. (2002). The contribution of VHL substrate binding and HIF1-alpha to the phenotype of VHL loss in renal cell carcinoma. *Cancer Cell* **1**, 247–255.

Masson, N., Willam, C., Maxwell, P. H., Pugh, C. W., and Ratcliffe, P. J. (2001). Independent function of two destruction domains in hypoxia-inducible factor-alpha chains activated by prolyl hydroxylation. *EMBO J.* **20**, 5197–5206.

Maxwell, P. H., Dachs, G. U., Gleadle, J. M., Nicholls, L. G., Harris, A. L., Stratford, I. J., Hankinson, O., Pugh, C. W., and Ratcliffe, P. J. (1997). Hypoxia-inducible factor-1 modulates gene expression in solid tumors and influences both angiogenesis and tumor growth. *Proc. Natl. Acad. Sci. USA* **94**, 8104–8109.

Maxwell, P. H., Wiesener, M. S., Chang, G. W., Clifford, S. C., Vaux, E. C., Cockman, M. E., Wykoff, C. C., Pugh, C. W., Maher, E. R., and Ratcliffe, P. J. (1999). The tumour suppressor protein VHL targets hypoxia-inducible factors for oxygen-dependent proteolysis. *Nature* **399**, 271–275.

Melillo, G., Musso, T., Sica, A., Taylor, L. S., Cox, G. W., and Varesio, L. (1995). A hypoxia-responsive element mediates a novel pathway of activation of the inducible nitric oxide synthase promoter. *J. Exp. Med.* **182,** 1683–1693.

Mori, A., Arii, S., Furutani, M., Hanaki, K., Takeda, Y., Moriga, T., Kondo, Y., Gorrin Rivas, M. J., and Imamura, M. (1999). Vascular endothelial growth factor-induced tumor angiogenesis and tumorigenicity in relation to metastasis in a HT1080 human fibrosarcoma cell model. *Int. J. Cancer* **80,** 738–743.

Ohh, M., Park, C. W., Ivan, M., Hoffman, M. A., Kim, T. Y., Huang, L. E., Pavletich, N., Chau, V., and Kaelin, W. G. (2000). Ubiquitination of hypoxia-inducible factor requires direct binding to the beta-domain of the von Hippel–Lindau protein. *Nat. Cell. Biol.* **2,** 423–427.

Ohh, M., Yauch, R. L., Lonergan, K. M., Whaley, J. M., Stemmer-Rachamimov, A. O., Louis, D. N., Gavin, B. J., Kley, N., Kaelin, W. G., Jr., and Iliopoulos, O. (1998). The von Hippel–Lindau tumor suppressor protein is required for proper assembly of an extracellular fibronectin matrix. *Mol. Cell* **1,** 959–968.

Okino, S. T., Chichester, C. H., and Whitlock, J. P., Jr. (1998). Hypoxia-inducible mammalian gene expression analyzed *in vivo* at a TATA-driven promoter and at an initiator-driven promoter. *J. Biol. Chem.* **273,** 23837–23843.

Palmer, L. A., Semenza, G. L., Stoler, M. H., and Johns, R. A. (1998). Hypoxia induces type II NOS gene expression in pulmonary artery endothelial cells via HIF-1. *Am. J. Physiol.* **274,** L212–L219.

Pause, A., Lee, S., Lonergan, K. M., and Klausner, R. D. (1998). The von Hippel–Lindau tumor suppressor gene is required for cell cycle exit upon serum withdrawal. *Proc. Natl. Acad. Sci. USA* **95,** 993–998.

Peng, J., Zhang, L., Drysdale, L., and Fong, G. H. (2000). The transcription factor EPAS-1/hypoxia-inducible factor 2alpha plays an important role in vascular remodeling. *Proc. Natl. Acad. Sci. USA* **97,** 8386–8391.

Pugh, C. W., O'Rourke, J. F., Nagao, M., Gleadle, J. M., and Ratcliffe, P. J. (1997). Activation of hypoxia-inducible factor-1; definition of regulatory domains within the alpha subunit. *J. Biol. Chem.* **272,** 11205–11214.

Pugh, C. W., Tan, C. C., Jones, R. W., and Ratcliffe, P. J. (1991). Functional analysis of an oxygen-regulated transcriptional enhancer lying 3' to the mouse erythropoietin gene. *Proc. Natl. Acad. Sci. USA* **88,** 10553–10557.

Risau, W. (1997). Mechanisms of angiogenesis. *Nature* **386,** 671–674.

Rolfs, A., Kvietikova, I., Gassmann, M., and Wenger, R. H. (1997). Oxygen-regulated transferrin expression is mediated by hypoxia-inducible factor-1. *J. Biol. Chem.* **272,** 20055–20062.

Ryan, H. E., Lo, J., and Johnson, R. S. (1998). HIF-1 alpha is required for solid tumor formation and embryonic vascularization. *Embo. J.* **17,** 3005–3015.

Ryan, H. E., Poloni, M., McNulty, W., Elson, D., Gassmann, M., Arbeit, J. M., and Johnson, R. S. (2000). Hypoxia-inducible factor-1alpha is a positive factor in solid tumor growth. *Cancer Res.* **60,** 4010–4015.

Salceda, S., and Caro, J. (1997). Hypoxia-inducible factor 1alpha (HIF-1alpha) protein is rapidly degraded by the ubiquitin-proteasome system under normoxic conditions. Its stabilization by hypoxia depends on redox-induced changes. *J. Biol. Chem.* **272,** 22642–22647.

Schipani, E., Ryan, H. E., Didrickson, S., Kobayashi, T., Knight, M., and Johnson, R. S. (2001). Hypoxia in cartilage: HIF-1alpha is essential for chondrocyte growth arrest and survival. *Genes Dev.* **15,** 2865–2876.

Scortegagna, M., Ding, K., Oktay, Y., Gaur, A., Thurmond, F., Yan, L. J., Marck, B. T., Matsumoto, A. M., Shelton, J. M., Richardson, J. A., Bennett, M. J., and Garcia, J. A.

(2003a). Multiple organ pathology, metabolic abnormalities and impaired homeostasis of reactive oxygen species in Epas1(−/−) mice. *Nat. Genet.* **203,** 331–340.

Scortegagna, M., Morris, M. A., Oktay, Y., Bennett, M., and Garcia, J. A. (2003b). The HIF family member EPAS1/HIF-2{alpha} is required for normal hematopoiesis in mice. *Blood.*

Seagroves, T. N., Ryan, H. E., Lu, H., Wouters, B. G., Knapp, M., Thibault, P., Laderoute, K., and Johnson, R. S. (2001). Transcription factor HIF-1 is a necessary mediator of the pasteur effect in mammalian cells. *Mol. Cell Biol.* **21,** 3436–3444.

Semenza, G. L. (2000). HIF-1 and human disease: One highly involved factor. *Genes Dev.* **14,** 1983–1991.

Semenza, G. L., Jiang, B. H., Leung, S. W., Passantino, R., Concordet, J. P., Maire, P., and Giallongo, A. (1996). Hypoxia response elements in the aldolase A, enolase 1, and lactate dehydrogenase A gene promoters contain essential binding sites for hypoxia-inducible factor 1. *J. Biol. Chem.* **271,** 32529–32537.

Semenza, G. L., Nejfelt, M. K., Chi, S. M., and Antonarakis, S. E. (1991). Hypoxia-inducible nuclear factors bind to an enhancer element located 3′ to the human erythropoietin gene. *Proc. Natl. Acad. Sci. USA* **88,** 5680–5684.

Semenza, G. L., Roth, P. H., Fang, H.-M., and Wang, G. L. (1994). Transcriptional regulation of genes encoding glycolytic enzymes by hypoxia-inducible factor 1. *J. Biol. Chem.* **269,** 23757–23769.

Shalaby, F., Ho, J., Stanford, W. L., Fischer, K. D., Schuh, A. C., Schwartz, L., Bernstein, A., and Rossant, J. (1997). A requirement for Flk1 in primitive and definitive hematopoiesis and vasculogenesis. *Cell* **89,** 981–990.

Shalaby, F., Rossant, J., Yamaguchi, T. P., Gertsenstein, M., Wu, X. F., Breitman, M. L., and Schuh, A. (1995). Failure of blood-island formation and vasculogenesis in *Flk*-1 deficient mice. *Nature* **376,** 62–66.

Tacchini, L., Bianchi, L., Bernelli-Zazzera, A., and Cairo, G. (1999). Transferrin receptor induction by hypoxia. HIF-1-mediated transcriptional activation and cell-specific post-transcriptional regulation. *J. Biol. Chem.* **274,** 24142–24146.

Talks, K. L., Turley, H., Gatter, K. C., Maxwell, P. H., Pugh, C. W., Ratcliffe, P. J., and Harris, A. L. (2000). The expression and distribution of the hypoxia-inducible factors HIF-1alpha and HIF-2alpha in normal human tissues, cancers, and tumor-associated macrophages. *Am. J. Pathol.* **157,** 411–421.

Tanimoto, K., Makino, Y., Pereira, T., and Poellinger, L. (2000). Mechanism of regulation of the hypoxia-inducible factor-1 alpha by the von Hippel–Lindau tumor suppressor protein. *Embo. J.* **19,** 4298–4309.

Tian, H., Hammer, R. E., Matsumoto, A. M., Russell, D. W., and McKnight, S. L. (1998). The hypoxia-responsive transcription factor EPAS1 is essential for catecholamine homeostasis and protection against heart failure during embryonic development. *Genes Dev.* **12,** 3320–3324.

Tian, H., McKnight, S. L., and Russell, D. W. (1997). Endothelial PAS domain protein 1 (EPAS1), a transcription factor selectively expressed in endothelial cells. *Genes Dev.* **11,** 72–82.

Turner, K. J., Moore, J. W., Jones, A., Taylor, C. F., Cuthbert-Heavens, D., Han, C., Leek, R. D., Gatter, K. C., Maxwell, P. H., Ratcliffe, P. J., Cranston, D., and Harris, A. L. (2002). Expression of hypoxia-inducible factors in human renal cancer: Relationship to angiogenesis and to the von Hippel–Lindau gene mutation. *Cancer Res.* **62,** 2957–2961.

Wang, G. L., Jiang, B.-H., Rue, E. A., and Semenza, G. L. (1995). Hypoxia-inducible factor 1 is a basic-helix-loop-helix-PAS heterodimer regulated by cellular O_2 tension. *Proc. Natl. Acad. Sci. USA* **92,** 5510–5514.

Wang, G. L., and Semenza, G. L. (1993). Characterization of hypoxia-inducible factor 1 and regulation of DNA binding activity by hypoxia. *J. Biol. Chem.* **268,** 21513–21518.

Wenger, R. H., and Gassmann, M. (1997). Oxygen(es) and the hypoxia-inducible factor-1. *Biol. Chem.* **378**, 609–616.

Wiesener, M. S., Jurgensen, J. S., Rosenberger, C., Scholze, C. K., Horstrup, J. H., Warnecke, C., Mandriota, S., Bechmann, I., Frei, U. A., Pugh, C. W., Ratcliffe, P. J., Bachmann, S., Maxwell, P. H., and Eckardt, K. U. (2003). Widespread hypoxia-inducible expression of HIF-2alpha in distinct cell populations of different organs. *Faseb. J.* **17**, 271–273.

Williams, K. J., Telfer, B. A., Airley, R. E., Peters, H. P., Sheridan, M. R., van der Kogel, A. J., Harris, A. L., and Stratford, I. J. (2002). A protective role for HIF-1 in response to redox manipulation and glucose deprivation: Implications for tumorigenesis. *Oncogene* **21**, 282–290.

Wood, S. M., Gleadle, J. M., Pugh, C. W., Hankinson, O., and Ratcliffe, P. J. (1996). The role of aryl hydrocarbon receptor nuclear translocator (ARNT) in hypoxia induction of gene expression. *J. Biol. Chem.* **271**, 15117–15123.

Wood, S. M., Wiesener, M. S., Yeates, K. M., Okada, N., Pugh, C. W., Maxwell, P. H., and Ratcliffe, P. J. (1998). Selection and analysis of a mutant cell line defective in the hypoxia-inducible factor-1 alpha-subunit (HIF-1alpha). Characterization of hif-1alpha-dependent and -independent hypoxia-inducible gene expression. *J. Biol. Chem.* **273**, 8360–8368.

Wu, H., Lee, S. H., Gao, J., Liu, X., and Iruela-Arispe, M. L. (1999). Inactivation of erythropoietin leads to defects in cardiac morphogenesis. *Development* **126**, 3597–3605.

Wu, H., Liu, X., Jaenisch, R., and Lodish, H. F. (1995). Generation of committed erythroid BFU-E and CFU-E progenitors does not require erythropoietin or the erythropoietin receptor. *Cell* **83**, 59–67.

Yu, A. Y., Shimoda, L. A., Iyer, N. V., Huso, D. L., Sun, X., McWilliams, R., Beaty, T., Sham, J. S., Wiener, C. M., Sylvester, J. T., and Semenza, G. L. (1999). Impaired physiological responses to chronic hypoxia in mice partially deficient for hypoxia-inducible factor 1alpha. *J. Clin. Invest.* **103**, 691–696.

Yu, F., White, S. B., Zhao, Q., and Lee, F. S. (2001a). Dynamic, site-specific interaction of hypoxia-inducible factor-1alpha with the von Hippel–Lindau tumor suppressor protein. *Cancer Res.* **61**, 4136–4142.

Yu, F., White, S. B., Zhao, Q., and Lee, F. S. (2001b). HIF-1alpha binding to VHL is regulated by stimulus-sensitive proline hydroxylation. *Proc. Natl. Acad. Sci. USA* **98**, 9630–9635.

Zagzag, D., Zhong, H., Scalzitti, J. M., Laughner, E., Simons, J. W., and Semenza, G. L. (2000). Expression of hypoxia-inducible factor 1alpha in brain tumors: Association with angiogenesis, invasion, and progression. *Cancer* **88**, 2606–2618.

Zhang, H. T., Craft, P., Scott, P. A., Ziche, M., Weich, H. A., Harris, A. L., and Bicknell, R. (1995). Enhancement of tumor growth and vascular density by transfection of vascular endothelial cell growth factor into MCF-7 human breast carcinoma cells. *J. Natl. Cancer Inst.* **87**, 213–219.

Zhong, H., Agani, F., Baccala, A. A., Laughner, E., Rioseco-Camacho, N., Isaacs, W. B., Simons, J. W., and Semenza, G. L. (1998). Increased expression of hypoxia inducible factor-1alpha in rat and human prostate cancer. *Cancer Res.* **58**, 5280–5284.

Zhong, H., De Marzo, A. M., Laughner, E., Lim, M., Hilton, D. A., Zagzag, D., Buechler, P., Isaacs, W. B., Semenza, G. L., and Simons, J. W. (1999). Overexpression of hypoxia-inducible factor 1alpha in common human cancers and their metastases. *Cancer Res.* **59**, 5830–5835.

3
Blood Vessel Patterning at the Embryonic Midline

Kelly A. Hogan* and Victoria L. Bautch*,[†],[‡]
*Department of Biology
[†]Carolina Cardiovascular Biology Center
[‡]Curriculum in Genetics and Molecular Biology
University of North Carolina at Chapel Hill
Chapel Hill, North Carolina 27599

I. Introduction
II. Vascular Development and Patterning
   A. Overview
   B. Blood Vessel Formation
   C. Blood Vessel Patterning
   D. Vascular Patterning Information from Different Model Organisms
III. Signaling Pathways Implicated in Vascular Patterning
   A. Overview
   B. Signaling Pathways Implicated in Vascular Patterning
IV. Axial Structures Implicated in Vascular Patterning
   A. Hypochord
   B. Endoderm
   C. Notochord
   D. Neural Tube
V. Conclusions and Future Directions
   Acknowledgments
   References

The reproducible pattern of blood vessels formed in vertebrate embryos has been described extensively, but only recently have we obtained the genetic and molecular tools to address the mechanisms underlying these processes. This review describes our current knowledge regarding vascular patterning around the vertebrate midline and presents data derived from frogs, zebrafish, avians, and mice. The embryonic structures implicated in midline vascular patterning, the hypochord, endoderm, notochord, and neural tube, are discussed. Moreover, several molecular signaling pathways implicated in vascular pattering, VEGF, Tie/tek, Notch, Eph/ephrin, and Semaphorin, are described. Data showing that VEGF is critical to patterning the dorsal aorta in frogs and zebrafish, and to patterning the vascular plexus that forms around the neural tube in amniotes, is presented. A more complete knowledge of vascular patterning is likely to come from the next generation of experiments using ever more sophisticated tools, and these results promise to directly impact on clinically important issues such as forming new vessels in the human body and/or in bioreactors. © 2004, Elsevier Inc.

I. Introduction

Developmental vascular biology presents an interesting paradox. Blood vessels are easy to see because of the blood within. In 1672, Marcello Malpighi first described that blood coursed through specific tubes in chick embryos (Gilbert, 2003), and much subsequent embryology described the development and elaboration of blood vessels. These early studies reached an apex with the publication of careful, descriptive studies of blood vessel formation by Herbert Evans, E. R. Clark, and Florence Sabin (Clark, 1918; Evans, 1909; Sabin, 1917, 1920). Yet until very recently, developmental blood vessel formation has been understudied relative to other developmental processes such as limb formation and neural development. This was partially due to the ubiquitous presence of blood vessels in almost all tissues, which prevented extensive molecular analysis until these studies could be carried out at the single-cell level. Vascular pattern formation has been even more refractory to mechanistic analysis, even though these patterns have been described for hundreds of years. However, the recent surge in interest in vascular patterning has resulted in much new information and models for further testing. Moreover, beyond the basic developmental questions are applications to diseases and therapies that also motivate investigations of vascular pattern formation. For example, if we understand how the embryo coordinates the pattern of vessels with the development of other organs and tissues, we may be able to apply this information to the reconstruction of functional vasculature in the adult or even in an artificial setting. It is an exciting time to work in this field. The genetic and analytic tools are available and the questions are compelling.

II. Vascular Development and Patterning

A. Overview

The embryonic vasculature is formed via the coordination of multiple cellular processes. These include the specification of mesodermal precursor cells called angioblasts, their differentiation into endothelial cells, and the migration and assembly of angioblasts and endothelial cells into vessels (reviews: Cleaver and Krieg, 1999; Daniel and Abrahamson, 2000; Drake and Little,

1999; Jain, 2003; Folkman, 2003; Risau, 1997; Poole et al., 2001; Weinstein, 2002; Yancopoulos et al., 2000). These processes must be synchronized within the vascular lineage, and they must also interface with the developmental programs of other embryonic lineages. This coordination is called vascular patterning, and it results in a primary vessel network that is reproducible in both time and space. Signals produced by other tissues impinge on angioblasts and endothelial cells to pattern the embryonic vasculature (Coffin and Poole, 1991; Noden, 1988; Poole and Coffin, 1989). However, vascular patterning signals have been identified only recently, and the list is very incomplete. Moreover, little is known about where and how these signals act to pattern vessels.

This review combines information from zebrafish, frogs, avians, and mice and focuses on the migration and assembly of vessels guided by axial structures that straddle the embryonic midline. We describe the molecular signaling pathways implicated in vascular patterning, then describe the evidence for involvement of specific axial structures in vessel patterning around the midline: the hypochord, endoderm, notochord, and neural tube. Due to space constraints, we will not discuss in detail the short-range patterning of vessels that occurs in the limb and retina (Mukouyama et al., 2002; Otani et al., 2002; Stone et al., 1995; Zhang et al., 1999). Likewise, other interesting aspects of patterning, such as how endothelial cells signal to tissues (reviewed in Cleaver and Melton, 2003) and arterial/venous differentiation (reviewed in Lawson and Weinstein, 2002a), are the subject of recent excellent reviews and will not be covered in detail here.

B. Blood Vessel Formation

Blood vessels in the embryo form through a combination of two developmental processes—vasculogenesis and angiogenesis. The coalescence and differentiation of mesodermal precursor cells to form vessels de novo is termed vasculogenesis. Angiogenesis involves the migration and division of already differentiated endothelial cells to form new vessels. The work of several labs, including ours, demonstrates that many vessels in the embryo form by a combination of both processes (Ambler et al., 2001; Brand-Saberi et al., 1995; Childs et al., 2002; Feinberg and Nolden, 1991). Historically, both normal and pathological neovascularization were thought to occur solely by angiogenic processes in the adult. However, recent studies show that bone-marrow-derived circulating endothelial cells contribute to adult neoangiogenesis, suggesting that both vasculogenesis and angiogenesis occur throughout the life of an organism (Asahara et al., 1997, 1999; Otani et al., 2002).

C. Blood Vessel Patterning

Early descriptive studies of blood vessel patterning were based on live observations of vessel development in chick embryos and in the tails of frog tadpoles, complemented by analysis of ink injections and histological sections of embryos at different developmental stages (Clark, 1918; Evans, 1909; Sabin, 1917, 1920). These early studies were built on by the Clarks, who elegantly described vessel formation and remodeling in frog tails by examining living specimens (Clark and Clark, 1939). Subsequently, several groups carefully described vessel formation and patterning in the developing quail using a vascular cell-specific antibody (Coffin and Poole, 1988; Pardanaud et al., 1987). These studies in aggregate led to a model of blood vessel patterning in which an early primitive vascular plexus is first formed by either vasculogenesis or angiogenesis at the site of a future vessel or vessel bed, and it is then extensively remodeled to form the final vascular pattern (review: Drake et al., 1998). Experimental analysis of the migratory behavior and origin of angioblasts was then carried out by several groups using quail transplants in chick hosts (Noden, 1989; Pardanaud et al., 1989; Poole and Coffin, 1989). Noden determined that most mesodermal tissues, with the exception of the prechordal plate, contain cells with angiogenic potential that can migrate large distances to form vessels in the embryo (Noden, 1989). It is important that Noden as well as and Poole and Coffin, observed that angioblasts that normally form vessels in the trunk were able to participate in the formation of blood vessels unique and appropriate to the head (Noden, 1989; Poole and Coffin, 1989). These observations continue to influence the way we think about vascular patterning, and subsequent work supports a model in which the pattern of blood vessels is guided by environmental cues rather than intrinsic to endothelial cells or their progenitors. The identification of an embryonic structure as the source of a vascular patterning signal came from Cleaver and Krieg, who first showed that the vertebrate midline produced a vascular patterning signal for dorsal aorta formation and suggested that hypochord-derived VEGF was important for this signal (Cleaver and Krieg, 1998; Cleaver et al., 1997). These landmark studies have set up important questions in vascular patterning that are currently under investigation, such as: What are additional embryonic sources of vascular patterning cues? What is the molecular composition of the signals that emanate from these sources? How are these signals coordinated with ongoing development and patterning in the rest of the embryo?

D. Vascular Patterning Information from Different Model Organisms

An interesting evolutionary question is: When did blood vessels first arise? A bona fide vasculature is associated with vertebrates, although many

3. Midline Vessel Patterning 59

productive analogies have been made between vertebrate blood vessel formation and *Drosophila* tracheal development (Ghabrial *et al.*, 2003). Current models of embryonic vascular patterning rely on information derived from several different vertebrate model organisms, and each of these different model organisms has provided unique opportunities to dissect aspects of vascular patterning.

1. *Xenopus laevis* (tropical frog)

Blood and lymphatic vessels in frog tails were some of the earliest specimens to be analyzed by live observation. The modern advantages of this well-used embryological model are the large size and free-living status of the early embryo. This allows for molecular manipulation via injection of individual early cells with RNAs that misexpress proteins or morpholinos that block expression. Unfortunately, significant vascular patterning occurs later in development when molecular manipulation of the *Xenopus* embryo is more difficult. Moreover, there are few markers for *Xenopus* vascular cells. However, the first studies defining midline vascular patterning signals were carried out by Cleaver and Krieg in the *Xenopus* embryo (Cleaver and Krieg, 1998; Cleaver *et al.*, 1997). Recently, vascular development in *Xenopus* was documented by visualization of vessels using injection of DiI-Ac-LDL, a compound that selectively binds to endothelial cells in many organisms (Levine *et al.*, 2003).

2. Zebrafish (*Danio rerio*)

There are several advantages to this rather new model of vertebrate development. The embryo is transparent and free-living, facilitating the acquisition of descriptive information. The relatively short life cycle and small size permit forward genetic screens to uncover novel genes important in vascular development and patterning. The recent finding that morpholino injection can provide information on the phenotypic consequences of reduced function of specific genes has added an important tool to analysis of zebrafish development. These advantages have been exploited by a number of investigators to study aspects of vascular patterning, such as arterial-venous differentiation (Lawson *et al.*, 2001, 2002; Zhong *et al.*, 2001) and sprouting of intersomitic vessels (Childs *et al.*, 2002). Moreover, analysis of the *gridlock* mutation in zebrafish showed that only vessels in specific parts of the embryo were compromised (Weinstein *et al.*, 1995; Zhong *et al.*, 2000), suggesting that local cues from surrounding tissues pattern vessels and supporting the work of Noden (described in Section II.C). The midline structures of the zebrafish embryo are well characterized and amenable to disruption by mutations. Recently, Weinstein and colleagues injected vessels of zebrafish embryos with fluorescent microspheres (microangiography), and they subsequently generated transgenic zebrafish that express green fluorescent protein (GFP) in the developing vasculature. They then used state-of-the-art live imaging to describe vascular development in the zebrafish embryo (Isogai et al., 2001; Lawson and Weinstein, 2002b). These studies have significantly increased our understanding of vessel development and patterning and promise to further our knowledge even more in the future.

3. Avians (chick and quail)

The avian embryo has historically had a central place in investigations of vascular patterning. Early investigators used the chick embryo for live observation and ink injections, so by 1920 it was the best-described embryological model of vascular development. Subsequently, the accessibility of the avian embryo was exploited for surgical manipulations, and the observation that quail cells could be distinguished from chick cells by nuclear morphology was used in early graft experiments to follow angioblast migration and patterning. An important refinement was development of the QH1 antibody that recognizes an epitope specific to quail endothelial cells and progenitors, but does not recognize chick endothelial cells (Pardanaud et al., 1987). A series of elegant studies analyzed quail grafts placed into chick hosts and provided a rich source of data that led to our current model of vascular patterning around the avian midline (Klessinger and Christ, 1996; Pardanaud and Dieterlen-Lievre, 1995; Pardanaud et al., 1996; Wilting et al., 1995). Others have used the quail to experimentally manipulate molecules implicated in vessel formation and patterning such as VEGF, bFGF, and integrins (Cox and Poole, 2001; Drake and Little, 1995; Drake et al., 2000; Finkelstein and Poole, 2003). Recently, Little and colleagues have devised protocols for dynamic image analysis of vessel development and patterning in the avian embryo, and analytical tools for quantitative assessments of vessel behavior (Rupp et al., 2003).

4. Mouse

The mouse is the least tractable model for visual and surgical manipulations, but in contrast, both gain- and loss-of-function genetic experiments in the mouse have shed light on genetic pathways important in vessel development and patterning in mammals. The earliest manipulations were of the VEGF and Tie/tek signaling pathways, and both are clearly crucial to vessel formation (Carmeliet et al., 1996; Dumont et al., 1994; Ferrara et al., 1996; Fong et al., 1995; Sato et al., 1995; Shalaby et al., 1995). However, the early and profound effects of mutations in these and a number of other pathways have somewhat hindered analysis of effects on vascular patterning in mammals.

3. Midline Vessel Patterning

The recent explosion in analysis of mutations of genes in specific cell types of the mouse by tissue-specific excision using the Cre-lox system promises to circumvent this problem and lead to a more sophisticated understanding of vascular patterning in mammals. Our group has attempted to utilize the advantages of both avian and mouse models by analyzing mouse grafts placed into avian hosts (Ambler et al., 2001, 2003; Hogan et al., 2004). This work will be described in more detail in Section IV.D.

The analysis of vascular patterning events and mechanisms in the models described above has been useful in revealing universal aspects of vascular patterning while defining differences among different organisms. Examples of universal features of vascular pattern formation are: (1) the use of both vasculogenesis and angiogenesis to form the embryonic vasculature; (2) the midline aggregation of vascular cells to form the dorsal aorta; and (3) the importance of the VEGF signaling pathway in vascular development. Examples of differences are: (1) the formation of large vessels by remodeling of an initial plexus in amniotes versus formation of the large vessels de novo in zebrafish and (2) formation of the dorsal aorta via signals from the hypochord, a VEGF-secreting structure at the midline of frogs and zebrafish but lacking in avians and mammals.

III. Signaling Pathways Implicated in Vascular Patterning

A. Overview

There are numerous signaling pathways thought to be important in aspects of vascular patterning. This has been one of the most exciting areas of research recently; the genetic and molecular tools available have permitted not only the testing of hypotheses about specific pathways, but the identification of mutations that affect patterning. Here we describe only the subset of pathways for which evidence exists for a role in vessel patterning. We do not discuss in detail interesting mutations, such as *out-of-bounds,* that have not yet been cloned. We also do not describe important signaling pathways whose primary effect is thought to be at the level of vessel stability and remodeling, such as the SIP/EDG, TGF beta superfamily, and PDGF pathways.

B. Signaling Pathways Implicated in Vascular Patterning

1. VEGF (Vascular endothelial growth factor)

The VEGF family of ligands and their receptors play a central role in many aspects of blood vessel formation, including vascular patterning. The preponderance of evidence for effects on vascular patterning is restricted to the most studied family member, VEGF-A. Thus we will discuss the role of VEGF-A (VEGF) and its receptors flk-1 (VEGFR-2) and flt-1 (VEGFR-1) in vascular patterning. An interacting set of coreceptors, neuropilin-1 and neuropilin-2, will be covered in Section III.B.5, as they were initially identified as Semaphorin receptors.

VEGF-A is a multi-functional protein involved in differentiation, proliferation, and migration of endothelial cells (reviews: Cross et al., 2003; Ferrara et al., 2003). It is expressed early in development in vertebrates, and its expression coincides temporally and spatially with blood vessel formation at numerous embryonic sites (Cleaver et al., 1997; Dumont et al., 1995; Flamme et al., 1995; Miquerol et al., 1999). A requirement for VEGF in vascular development is demonstrated by the paucity of vasculature and the embryonic lethality when one VEGF-A allele is deleted in mice, and the almost complete lack of vessels in embryos and embryonic stem cells lacking VEGF-A (Bautch et al., 2000; Carmeliet et al., 1996; Ferrara et al., 1996). In zebrafish, morpholino knockdown of VEGF indicates that the initial establishment of axial vasculature does not require VEGF-A, although it is required for patterning intersegmental vessels (Nasevicius et al., 2000). In addition, modest increases or decreases in VEGF-A levels in mice also disrupt vessel development and lead to embryonic lethality (Damert et al., 2002; Miquerol et al., 2000), indicating that tight dosage control of the VEGF-A signal is important.

However, as mentioned in Section II.D.4, the pleiotropic effects of VEGF-A on vascular development have impeded a direct assessment of its role in vascular patterning until quite recently (see Section IV.A, D). Gain-of-function experiments indicate that VEGF-A is involved in patterning, since injection of VEGF or placement of VEGF-coated beads into avian embryos results in ectopic and mis-patterned vessels (Bates et al., 2003; Drake and Little, 1995; Drake et al., 2000; Flamme et al., 1995; Finkelstein and Poole, 2003). VEGF-A RNA is alternatively spliced to produce three major isoforms of 120, 164, and 188 amino acids (Park et al., 1993). These isoforms have different biochemical properties, suggesting that they may be differentially deposited in the embryo: VEGF120 is predicted to be freely diffusible; VEGF188 is predicted to be matrix bound; and VEGF164 is predicted to be intermediate in these properties. Mice that are genetically engineered to express only VEGF120 or VEGF188 have consistent differences in the caliber and branching of vessels, suggesting that isoform composition affects local assembly and patterning of the vascular plexus (Carmeliet et al., 1999; Gerhardt et al., 2003; Ng et al., 2001; Ruhrberg et al., 2002; Stalmans et al., 2002).

The VEGF receptor tyrosine kinases, flk-1 and flt-1, both bind VEGF-A with high affinity. VEGF-A binding to flk-1 induces receptor tyrosine phosphorylation, and endothelial cells respond with downstream signaling that leads to proliferation, migration, survival, and permeability changes (reviews: Cross et al., 2003; Ferrara et al., 2003). Deletion of *flk-1* in mice or embryonic stem cells is embryonic lethal with lack of organized blood vessels (Schuh et al., 1999; Shalaby et al., 1995, 1997). The similarity between the loss-of-function phenotypes for VEGF-A and flk-1 suggest that most VEGF-A signaling is mediated by the flk-1 receptor. Evidence that signaling through the flk-1 receptor is important in vascular patterning around the amniote axial midline comes from analysis of embryonic stem-cell-derived embryoid bodies placed into the presomitic mesoderm cavity of quail hosts (Ambler et al., 2003). Although wild-type angioblasts migrated and patterned properly in this model, angioblasts genetically deleted for *flk-1* did not respond to avian patterning cues to migrate to specific embryonic locations. Moreover, in zebrafish, a *flk-1* mutation that severely down regulates flk mRNA did not disrupt initial vasculogenesis, but prevented sprouting of intersomitic and other vessels (Habeck et al., 2002). Taken together, these results strongly indicate that VEGF signaling through flk-1 mediates vascular patterning around the embryonic midline.

The role of the flt-1 receptor in vascular development and patterning appears more complex, and it has been somewhat controversial. Flt-1 is clearly necessary for proper vascular development, since deletion of *flt-1* leads to vessel overgrowth and embryonic lethality (Fong et al., 1995). It was subsequently suggested that flt-1 normally modulates the cell fate decision that induces hemangioblasts, progenitor cells capable of giving rise to both hematopoietic and endothelial cells, to form from mesoderm (Fong et al., 1999). However, corroboration of that model has been lacking. Recently, we showed that vascular cells lacking flt-1 have a higher rate of cell division than wild-type controls, suggesting that flt-1 normally negatively modulates cell division (Kearney et al., 2002). This is consistent with a model of flt-1 action in which its ability to act as a sink for the VEGF-A ligand is important developmentally, likely through a soluble form of the receptor that is generated via alternative splicing (Kendall and Thomas, 1993; Kendall et al., 1996). However, these basic questions regarding the cellular phenotype and mechanism of flt-1 action have precluded extensive analysis of its role in vascular patterning. We recently used dynamic image analysis of embryonic stem-cell-derived vessels to show that flt-1 is a positive modulator of vascular sprout formation and branching; that is to say, in the absence of flt-1, the vascular plexus is less branched (Kearney et al., 2004). The defect of the *flt-1* mutants could be largely rescued with a transgene that only expressed the soluble form of flt-1, suggesting that the morphogenetic effect is largely mediated by soluble flt-1. These findings suggest that flt-1 does affect vascular patterning at the local level. It will be interesting to determine if flt-1 is also involved in modulation of midline vascular patterning mediated by VEGF-A.

2. Tie/Tek Pathway

Tie2 (also called tek) and Tie1 (also called Tie) are receptor tyrosine kinases expressed on endothelial cells during development (reviews: Loughna and Sato 2001a; Ward and Dumont 2002). Angiopoietins (Ang) are the ligands for Tie2, while the Tie1 ligand(s) is unknown. Deletion of *Tie2* in mice is embryonic lethal at midgestation, with extensive vascular defects that include decreased sprouting, simplified vessel branching, and lack of pericyte recruitment (Dumont *et al.*, 1994; Patan, 1998; Puri *et al.*, 1999; Sato *et al.*, 1995). *Tie1* mutant mice also die as embryos, but at later stages and with reduced vessel integrity (Puri *et al.*, 1995; Sato *et al.*, 1995). Deletion of *Ang1* is also embryonic lethal, but the vasculature is less affected than in embryos lacking Tie2, suggesting that other ligands for Tie2 exist *in vivo* (Suri *et al.*, 1996). Ang2 has complex effects on vessel development that appear to be context dependent. That is to say, Ang2 can act as either an agonist or an antagonist of Tie2 signaling in different situations. Although most evidence suggests that Tie/tek signaling is involved in local patterning and remodeling, an interesting study suggests that it may impinge on global patterning. Loughna and Sato generated embryos that were deleted for both Ang1 and Tie1, the receptor without identified ligands (Loughna and Sato, 2001b). These double knockout mice had disruption of the venous system only on the righthand side of the embryo. This phenotype correlated with the expression of Ang1 on the right side veins at this time, although Tie1 is more generally expressed in vessels. These findings suggest that these pathways intersect to specify veins in a particular location of the embryo.

3. Notch Pathway

Notch proteins are large trans-membrane receptors that are important to a variety of developmental processes, including blood vessel formation (review: Iso *et al.*, 2003). Their ligands, Delta-like and Jagged proteins, are also largely membrane localized, leading to models of Notch signaling that involve neighboring cell types. Mice have four *Notch* genes, and deletion of *Notch1* or hypomorphism for *Notch2* in mice results in complex phenotypes with vascular abnormalities (McCright *et al.*, 2001; Swiatek *et al.*, 1994). Deletion of *Notch4* produces normal mice, but *Notch1–Notch4* double mutants have more severe vascular defects than *Notch1* mutant mice, suggesting functional overlap among Notch family members (Krebs *et al.*, 2000). Further evidence that Notch signaling is important in vascular patterning comes from analysis of transgenic mice that express an activated Notch4 in the vasculature (Uyttendaele *et al.*, 2001). The mutant embryos have numerous vascular patterning defects, including disorganized vessel networks and fewer smaller vessels. Mutations in *Notch3* in humans lead to a vascular defect associated with adult stroke and dementia called

3. Midline Vessel Patterning   65

CADASIL (Joutel et al., 1996; Salloway and Hong, 1998). Deletion of Notch ligands *Jag1* or *Dll1* in mice results in vascular defects and hemorrhage in the head and yolk sac and further supports a role for the Notch pathway in vessel patterning (Hrabe de Angelis et al., 1997; Xue et al., 1999). Thus far, the phenotypes generated in mice via targeted mutations are complex, and it is not completely clear what role Notch signaling plays in vascular patterning, although vessel morphogenesis appears to be affected by some genetic manipulations. It will be interesting to determine the phenotypes of Notch mutations localized to the vasculature.

Notch signaling is also important for arterial differentiation in zebrafish. Embryos lacking Notch activity fail to express artery-specific markers in the dorsal aorta, and mutation of a downstream target of Notch signaling, *gridlock*, results in defective dorsal aorta patterning (Lawson et al., 2001; Weinstein et al., 1995; Zhong et al., 2000, 2001). Further studies suggest a signaling cascade in which Shh can activate VEGF, which can in turn activate Notch signaling and arterial differentiation (Lawson et al., 2002). The zebrafish studies suggest that one role for Notch signaling in vascular development is to control cell fate decisions, a model consistent with how Notch signaling affects other developmental processes. However, the data in mice also suggest that Notch signaling is important, at least in vessel morphology, and it may be important in how neighboring tissues pattern vessels locally.

4. Ephrins/Eph Receptor Pathway

Eph receptor tyrosine kinases and their membrane-bound ligands, ephrins, are required for the proper placement of tissues developmentally. For example, neural crest cells that emanate from the neural tube migrate from the hindbrain in specific stripes that are defined by ephrin/Eph expression patterns (reviews: Adams, 2002; Himanen and Nikolov, 2003). In mice, ephrin-B2 is expressed by arterial endothelial cells, and its receptor, EphB4, is expressed predominantly on veins (Adams et al., 1999; Gerety et al., 1999; Wang et al., 1998). Deletion of either gene in mice results in defective angiogenic remodeling in veins and arteries of the yolk sac and head, with mid-gestational embryonic lethality. These data demonstrate a genetic component to the distinction between arteries and veins and suggest that disruption of the arterial/venous boundaries have severe consequences for vascular patterning. In many cases, cells that express Eph receptors and those that express ephrins are prevented from mixing, and endothelial cells clearly use this pathway (and likely others as well) to distinguish arteries from veins (review: Wilkinson, 2000). Although intersomitic vessels normally do not enter somites, the intersomitic vessels of the *ephrin-B2* null embryos often invade somites (Adams et al., 1999). The role of ephrins and their receptors in vascular development is likely to be complex, as other family members such as EphB3 and ephrin-B1 are also expressed in the vasculature. Moreover, some family members, such as EphB2, are also expressed in the mesenchyme and likely function in endothelial–mesenchymal cell signaling, which could affect vascular patterning (Adams et al., 1999). Gene targeting of ephrins in the vascular compartment will likely resolve some of these issues. However, current evidence clearly indicates that vessel patterns as well as vessel identity are influenced by ephrin-Eph receptor interactions, and further studies will likely uncover more important specific roles for these signals in vascular patterning.

5. Semaphorin/Neuropilin/Plexin Pathway

Semaphorins are another group of signaling molecules that have recently been implicated in the restriction of angioblast/endothelial cell migration. Semaphorins were first identified as guidance molecules for neurons, and the family includes both secreted and transmembrane signaling proteins (reviews: Goshima et al., 2002; Kolodkin, 1998; Tessier-Lavigne and Goodman 1996). Semaphorins bind to two different transmembrane-receptor families, neuropilin (NP) and plexin. It is now thought that NPs provide a ligand binding site and plexin proteins play a signaling role (Goshima et al., 2002). It is interesting that Sema3A binds NP1, which is also a coreceptor for the 165 isoform of VEGF-A expressed on endothelial cells (Soker et al., 1998). *In Vitro*, NP1/VEGF-A interactions appear to strengthen the ability of VEGF-A to promote chemotaxis of endothelial cells (Soker et al., 1998). In contrast, NP1/Sema3A interactions inhibit the motility of endothelial cells expressing NP1 (Miao et al., 1999). These findings suggest that differential or competitive binding of Sema3A or VEGF-A165 to NP1 might be involved in modulation of vascular patterning mediated by VEGF-A. NP2 is a second co-receptor that binds Sema3F with high affinity, and it is also expressed by endothelial cells and binds VEGF165 (Gluzman-Poltorak et al., 2000). A recent report showed that Sema3F effectively blocks tumor neoangiogenesis and interferes with VEGF signaling (Kessler et al., 2004). Thus, both Sema3A/NP1 and Sema3F/NP2 interactions may impact vascular patterning.

In Vivo experiments demonstrate the importance of this pathway and support a model whereby both NP coreceptors affect vascular patterning. Sema3A bead implantations into E4.5 avian forelimbs caused failure of blood vessel formation and formed vessels deviate away from the bead, while overexpression of Sema3A or Sema3F in the perineural vascular network of the head of chick embryos led to defects in vascular remodeling (Bates et al., 2003; Serini et al., 2003). In addition, Sema3A and Sema3C knockout mice both have cardiovascular defects (Feiner et al., 2001; Serini *et al.*, 2003). Overexpression or deletion of NP1 in mice leads to numerous vascular defects (Kawasaki *et al.*, 1999; Kitsukawa *et al.*, 1995), while deletion of NP2 does not (Chen *et al.*, 2000; Giger *et al.*, 2000). However, the combined deletion of NP1 and NP2 results in severe defects, resembling the *Vegfa* and *flk-1* knockouts (Takashima *et al.*, 2002). Recently, Kolodkin and Ginty and colleagues reported a set of elegant studies showing that VEGF/NP1 but not Sema/NP1 signaling was required for general vascular development (Gu *et al.*, 2003). To determine this, they first conditionally ablated NP1 in endothelial cells and found severe vascular defects. They then generated mice that expressed a mutant NP1 from the NP locus that could bind to VEGF but not to Sema, and bred it to the NP2-deficient background so that mutant NP1 was the sole NP. Vascular defects were not seen in these mice, showing that the vascular abnormalities were specific to the lack of VEGF/NP1 signaling in endothelial cells. Thus, the effects of the semaphorins on vascular development and endothelial function described above could result from competitive inhibition of VEGF/NP signaling, since although Semas and VEGF have distinct binding sites, it is thought that steric hindrance may prevent simultaneous binding.

Data from zebrafish also demonstrate an important role for this pathway. NP1 antisense morpholinos injected into zebrafish produce defects in the intersegmental vessels, and knockdown or ubiquitous expression of Sema3a1 impairs dorsal aorta formation (Lee *et al.*, 2002; Shoji *et al.*, 2003). Thus, the evidence to date implicates semaphorins in vascular patterning events, but exactly how and where semaphorins affect patterning remains to be elucidated.

IV. Axial Structures Implicated in Vascular Patterning

As mentioned in Section II, this review focuses on the axial midline structures implicated in vascular patterning, since these structures are presumably the source of vascular patterning signals described in Section III. Figure 1 shows a schematized cross section of an amphibian/zebrafish embryo (A) and an avian/mammalian embryo (B) to highlight the similarities and differences in the different model organisms. Both types of embryo have a dorsal neural tube and a notochord immediately ventral to the neural tube. Both also have a major axial artery(s) called the dorsal aorta, and a major axial vein(s) called the cardinal vein. However, in amphibians and zebrafish these structures are single vessels, while in avians and mammals the dorsal aorta initially forms as a set of paired vessels on either side of the midline (shown in Fig. 1), then fuses to a single vessel only in the mid-trunk region of the embryo. The cardinal veins form as paired vessels on either side of the midline and remain that way in avians and mammals. A second major difference is the formation

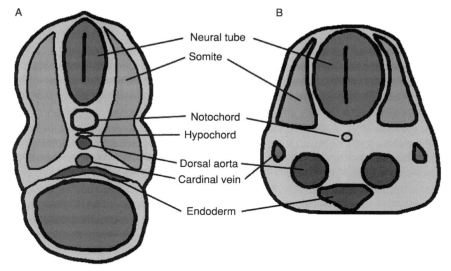

Figure 1 Schematic cross section through the trunk of midgestation vertebrate embryos at forelimb level. (A) Representative amphibian/zebrafish embryo. (B) Representative avian/mammalian embryo. (See Color Insert.)

of the hypochord in amphibians and zebrafish that is lacking in amniotes. Finally, all embryos have endoderm ventral to the other structures that will form the gut and other endodermally derived organs. We now discuss each of the major non-vascular axial midline structures.

A. Hypochord

The hypochord is a temporary, rodlike structure in amphibian and fish embryos located just ventral to the notochord and in close association with the dorsal aorta (Lofberg and Collazo, 1997) (see Fig. 1). The hypochord is thought to derive from the endoderm in frogs, but fate mapping shows it to be a mesodermal derivative in zebrafish (Cleaver et al., 2000; Latimer et al., 2002). Cleaver and Krieg demonstrated that the hypochord transiently expresses high levels of VEGF, and that this expression correlates with the formation of the dorsal aorta in *Xenopus* (Cleaver et al., 1997; Cleaver and Krieg, 1998). They showed that angioblasts originating in the lateral plate mesoderm migrate to the midline to form the dorsal aorta. Their data further suggests that a long-range diffusible form of VEGF (the 121 isoform) acts as a midline chemoattractant for the migrating angioblasts. It is unclear whether VEGF is expressed by the zebrafish hypochord (Eriksson and Lofberg, 2000; Liang *et al.*, 2001; Weinstein, 2002). However, fish hypochord expresses Ang1, a ligand of Tie/tek receptors that is important in vessel formation, and other genes such as *radar*, which are required for vascular integrity (Eriksson and Lofberg, 2000; Hall *et al.*, 2002; Pham *et al.*, 2001). Cleaver and Krieg suggest that the notochord may be responsible for induction of the hypochord that, in turn, patterns the dorsal aorta (Cleaver *et al.*, 2000). It should be possible to test this model in zebrafish, as mutants exist in which notochord is formed at the expense of hypochord, and vice versa (Latimer *et al.*, 2002). Regardless of how these structures affect axial vascular patterning in amphibians and fish, only these model organisms possess a hypochord, so another structure must guide dorsal aorta patterning in avians and mammals. This may be the endoderm.

B. Endoderm

There is evidence that endoderm guides the assembly of the axial vein in zebrafish. One-eyed pinhead (*oep*) mutant embryos lack several tissues, including most endoderm and endoderm derivatives (Hammerschmidt *et al.*, 1996; Schier *et al.*, 1996, 1997; Strahle *et al.*, 1997). It is interesting that these embryos only have a single vessel formed immediately ventral to the notochord, which is most likely the dorsal aorta (Brown *et al.*, 2000). Other zebrafish mutants with similar defects, yet with some endoderm (*sqt* and *cyc*), still retain axial veins. These data suggest that endoderm, which is found in close proximity to the axial vein, actually signals to migrating angioblasts to form the axial vein. Sonic hedgehog (Shh) is one candidate molecular mediator of the endoderm signal, since it is expressed in endoderm and is known to act as an induction/patterning signal at other embryonic sites (Krauss *et al.*, 1993; Strahle *et al.*, 1996). *Sonic-you* is a mutation in the zebrafish homolog of Shh, and these embryos resemble mice lacking Shh in many respects (Schauerte *et al.*, 1998). However, unlike $Shh^{-/-}$ mutant mice, *sonic-you* zebrafish embryos do not form either the dorsal aorta or the axial vein. Because the other zebrafish mutants that lack endoderm lack the axial vein but not the dorsal aorta, the lack of an axial vein in the *sonic-you* embryos is likely due to a loss of Shh expression specifically from the endoderm (Brown *et al.*, 2000).

The endoderm may also be important for dorsal aorta formation in avians and mammals. Historically, endoderm was implicated as the source of an angioblast induction signal. However, recently reported experiments show that removal of endoderm from frog or avian embryos does not alter specification of angioblasts, but angioblasts do not coalesce into tubes without endoderm (Vokes and Krieg, 2002). These data are consistent with a role for endoderm in dorsal aorta formation. Moreover, endoderm is a potent site of VEGF expression in avians and mammals (Aitkenhead et al., 1998; Dumont et al., 1995; Flamme et al., 1995; Miquerol et al., 1999). Thus, VEGF expression from the definitive endoderm in avians and mammals may be analogous to the strong VEGF midline expression from the hypochord in amphibians. In addition, *VegfA* mutant embryos have abnormal dorsal aorta formation (Carmeliet et al., 1996; Ferrara et al., 1996). However, since the expression data is correlative and the genetic experiments removed VEGF globally, targeted deletion of VEGF in the endoderm will be required to determine if endoderm-derived VEGF is critical to dorsal aorta formation in mammals.

C. Notochord

The notochord is a transient structure that lies ventral to the neural tube in all vertebrates, and it has a critical role in organizing the midline structures. Notochord signals induce differentiation of the floor plate and motor neurons of the neural tube, the sclerotome of the somites, and oligodendrocytes (Brand-Saberi et al., 1993; Fan et al., 1995; Halpern et al., 1993; Placzek et al., 1990; Trousse et al., 1995; Yamada et al., 1993). The notochord also patterns the endoderm (Cleaver et al., 2000). Despite these known patterning interactions, the role of the notochord in patterning the vasculature is poorly understood. In fact, there is data to support a role for the notochord as a source of both positive and negative vascular patterning signals.

1. The Notochord as the Source of Positive Vascular Patterning Signal(s)

One area of recent focus has been the role of notochord in assembly of the dorsal aorta, which comes to lie ventral to the notochord. In zebrafish, angioblasts assemble at the midline by the 14 somite stage to form the dorsal aorta, and other angioblasts begin forming the axial vein (also known as the posterior cardinal vein) ventral to the dorsal aorta by the 14–20 somite stage (Fouquet et al., 1997). Two zebrafish mutants have provided evidence that the notochord guides the initial assembly of angioblasts that will form the dorsal aorta at the midline. *Floating head* (*flh*) is a mutant that lacks notochord, has an un-patterned neural tube, and has fused somites (Halpern et al., 1995). *No tail* (*ntl*) is a mutant that retains some notochord precursors, a neural tube that consists of a wider than normal floor plate, and unfused but abnormally shaped somites (Halpern et al., 1993; Melby et al., 1996). Both mutants lack trunk circulation, and only a single vessel forms between the notochord and endoderm. This vessel is likely to be the axial vein based on its position in the *flh* mutants (Fouquet et al., 1997; Sumoy

3. Midline Vessel Patterning

et al., 1997). Mosaic embryos with both *flh* mutant cells and wild-type cells had patches of notochord and evidence of dorsal aorta assembly adjacent to these patches (Fouquet *et al.*, 1997). In sum, these findings strongly suggest that notochord signals are necessary for dorsal aorta formation, but do not rule out an indirect effect of the notochord in patterning other tissues that, in turn, relay directive signals to migrating angioblasts. For example, the *flh* mutants lack a hypochord as well as a notochord (Talbot *et al.*, 1995), and the hypochord seems critical to dorsal aorta development in amphibians (see Section IV.A). In addition, VEGF expression is missing in somites, and Sema3a1 is inappropriately expressed throughout the entire somite in *flh* mutants (Liang *et al.*, 2001; Shoji *et al.*, 1998). Reduced expression of either VEGF or Sema3a1 via morpholinos also leads to abnormalities in dorsal aorta development (Nasevicius *et al.*, 2000; Shoji *et al.*, 2003). More studies will be needed to determine whether the notochord acts directly or indirectly to pattern the angioblasts that form the dorsal aorta.

Our lab has recently examined the vasculature in mice mutant for *Brachyury*, which is a T-box gene that is the mammalian homologue of zebrafish *ntl*. The *Brachyury* mutant embryos (*T/T*) do not form a mature notochord in the posterior half of the embryo. It is interesting that the dorsal aorta forms throughout the A–P axis of these mutant embryos, although with more perturbations in the size and shape in the posterior of the embryo (Jenkins, T., K.A.H. and V.L.B., unpublished results). Thus, it appears that the mouse differs from the zebrafish in the requirement for the notochord for dorsal aorta assembly. These differences may reflect differences in the midline structure that is the source of important vascular patterning signals. Specifically, the endoderm in amniotes may provide the initial patterning signals for the dorsal aorta, and this role may be carried out by the notochord/hypochord in the amphibian and zebrafish.

The identity of notochord signals important in dorsal aorta formation is unknown. Both *flh* and *ntl* encode transcription factors, so their effects are likely to be indirect. Sonic hedgehog is the best-characterized signal emanating from the notochord, making it a prime candidate in the search for a notochord signal responsible for dorsal aorta assembly. In support of this hypothesis, the sonic hedgehog mutant, *sonic-you*, lacks both the dorsal aorta and the axial vein (Brown *et al.*, 2000). It is likely that lack of notochord-derived Shh is upstream of the dorsal aorta defect, while endoderm-derived Shh is upstream of the axial vein defect. These studies, once again, do not rule out an indirect effect of Shh signaling via other cell types that are no longer patterned when Shh is missing. Another member of the T-box family of transcription factors, *hrT*, is expressed in the dorsal aorta and heart, and mutants lacking *hrT* resemble *flh* mutants in that no dorsal aorta forms (Ahn *et al.*, 2000; Griffin *et al.*, 2000; Szeto *et al.*, 2002). The zebrafish hedgehog signaling mutant *you-too* lacks hrT expression, suggesting that *hrT* is a downstream effector of hedgehog signaling (Szeto *et al.*, 2002). Additional studies in zebrafish demonstrate that Shh has a role in arterial identity acting upstream of both the VEGF and the Notch pathways (Lawson *et al.*, 2002).

The effects of Shh on vascular development in mammalian models appear to be indirect. Transgenic overexpression of Shh in the dorsal neural tube of mice results in hypervascularization of the tissue (Rowitch *et al.*, 1999). Shh administration to aged mice induces neovascularization in an ischemic hind limb model, and the same treatment induces angiogenesis with large branching vessels in the murine cornea (Pola *et al.*, 2001). However, direct treatment of human umbilical vein endothelial vein cells (HUVECs) with Shh had no effect, although fibroblast cells responded to Shh by upregulating VEGF and angiopoietins (Kanda *et al.*, 2003; Pola *et al.*, 2001). Thus, it was hypothesized that Shh upregulates VEGF in non-vascular cells such as fibroblasts, and this upregulation in turn positively affects blood vessel formation. Nonetheless, Shh can induce capillary morphogenesis of HUVECs in Matrigel through the activation of PI-3 kinase (Kanda *et al.*, 2003).

2. The Notochord as the Source of Negative Vascular Patterning Signal(s)

In mouse and avian embryos, the notochord is associated with negative patterning signals in two ways. First, early in gastrulation, the embryonic midline is set up, and subsequently, angioblasts/endothelial cells are largely prevented from crossing the midline to colonize the contralateral side of the embryo. Second, surrounding the notochord is an avascular zone that presents a striking contrast to most other embryonic tissues and structures that are highly vascularized (Fig. 2A).

The midline barrier to angioblast–endothelial cell crossing was demonstrated experimentally through the use of quail chick chimeras. Somite and presomitic mesoderm grafts produced vascular cells that migrated cranially, caudally, and laterally, but never crossed the axial midline (Pardanaud *et al.*, 1996; Wilting *et al.*, 1995). However, removal of the notochord resulted in extensive midline crossing, showing that the notochord was required to maintain the midline barrier (Klessinger and Christ, 1996). Using mouse–avian chimeras, we demonstrated that mouse angioblast/endothelial cells also respect the midline barrier in the avian host (Ambler *et al.*, 2001). The molecular nature of this midline barrier is unknown.

The avascular zone surrounding the notochord has been observed but not experimentally manipulated. Initially, the notochord closely abuts both the neural tube dorsally and the dorsal aorta ventrally. However, the notochord soon becomes surrounded by cells, as the sclerotome cells of

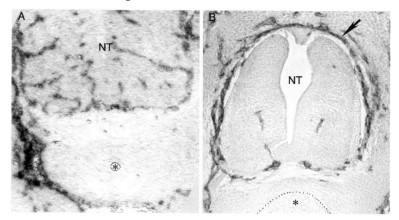

Figure 2 Vascular patterning at the embryonic midline of amniotes. (A) The perinotochordal area is avascular as visualized by *in situ* hybridization for flt-1 mRNA in an E11.5 mouse embryo. The notochord is denoted by an asterisk and surrounded by dotted lines, and the vessels are stained purple. (B) The peri-neural vascular plexus (PNVP) (arrow) surrounds the neural tube (NT) as shown in this HH stage 24 quail embryo. The vessels are reacted with QH1 antibody and appear brown. (See Color Insert.)

the somites migrate around the notochord to form the peri-notochordal mesenchyme (review: Christ *et al.*, 2000). As the intersomitic vessels sprout from the dorsal aorta and migrate dorsally, they hug but do not invade the avascular zone. We have begun to investigate the avascular zone of the peri-notochordal mesenchyme, and preliminary experiments suggest that the avascular zone requires the notochord, because surgical placement of the notochord into somites resulted in the formation of an avascular zone around the ectopic notochord (Ambler, C. A., and V.L.B., unpublished results).

How do we reconcile the data showing that the notochord positively affects dorsal aorta assembly but prevents midline crossing and sets up an avascular zone? First, it is quite possible that the notochord has different midline effects depending on the organism being studied. To date, the notochord effects on dorsal aorta assembly are exclusive to amphibians and zebrafish, and in these models the distinction between notochord and hypochord effects is not completely clear. There may also be stage-specific effects of the notochord on vessel development. For example, initially, the notochord may impart positive patterning signals to recruit the dorsal aorta, and at later stages the notochord may up regulate a negative vascular patterning signal. Finally, it is quite likely that some or even all of the effects may be indirect, that is to say, that notochord-derived signals may induce expression of distinct sets of genes in different target cells. These notochord-activated genes may encode proteins that mediate either positive or repulsive vascular patterning signals. Indeed, the tissues neighboring the notochord change dynamically during mid-gestation, suggesting a model whereby both positive and negative vascular patterning effects could result from the notochord-derived signals at different times as a result of indirect effects on adjacent cells. It will be interesting to dissect the role of the notochord in vascular patterning more completely.

D. Neural Tube

While the dorsal aorta is the first vessel formed in the embryo by midline signals, other vessels are also patterned at the midline at later stages. One such patterning event is the development of the perineural vascular plexus (PNVP), which comes to surround the neural tube (Fig. 2B). This plexus provides essential nutrients and oxygen to the developing neural tissue, and it is the source of vascular sprouts that subsequently invade and metabolically support the neural tissue. These PNVP-derived vessels go on to form the blood–brain barrier that is critical to proper CNS function in the adult (Bar, 1980; Bauer *et al.*, 1993; Risau and Wolburg, 1990). Although the invasion of angiogenic sprouts into neural tissue has been described, the developmental processes that result in the formation of the PNVP at the proper location have not been investigated. Noden noted from his quail chick chimera studies that the neural tube itself does not have vascular potential, and thus it must become vascularized by exogenous endothelial precursors (Noden, 1989). Both quail-chick and mouse–quail chimera analysis identified somite-derived precursor cells as an important source of endothelial cells that comprise the PNVP (Ambler *et al.*, 2001; Klessinger and Christ, 1996; Pardanaud *et al.*, 1996; Pardanaud and Dieterlen-Lievre, 1999; Wilting *et al.*, 1995). However, the source and nature of the signal(s) that act on somite-derived angioblasts to pattern the PNVP have only recently been investigated.

To formally prove that the neural tube is the source of a vascular patterning signal, our group placed mouse neural tubes ectopically in avian hosts and showed that a vascular plexus forms around the neural tube regardless of its position (Fig. 3A) (Hogan *et al.*, 2004). We then did a graft experiment in which a buffer of avian tissue separated the mouse graft (presomitic mesoderm) from the avian host neural tube. The finding that graft-derived endothelial cells were found in the PNVP indicated that the neural tube signal could act at a distance to recruit endothelial cells to the PNVP. We next tested the role of VEGF-A in neural tube patterning through the use of a novel explant assay in which vessels form from presomitic mesoderm grafts in a VEGF-dependent manner. The neural tube

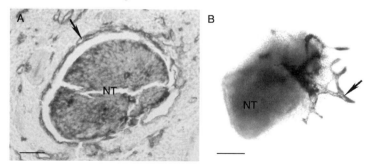

Figure 3 The neural tube directs PNVP formation. (A) Transverse section through the trunk of an HH stage 24 quail embryo containing a grafted *ROSA* +/− mouse neural tube, 3 days postsurgery. Chimeric embryos were whole-mount stained for β-galactosidase (blue), then sectioned and reacted with QH1 antibody (brown). The arrow points to a host-derived vascular plexus surrounding the grafted neural tube. NC, notochord; NT, neural tube. (B) *Flt1* +/− mouse presomitic mesoderm grafts from E8.5 mouse embryos were placed beside stage HH 10–13 quail neural tubes in collagen gels and cultured for 72 h in basal medium. Mouse vascular cells were visualized by β-galactosidase staining (blue). Arrow indicates vascular plexus. Scale bars are 50 μm in A and 200 μm in B. (See Color Insert.)

could replace the requirement for VEGF in this system (Fig. 3B), and both pharmacological inhibition and genetic ablation of VEGF-A signaling showed that this pathway is required for the neural tube to pattern vessels from the presomitic mesoderm (Hogan *et al.*, 2004). Thus, not only is VEGF-A implicated in the first midline patterning events that lead to dorsal aorta formation, it is also important in later midline patterning mediated by the neural tube.

V. Conclusions and Future Directions

As with many endeavors, the journey to understand the mechanisms that underlie vascular patterning events has provided more questions than answers so far. The initial hypothesis that angioblasts respond to extrinsic vascular patterning cues to migrate and assemble in the embryo has been solidified by numerous experiments. Clearly, the axial midline of the vertebrate embryo acts to organize the developing vasculature, as it does other structures and tissues in the organism. Moreover, it does so in complex ways. There are differences in time; for example, at early times the hypochord and endoderm appear critical to patterning the dorsal aorta, while at later times the neural tube provides positive vascular patterning cues to promote the migration and assembly of vessels. There are differences in which structures are responsible for certain effects. For example, the hypochord of amphibians and zebrafish is missing in avians and mammals, and the endoderm may provide similar patterning cues in the latter embryos. There are also differences in how the structures themselves affect patterning. The notochord seems to act as a source of positive vascular patterning cues in zebrafish that are critical to dorsal aorta formation, yet the notochord of amniotes prevents midline crossing of angioblasts and is surrounded by an extensive avascular zone. It is quite likely that further studies using the sophisticated molecular and genetic tools available will lead to a much more refined model of vascular patterning around the vertebrate midline.

There is no doubt that the VEGF signaling pathway presents an important nexus for vascular patterning events, as it is involved in almost every aspect of endothelial cell formation and function. To date, all clearly defined axial patterning events require VEGF signaling—the initial formation of the dorsal aorta, the specification of this vessel in zebrafish, and the formation of the plexus surrounding the neural tube in mouse and avian embryos. However, other signals and receptors are clearly important in vascular patterning, and it is quite likely that soon more studies will show defined roles for the Tie/tek, Notch, Eph/ephrin, and Semaphorin pathways in vascular patterning at the midline. The near future will likely see the development of models that effectively incorporate all of these molecular inputs to pattern vessels. The future will also see the integration of these signaling pathways with upstream inputs and downstream targets, such as transcription factors and cytoskeletal proteins.

We have also just begun to use the power of genetics to uncover novel pathways and genes involved in vascular patterning. In this realm, the zebrafish is clearly a star, as it is possible to perform forward genetic screens for novel genes. An interesting group of genes that affect vessel patterning, such as *out-of-bounds*, is sure to become larger with time and expand our knowledge of the molecular control of vascular patterning beyond the "usual suspects."

The future is very bright. The patterning of blood vessels presents a basic question of developmental biology: As embryos develop and form diverse cell types, how do these distinct groups of cells communicate with each other and influence each other's behavior? These processes have intrigued developmental biologists for hundreds of years, and we now have available the tools to address the mechanisms underlying vascular patterning events. In addition, understanding vascular patterning is relevant to potential therapeutic applications as never before. We now dream of inducing new vessels in the human body where and when we need them, and we even hope to be able to produce vessels *in vitro* for placement in the human body. Both of these clinical applications will move forward more quickly as we increase our knowledge of how the body regulates and patterns vessels.

Acknowledgments

We thank members of the Bautch Lab for fruitful discussion, and we apologize to colleagues whose work we could not cite due to space constraints. We thank Susan Whitfield for photography. This work was supported by National Institutes of Health grants R01 HL 43174 and R21 HL 71993 (V.L.B.) and AHA 0120572U and NIH F32 HL68484 (K.A.H).

References

Adams, R. H. (2002). Vascular patterning by Eph receptor tyrosine kinases and ephrins. *Semin. Cell Dev. Biol.* **13,** 55–60.

Adams, R. H., Wilkinson, G. A., Weiss, C., Diella, F., Gale, N. W., Deutsch, U., Risau, W., and Klein, R. (1999). Roles of ephrinB ligands and EphB receptors in cardiovascular development: demarcation of arterial/venous domains, vascular morphogenesis, and sprouting angiogenesis. *Genes Dev.* **13,** 295–306.

Ahn, D. G., Ruvinsky, I., Oates, A. C., Silver, L. M., and Ho, R. K. (2000). tbx20, a new vertebrate T-box gene expressed in the cranial motor neurons and developing cardiovascular structures in zebrafish. *Mech. Dev.* **95,** 253–258.

Aitkenhead, M., Christ, B., Eichmann, A., Feucht, M., Wilson, D. J., and Wilting, J. (1998). Paracrine and autocrine regulation of vascular endothelial growth factor during tissue differentiation in the quail. *Dev. Dyn.* **212,** 1–13.

Ambler, C. A., Nowicki, J. L., Burke, A. C., and Bautch, V. L. (2001). Assembly of trunk and limb blood vessels involves extensive migration and vasculogenesis of somite-derived angioblasts. *Dev. Biol.* **234,** 352–364.

Ambler, C. A., Schmunk, G. A., and Bautch, V. L. (2003). Stem cell–derived endothelial cells/progenitors migrate and pattern in the embryo using the VEGF signaling pathway. *Dev. Biol.* **257,** 205–219.

Asahara, T., Masuda, H., Takahashi, T., Kalka, C., Pastore, C., Silver, M., Kearne, M., Magner, M., and Isner, J. M. (1999). Bone marrow origin of endothelial progenitor cells responsible for postnatal vasculogenesis in physiological and pathological neovascularization. *Circ. Res.* **85,** 221–228.

Asahara, T., Murohara, T., Sullivan, A., Silver, M., van der Zee, R., Li, T., Witzenbichler, B., Schatteman, G., and Isner, J. M. (1997). Isolation of putative progenitor endothelial cells for angiogenesis. *Science* **275,** 964–967.

Bar, T. (1980). The vascular system of the cerebral cortex. *Adv. Anat. Embryol. Cell Biol.* **59,** 1–62.

Bates, D., Taylor, G. I., Minichiello, J., Farlie, P., Cichowitz, A., Watson, N., Klagsbrun, M., Mamluk, R., and Newgreen, D. F. (2003). Neurovascular congruence results from a shared patterning mechanism that utilizes Semaphorin3A and Neuropilin-1. *Dev. Biol.* **255,** 77–98.

Bauer, H. C., Bauer, H., Lametschwandtner, A., Amberger, A., Ruiz, P., and Steiner, M. (1993). Neovascularization and the appearance of morphological characteristics of the blood–brain barrier in the embryonic mouse central nervous system. *Brain Res. Dev. Brain Res.* **75,** 269–278.

Bautch, V. L., Redick, S. D., Scalia, A., Harmaty, M., Carmeliet, P., and Rapoport, R. (2000). Characterization of the vasculogenic block in the absence of vascular endothelial growth factor-A. *Blood* **95,** 1979–1987.

Brand-Saberi, B., Ebensperger, C., Wilting, J., Balling, R., and Christ, B. (1993). The ventralizing effect of the notochord on somite differentiation in chick embryos. *Anat. Embryol.* **188,** 239–245.

Brand-Saberi, B., Seifert, R., Grim, M., Wilting, J., Kuhlewein, M., and Christ, B. (1995). Blood vessel formation in the avian limb bud involves angioblastic and angiotrophic growth. *Dev. Dyn.* **202**, 181–194.

Brown, L. A., Rodaway, A. R., Schilling, T. F., Jowett, T., Ingham, P. W., Patient, R. K., and Sharrocks, A. D. (2000). Insights into early vasculogenesis revealed by expression of the ETS-domain transcription factor Fli-1 in wild-type and mutant zebrafish embryos. *Mech. Dev.* **90**, 237–252.

Carmeliet, P., Ferreira, V., Breier, G., Pollefeyt, S., Kieckens, L., Gertsenstein, M., Fahrig, M., Vandenhoeck, A., Harpal, K., Eberhardt, C., Declercq, C., Pawling, J., Moons, L., Collen, D., Risau, W., and Nagy, A. (1996). Abnormal blood vessel development and lethality in embryos lacking a single VEGF allele. *Nature* **380**, 435–439.

Carmeliet, P., Ng, Y. S., Nuyens, D., Theilmeier, G., Brusselmans, K., Cornelissen, I., Ehler, E., Kakkar, V. V., Stalmans, I., Mattot, V., Perriard, J. C., Dewerchin, M., Flameng, W., Nagy, A., Lupu, F., Moons, L., Collen, D., D'Amore, P. A., and Shima, D. T. (1999). Impaired myocardial angiogenesis and ischemic cardiomyopathy in mice lacking the vascular endothelial growth factor isoforms VEGF164 and VEGF188. *Nat. Med.* **5**, 495–502.

Chen, H., Bagri, A., Zupicich, J. A., Zou, Y., Stoeckli, E., Pleasure, S. J., Lowenstein, D. H., Skarnes, W. C., Chedotal, A., and Tessier-Lavigne, M. (2000). Neuropilin-2 regulates the development of selective cranial and sensory nerves and hippocampal mossy fiber projections. *Neuron* **25**, 43–56.

Childs, S., Chen, J. N., Garrity, D. M., and Fishman, M. C. (2002). Patterning of angiogenesis in the zebrafish embryo. *Development* **129**, 973–982.

Christ, B., Huang, R., and Wilting, J. (2000). The development of the avian vertebral column. *Anat. Embryol.* **202**, 179–194.

Clark, E. R. (1918). Studies on the growth of blood vessels in the tail of the frog larvae. *Am. J. Anat.* **23**, 37–88.

Clark, E. R., and Clark, E. L. (1939). Microscopic observations on the growth of blood capillaries in the living mammal. *Am. J. Anat.* **64**, 251–299.

Cleaver, O., and Krieg, P. (1998). VEGF mediates angioblast migration during development of the dorsal aorta in *Xenopus*. *Development* **125**, 3905–3914.

Cleaver, O., and Krieg, P. A. (1999). Molecular mechanisms of vascular development. *In* "Heart Development" (R. P. Harvey and N. Rosenthal, Eds.), pp. 221–252. Academic Press, New York.

Cleaver, O., and Melton, D. A. (2003). Endothelial signaling during development. *Nat. Med.* **9**, 661–668.

Cleaver, O., Seufert, D. W., and Krieg, P. A. (2000). Endoderm patterning by the notochord: development of the hypochord in *Xenopus*. *Development* **125**, 869–879.

Cleaver, O., Tonissen, K. F., Saha, M. S., and Krieg, P. A. (1997). Neovascularization of the *Xenopus* embryo. *Dev. Dyn.* **210**, 66–77.

Coffin, J. D., and Poole, T. J. (1988). Embryonic vascular development: Immunohistochemical identification of the origin and subsequent morphogenesis of the major vessel primordia in quail embryos. *Development* **102**, 735–748.

Coffin, J. D., and Poole, T. J. (1991). Endothelial cell origin and migration in embryonic heart and cranial blood vessel development. *Anat. Rec.* **231**, 383–395.

Cox, C. M., and Poole, T. J. (2001). Angioblast differentiation is influenced by the local environment: FGF-2 induces angioblasts and patterns vessel formation in the quail embryo. *Dev. Dyn.* **218**, 371–382.

Cross, M. J., Dixelius, J., Matsumoto, T., and Claesson-Welsh, L. (2003). VEGF-receptor signal transduction. *Trends. Biochem. Sci.* **28**, 488–494.

Damert, A., Miquerol, L., Gertsenstein, M., Risau, W., and Nagy, A. (2002). Insufficient VEGFA activity in yolk sac endoderm compromises haematopoietic and endothelial differentiation. *Development* **129**, 1881–1892.

Daniel, T. O., and Abrahamson, D. (2000). Endothelial signal integration in vascular assembly. *Ann. Rev. Physiol.* **62,** 649–671.

Drake, C. J., Hungerford, J. E., and Little, C. D. (1998). Morphogenesis of the first blood vessels. *In* "Morphogenesis: Cellular Interactions" (R. Fleischmajer, R. Timpl, and Z. Werb, Eds.), Vol. 857, pp. 155–180. New York Academy of Sciences, New York.

Drake, C. J., and Little, C. D. (1995). Exogenous vascular endothelial growth factor induces malformed and hyperfused vessels during embryonic neovascularization. *Proc. Natl. Acad. Sci. USA* **92,** 7657–7661.

Drake, C. J., and Little, C. D. (1999). VEGF and vascular fusion: Implications for normal and pathological vessels. *J. Histochem. Cytochem.* **47,** 1351–1355.

Drake, C. J., LaRue, A., Ferrara, N., and Little, C. D. (2000). VEGF regulates cell behavior during vasculogenesis. *Dev. Biol.* **224,** 178–188.

Dumont, D. J., Gradwohl, G., Fong, G. H., Puri, M. C., Gertsenstein, M., Auerbach, A., and Breitman, M. L. (1994). Dominant-negative and targeted null mutations in the endothelial receptor tyrosine kinase, tek, reveal a critical role in vasculogenesis of the embryo. *Genes Dev.* **8,** 1897–1909.

Dumont, D. J., Fong, G. H., Puri, M. C., Gradwohl, G., Alitalo, K., and Breitman, M. L. (1995). Vascularization of the mouse embryo: A study of flk-1, tek, tie, and vascular endothelial growth factor expression during development. *Dev. Dyn.* **203,** 80–92.

Eriksson, J., and Lofberg, J. (2000). Development of the hypochord and dorsal aorta in the zebrafish embryo (Danio rerio). *J. Morphol.* **244,** 167–176.

Evans, H. M. (1909). On the development of the aortae, cardinal and umbilical veins, and other blood vessels of the vertebrate embryos from vertebrates. *Anat. Rec.* **3,** 498–518.

Fan, C. M., Porter, J. A., Chiang, C., Chang, D. T., Beachy, P. A., and Tessier-Lavigne, M. (1995). Long-range sclerotome induction by sonic hedgehog: Direct role of the amino-terminal cleavage product and modulation by the cyclic AMP signaling pathway. *Cell* **81,** 457–465.

Feinberg, R. N., and Noden, D. M. (1991). Experimental analysis of blood vessel development in the avian wing bud. *Anat. Rec.* **231,** 136–144.

Feiner, L., Webber, A. L., Brown, C. B., Lu, M. M., Jia, L., Feinstein, P., Mombaerts, P., Epstein, J. A., and Raper, J. A. (2001). Targeted disruption of semaphorin 3C leads to persistent truncus arteriosus and aortic arch interruption. *Development* **128,** 3061–3070.

Ferrara, N., Carver-Moore, K., Chen, H., Dowd, M., Lu, L., O'Shea, K. S., Powell-Braxton, L., Hillan, K. J., and Moore, M. W. (1996). Heterozygous embryonic lethality induced by targeted inactivation of the VEGF gene. *Nature* **380,** 439–442.

Ferrara, N., Gerber, H. P., and LeCouter, J. (2003). The biology of VEGF and its receptors. *Nat. Med.* **9,** 669–676.

Finkelstein, E. B., and Poole, T. J. (2003). Vascular endothelial growth factor: A regulator of vascular morphogenesis in the Japanese quail embryo. *Anat. Rec.* **272A,** 403–414.

Flamme, I., Breier, G., and Risau, W. (1995). Vascular endothelial growth factor (VEGF) and VEGF receptor 2 (flk-1) are expressed during vasculogenesis and vascular differentiation in the quail embryo. *Dev. Biol.* **169,** 699–712.

Flamme, I., von Reutern, M., Drexler, H. C., Syed-Ali, S., and Risau, W. (1995). Overexpression of vascular endothelial growth factor in the avian embryo induces hypervascularization and increased vascular permeability without alterations of embryonic pattern formation. *Dev. Biol.* **171,** 399–414.

Folkman, J. (2003). Fundamental concepts of the angiogenic process. *Curr. Mol. Med.* **3,** 643–651.

Fong, G. H., Rossant, J., Gertsenstein, M., and Breitman, M. (1995). Role of the Flt-1 receptor tyrosine kinase in regulating the assembly of vascular endothelium. *Nature* **376,** 66–70.

Fong, G. H., Zhang, L., Bryce, D. M., and Peng, J. (1999). Increased hemangioblast commitment, not vascular disorganization, is the primary defect in flt-1 knock-out mice. *Development* **126,** 3015–3025.

Fouquet, B., Weinstein, B. M., Serluca, F. C., and Fishman, M. C. (1997). Vessel patterning in the embryo of the zebrafish: Guidance by notochord. *Dev. Biol.* **183**, 37–48.

Gerety, S. S., Wang, H. U., Chen, Z. F., and Anderson, D. J. (1999). Symmetrical mutant phenotypes of the receptor EphB4 and its specific transmembrane ligand ephrin-B2 in cardiovascular development. *Mol. Cell* **4**, 403–414.

Gerhardt, H., Golding, M., Fruttiger, M., Ruhrberg, C., Lundkvist, A., Abramsson, A., Jeltsch, M., Mitchell, C., Alitalo, K., Shima, D., and Betsholtz, C. (2003). VEGF guides angiogenic sprouting utilizing endothelial tip cell filopodia. *J. Cell Biol.* **161**, 1163–1177.

Ghabrial, A., Luschnig, S., Metzstein, M. M., and Krasnow, M. A. (2003). Branching morphogenesis of the *Drosophila* tracheal system. *Ann. Rev. Cell Dev. Biol.* **19**, 623–647.

Giger, R. J., Cloutier, J. F., Sahay, A., Prinjha, R. K., Levengood, D. V., Moore, S. E., Pickering, S., Simmons, D., Rastan, S., Walsh, F. S., Kolodkin, A. L., Ginty, D. D., and Geppert, M. (2000). Neuropilin-2 is required *in vivo* for selective axon guidance responses to secreted semaphorins. *Neuron* **25**, 29–41.

Gilbert, S. F. (2003). "Developmental Biology." Sinauer, Sunderland, MA.

Gluzman-Poltorak, Z., Cohen, T., Herzog, Y., and Neufeld, G. (2000). Neuropilin-2 is a receptor for the vascular endothelial growth factor (VEGF) forms VEGF-145 and VEGF-165 [corrected]. *J. Biol. Chem.* **275**, 18040–18045.

Goshima, Y., Ito, T., Sasaki, Y., and Nakamura, F. (2002). Semaphorins as signals for cell repulsion and invasion. *J. Clin. Invest.* **109**, 993–998.

Griffin, K. J., Stoller, J., Gibson, M., Chen, S., Yelon, D., Stainier, D. Y., and Kimelman, D. (2000). A conserved role for H15-related T-box transcription factors in zebrafish and *Drosophila* heart formation. *Dev. Biol.* **218**, 235–247.

Gu, C., Rodriguez, E. R., Reimert, D. V., Shu, T., Fritzsch, B., Richards, L. J., Kolodkin, A. L., and Ginty, D. D. (2003). Neuropilin-1 conveys semaphorin and VEGF signaling during neural and cardiovascular development. *Dev. Cell* **5**, 45–57.

Habeck, H., Odenthal, J., Walderich, B., Maischein, H., and Schulte-Merker, S. (2002). Analysis of a zebrafish VEGF receptor mutant reveals specific disruption of angiogenesis. *Curr. Biol.* **12**, 1405–1412.

Hall, C. J., Flores, M. V. C., Davidson, A. J., Crosier, K. E., and Crosier, P. S. (2002). Radar is required for the establishment of vascular integrity in the zebrafish. *Dev. Biol.* **251**, 105–117.

Halpern, M. E., Ho, R. K., Walker, C., and Kimmel, C. B. (1993). Induction of muscle pioneers and floor plate is distinguished by the zebrafish no tail mutation. *Cell* **75**, 99–111.

Halpern, M. E., Thisse, C., Ho, R. K., Thisse, B., Riggleman, B., Trevarrow, B., Weinberg, E. S., Postlethwait, J. H., and Kimmel, C. B. (1995). Cell-autonomous shift from axial to paraxial mesodermal development in zebrafish floating head mutants. *Development* **121**, 4257–4264.

Hammerschmidt, M., Pelegri, F., Mullins, M. C., Kane, D. A., Brand, M., van Eeden, F. J., Furutani-Seiki, M., Granato, M., Haffter, P., Heisenberg, C. P., Jiang, Y. J., Kelsh, R. N., Odenthal, J., Warga, R. M., and Nusslein-Volhard, C. (1996). Mutations affecting morphogenesis during gastrulation and tail formation in the zebrafish, Danio rerio. *Development* **123**, 143–151.

Himanen, J. P., and Nikolov, D. B. (2003). Eph receptors and ephrins. *Int. J. Biochem. Cell Biol.* **35**, 130–134.

Hogan, K. A., Ambler, C. A., Chapman, D. L., Bautch, V. L. (2004). The neural tube patterns vessels developmentally using the VEGF signaling pathway. *Development* **131**, 1503–1513.

Hrabe de Angelis, M., McIntyre II, J., and Gossler, A. (1997). Maintenance of somite borders in mice requires the *Delta* homologue *Dll1*. *Nature* **386**, 717–721.

Iso, T., Hamamori, Y., and Kedes, L. (2003). Notch signaling in vascular development. *Arterioscler. Thromb. Vasc. Biol.* **23**, 543–553.

Isogai, S., Horiguchi, M., and Weinstein, B. M. (2001). The vascular anatomy of the developing zebrafish: an atlas of embryonic and early larval development. *Dev. Biol.* **230**, 278–301.

Jain, R. K. (2003). Molecular regulation of vessel maturation. *Nat. Med.* **9**, 685–693.
Joutel, A., Corpechot, C., Ducros, A., Vahedi, K., Chabriat, H., Mouton, P., Alamowitch, S., Domenga, V., Cecillion, M., Marechal, E., Maciazek, J., Vayssiere, C., Cruaud, C., Cabanis, E. A., Ruchoux, M. M., Weissenbach, J., Bach, J. F., Bousser, M. G., and Tournier-Lasserve, E. (1996). Notch3 mutations in CADASIL, a hereditary adult-onset condition causing stroke and dementia. *Nature* **383**, 707–710.
Kanda, S., Mochizuki, Y., Suematsu, T., Miyata, Y., Nomata, K., and Kanetake, H. (2003). Sonic hedgehog induces capillary morphogenesis by endothelial cells through phosphoinositide 3-kinase. *J. Biol. Chem.* **278**, 8244–8249.
Kawasaki, T., Kitsukawa, T., Bekku, Y., Matsuda, Y., Sanbo, M., Yagi, T., and Fujisawa, H. (1999). A requirement for neuropilin-1 in embryonic vessel formation. *Development* **126**, 4895–4902.
Kearney, J. B., Ambler, C. A., Monaco, K. A., Johnson, N., Rapoport, R. G., and Bautch, V. L. (2002). Vascular endothelial growth factor receptor Flt-1 negatively regulates developmental blood vessel formation by modulating endothelial cell division. *Blood* **99**, 2397–2407.
Kearney, J. B., Kappas, N. C., Ellerstrom, C., DiPaola, F. W., Bautch, V. L. (2004). The VEGF receptor flt-1 (VEGFR-1) is a positive modulator of vascular sprout formation and branching morphogenesis. *Blood* **103**, 4527–4535.
Kendall, R. L., and Thomas, K. A. (1993). Inhibition of vascular endothelial cell growth factor activity by an endogenously encoded soluble receptor. *Proc. Natl. Acad. Sci. USA* **90**, 10705–10709.
Kendall, R. L., Wang, G., and Thomas, K. A. (1996). Identification of a natural soluble form of the vascular endothelial growth factor receptor. FLT-1, and its heterodimerization with KDR. *Biochem. Biophys. Res. Comm.* **226**, 324–328.
Kessler, O., Shraga-Heled, N., Lange, T., Gutmann-Raviv, N., Sabo, E., Baruch, L., Machluf, M., and Neufeld, G. (2004). Semphorin-3F is an inhibitor of tumor angiogenesis. *Cancer Res.* **64**, 1008–1015.
Kitsukawa, T., Shimono, A., Kawakami, A., Kondoh, H., and Fujisawa, H. (1995). Overexpression of a membrane protein, neuropilin, in chimeric mice causes anomalies in the cardiovascular system, nervous system and limbs. *Development* **121**, 4309–4318.
Klessinger, S., and Christ, B. (1996). Axial structures control laterality in the distribution pattern of endothelial cells. *Anat. Embryol.* **193**, 319–330.
Kolodkin, A. L. (1998). Semaphorin-mediated neuronal growth cone guidance. *Prog. Brain Res.* **117**, 115–132.
Krauss, S., Concordet, J. P., and Ingham, P. W. (1993). A functionally conserved homolog of the *Drosophila* segment polarity gene hh is expressed in tissues with polarizing activity in zebrafish embryos. *Cell* **75**, 1431–1444.
Krebs, L. T., Xue, Y., Norton, C. R., Shutter, J. R., Maguire, M., Sundberg, J. P., Gallahan, D., Closson, V., Kitajewski, J., Callahan, R., Smith, G. H., Stark, K. L., and Gridley, T. (2000). Notch signaling is essential for vascular morphogenesis in mice. *Genes Dev.* **14**, 1343–1352.
Latimer, A. J., Dong, X., Markov, Y., and Appel, B. (2002). Delta-Notch signaling induces hypochord development in zebrafish. *Development* **129**, 2555–2563.
Lawson, N. D., Scheer, N., Pham, V. N., Kim, C. H., Chitnis, A. B., Campos-Ortega, J. A., and Weinstein, B. M. (2001). Notch signaling is required for arterial-venous differentiation during embryonic vascular development. *Development* **128**, 3675–3683.
Lawson, N. D., Vogel, A. M., and Weinstein, B. M. (2002). Sonic hedgehog and vascular endothelial growth factor act upstream of the notch pathway during arterial endothelial differentiation. *Dev. Cell* **3**, 127–136.
Lawson, N. D., and Weinstein, B. M. (2002a). Arteries and veins: Making a difference with zebrafish. *Nat. Rev. Genet.* **3**, 674–682.

Lawson, N. D., and Weinstein, B. M. (2002b). *In Vivo* imaging of embryonic vascular development using transgenic zebrafish. *Dev. Biol.* **248**, 307–318.

Lee, P., Goishi, K., Davidson, A. J., Mannix, R., Zon, L., and Klagsbrun, M. (2002). Neuropilin-1 is required for vascular development and is a mediator of VEGF-dependent angiogenesis in zebrafish. *Proc. Natl. Acad. Sci. USA* **99**, 10470–10475.

Levine, A. J., Munoz-Sanjuan, I., Bell, E., North, A. J., and Brivanlou, A. H. (2003). Fluorescent labeling of endothelial cells allows *in vivo*, continuous characterization of the vascular development of *Xenopus laevis*. *Dev. Biol.* **254**, 50–67.

Liang, D., Chang, J. R., Chin, A. J., Smith, A., Kelly, C., Weinberg, E. S., and Ge, R. (2001). The role of vascular endothelial growth factor (VEGF) in vasculogenesis, angiogenesis, and hematopoiesis in zebrafish development. *Mech. Dev.* **108**, 29–43.

Lofberg, J., and Collazo, A. (1997). Hypochord, an enigmatic embryonic structure: Study of the axolotl embryo. *J. Morphol.* **232**, 57–66.

Loughna, S., and Sato, T. N. (2001a). Angiopoietin and Tie signaling pathways in vascular development. *Matrix Biol.* **20**, 319–325.

Loughna, S., and Sato, T. N. (2001b). A combinatorial role of angiopoietin-1 and orphan receptor TIE1 pathways in establishing vascular polarity during angiogenesis. *Mol. Cell* **7**, 233–239.

McCright, B., Gao, X., Shen, L., Lozier, J., Lan, Y., Maguire, M., Herzlinger, D., Weinmaster, G., Jiang, R., and Gridley, T. (2001). Defects in development of the kidney, heart and eye vasculature in mice homozygous for a hypomorphic Notch2 mutation. *Development* **128**, 491–502.

Melby, A. E., Warga, R. M., and Kimmel, C. B. (1996). Specification of cell fates at the dorsal margin of the zebrafish gastrula. *Development* **122**, 2225–2237.

Miao, H. Q., Soker, S., Feiner, L., Alonso, J. L., Raper, J. A., and Klagsbrun, M. (1999). Neuropilin-1 mediates collapsin-1/semaphorin III inhibition of endothelial cell motility: functional competition of collapsin-1 and vascular endothelial growth factor-165. *J. Cell Biol.* **146**, 233–242.

Miquerol, L., Gertenstein, M., Harpal, K., Rossant, J., and Nagy, A. (1999). Multiple developmental roles of VEGF suggested by a lacZ-tagged allele. *Dev. Biol.* **212**, 307–322.

Miquerol, L., Langille, B. L., and Nagy, A. (2000). Embryonic development is disrupted by modest increases in vascular endothelial growth factor gene expression. *Development* **127**, 3941–3946.

Mukouyama, Y. S., Shin, D., Britsch, S., Taniguchi, M., and Anderson, D. J. (2002). Sensory nerves determine the pattern of arterial differentiation and blood vessel branching in the skin. *Cell* **109**, 693–705.

Nasevicius, A., Larson, J., and Ekker, S. C. (2000). Distinct requirements for zebrafish angiogenesis revealed by a VEGF-A morphant. *Yeast* **17**, 294–301.

Ng, Y. S., Rohan, R., Sunday, M. E., Demello, D. E., and D'Amore, P. A. (2001). Differential expression of VEGF isoforms in mouse during development and in the adult. *Dev. Dyn.* **220**, 112–121.

Noden, D. M. (1988). Interactions and fates of avian craniofacial mesenchyme. *Development* **103**, 121–140.

Noden, D. M. (1989). Embryonic origins and assembly of blood vessels. *Am. Rev. Respir. Disease* **140**, 1097–1103.

Otani, A., Kinder, K., Ewalt, K., Otero, F. J., Schimmel, P., and Friedlander, M. (2002). Bone marrow-derived stem cells target retinal astrocytes and can promote or inhibit retinal angiogenesis. *Nat. Med.* **8**, 1004–1010.

Pardanaud, L., Altmann, C., Kitos, P., Dieterlen-Lievre, F., and Buck, C. A. (1987). Vasculogenesis in the early quail blastodisc as studied with a monoclonal antibody recognizing endothelial cells. *Development* **100**, 339–349.

Pardanaud, L., and Dieterlen-Lievre, F. (1995). Does the paraxial mesoderm of the avian embryo have hemangioblastic capacity? *Anat. Embryol.* **192**, 301–308.
Pardanaud, L., and Dieterlen-Lievre, F. (1999). Manipulation of the angiopoietic/hemangiopoietic commitment in the avian embryo. *Development* **126**, 617–627.
Pardanaud, L., Luton, D., Prigent, M., Bourcheix, L. M., Catala, M., and Dieterlen-Lievre, F. (1996). Two distinct endothelial lineages in ontogeny, one of them related to hemopoiesis. *Development* **122**, 1363–1371.
Pardanaud, L., Yassine, F., and Dieterlen-Lievre, F. (1989). Relationship between vasculogenesis, angiogenesis and haemopoiesis during avian ontogeny. *Development* **105**, 473–485.
Park, J. E., Keller, G. A., and Ferrara, N. (1993). The vascular endothelial growth factor (VEGF) isoforms: Differential deposition into the subepithelial extracellular matrix and bioactivity of extracellular matrix-bound VEGF. *Mol. Biol. Cell* **4**, 1317–1326.
Patan, S. (1998). Requisite role of angiopoietin-1, a ligand for the TIE2 receptor, during embryonic angiogenesis. *Microvasc. Res.* **56**, 1–21.
Pham, V. N., Roman, B. L., and Weinstein, B. M. (2001). Isolation and expression analysis of three Zebrafish angiopoietin genes. *Dev. Dyn.* **221**, 470–474.
Placzek, M., Tessier-Lavigne, M., Yamada, T., Jessell, T., and Dodd, J. (1990). Mesodermal control of neural cell identity: Floor plate induction by the notochord. *Science* **250**, 985–988.
Pola, R., Ling, L. E., Silver, M., Corbley, M. J., Kearney, M. R., Blake, P., Shapiro, R., Taylor, F. R., Baker, D. P., Asahara, T., and Isner, J. M. (2001). The morphogen Sonic hedgehog is an indirect angiogenic agent upregulating two families of angiogenic growth factors. *Nat. Med.* **7**, 706–711.
Poole, T. J., and Coffin, J. D. (1989). Vasculogenesis and angiogenesis: Two distinct morphogenetic mechanisms establish embryonic vascular pattern. *J. Exp. Zool.* **251**, 224–231.
Poole, T. J., Finkelstein, E. B., and Cox, C. M. (2001). The role of FGF and VEGF in angioblast induction and migration during vascular development. *Dev. Dyn.* **220**, 1–17.
Puri, M. C., Rossant, J., Alitalo, K., Bernstein, A., and Partanen, J. (1995). The receptor tyrosine kinase TIE is required for integrity and survival of vascular endothelial cells. *Embo. J.* **14**, 5884–5891.
Puri, M. C., Partanen, J., Rossant, J., Bernstein, A., and Alitalo, K. (1999). Interaction of the TEK and TIE receptor tyrosine kinases during cardiovascular development. *Development* **126**, 4569–4580.
Risau, W. (1997). Mechanisms of angiogenesis. *Nature* **386**, 671–674.
Risau, W., and Wolburg, H. (1990). Development of the blood–brain barrier. *Trends Neurosci.* **13**, 174–178.
Rowitch, D. H., St.-Jacques, B., Lee, S. M., Flax, J. D., Snyder, E. Y., and McMahon, A. P. (1999). Sonic hedgehog regulates proliferation and inhibits differentiation of CNS precursor cells. *J. Neurosci.* **19**, 8954–8965.
Ruhrberg, C., Gerhardt, H., Golding, M., Watson, R., Ioannidou, S., Fujisawa, H., Betsholtz, C., and Shima, D. T. (2002). Spatially restricted patterning cues provided by heparin-binding VEGF-A control blood vessel branching morphogenesis. *Genes Dev.* **16**, 2684–2698.
Rupp, P. A., Czirok, A., and Little, C. D. (2003). Novel approaches for the study of vascular assembly and morphogenesis in avian embryos. *Trends Cardiovasc. Med.* **13**, 283–288.
Sabin, F. R. (1917). Origin and development of the primitive vessels of the chick and of the pig. *Contrib. Embryol. Carnegie Inst.* **6**, 61–124.
Sabin, F. R. (1920). Studies on the origin of the blood vessels and of red blood corpuscules as seen in the living blastoderm of chick during the second day of incubation. *Contrib. Embryol. Carnegie. Inst.* **9**, 215–262.
Salloway, S., and Hong, J. (1998). CADASIL syndrome: a genetic form of vascular dementia. *J. Geriatr. Psychiatry Neurol.* **11**, 71–77.

Sato, T. N., Tozawa, Y., Deutsch, U., Wolburg-Buchholz, K., Fujiwara, Y., Gendron-Maguire, M., Gridley, T., Wolburg, H., Risau, W., and Qin, Y. (1995). Distinct roles of the receptor tyrosine kinases Tie-1 and Tie-2 in blood vessel formation. *Nature* **376**, 70–74.

Schauerte, H. E., van Eeden, F. J., Fricke, C., Odenthal, J., Strahle, U., and Haffter, P. (1998). Sonic hedgehog is not required for the induction of medial floor plate cells in the zebrafish. *Development* **125**, 2983–2993.

Schier, A. F., Neuhauss, S. C., Harvey, M., Malicki, J., Solnica-Krezel, L., Stainier, D. Y., Zwartkruis, F., Abdelilah, S., Stemple, D. L., Rangini, Z., Yang, H., and Driever, W. (1996). Mutations affecting the development of the embryonic zebrafish brain. *Development* **123**, 165–178.

Schier, A. F., Neuhauss, S. C., Helde, K. A., Talbot, W. S., and Driever, W. (1997). The one-eyed pinhead gene functions in mesoderm and endoderm formation in zebrafish and interacts with no tail. *Development* **124**, 327–342.

Schuh, A. C., Faloon, P., Hu, Q. L., Bhimani, M., and Choi, K. (1999). In Vitro hematopoietic and endothelial potential of flk-1 −/− embryonic stem cells and embryos. *Proc. Natl. Acad. Sci. USA* **96**, 2159–2164.

Serini, G., Valdembri, D., Zanivan, S., Morterra, G., Burkhardt, C., Caccavari, F., Zammataro, L., Primo, L., Tamagnone, L., Logan, M., Tessier-Lavigne, M., Taniguchi, M., Puschel, A. W., and Bussolino, F. (2003). Class 3 semaphorins control vascular morphogenesis by inhibiting integrin function. *Nature* **424**, 391–397.

Shalaby, F., Ho, J., Stanford, W. L., Fischer, K. D., Schuh, A. C., Schwartz, L., Bernstein, A., and Rossant, J. (1997). A requirement for Flk1 in primitive and definitive hematopoiesis and vasculogenesis. *Cell* **89**, 981–990.

Shalaby, F., Rossant, J., Yamaguchi, T. P., Gertsenstein, M., Wu, X. F., Breitman, M. L., and Schuh, A. C. (1995). Failure of blood-island formation and vasculogenesis in Flk-1-deficient mice. *Nature* **376**, 62–66.

Shoji, W., Isogai, S., Sato-Maeda, M., Obinata, M., and Kuwada, J. Y. (2003). Semaphorin3a1 regulates angioblast migration and vascular development in zebrafish embryos. *Development* **130**, 3227–3236.

Shoji, W., Yee, C. S., and Kuwada, J. Y. (1998). Zebrafish semaphorin Z1a collapses specific growth cones and alters their pathway *in vivo*. *Development* **125**, 1275–1283.

Soker, S., Takashima, S., Miao, H. Q., Neufeld, G., and Klagsbrun, M. (1998). Neuropilin-1 is expressed by endothelial and tumor cells as an isoform-specific receptor for vascular endothelial growth factor. *Cell* **92**, 735–745.

Stalmans, I., Ng, Y. S., Rohan, R., Fruttiger, M., Bouche, A., Yuce, A., Fujisawa, H., Hermans, B., Shani, M., Jansen, S., Hicklin, D., Anderson, D. J., Gardiner, T., Hammes, H. P., Moons, L., Dewerchin, M., Collen, D., Carmeliet, P., and D'Amore, P. A. (2002). Arteriolar and venular patterning in retinas of mice selectively expressing VEGF isoforms. *J. Clin. Invest.* **109**, 327–336.

Stone, J., Itin, A., Alon, T., Pe'er, J., Gnessin, H., Chan-Ling, T., and Keshet, E. (1995). Development of retinal vasculature is mediated by hypoxia-induced vascular endothelial growth factor (VEGF) expression by neuroglia. *J. Neurosci.* **15**, 4738–4747.

Strahle, U., Blader, P., and Ingham, P. W. (1996). Expression of axial and sonic hedgehog in wildtype and midline defective zebrafish embryos. *Int. J. Dev. Biol.* **40**, 929–940.

Strahle, U., Jesuthasan, S., Blader, P., Garcia-Villalba, P., Hatta, K., and Ingham, P. W. (1997). one-eyed pinhead is required for development of the ventral midline of the zebrafish (Danio rerio) neural tube. *Genes Funct.* **1**, 131–148.

Sumoy, L., Keasey, J. B., Dittman, T. D., and Kimelman, D. (1997). A role for notochord in axial vascular development revealed by analysis of phenotype and the expression of VEGR-2 in zebrafish flh and ntl mutant embryos. *Mech. Dev.* **63**, 15–27.

Suri, C., Jones, P. F., Patan, S., Bartunkova, S., Maisonpierre, P. C., Davis, S., Sato, T. N., and Yancopoulos, G. D. (1996). Requisite role of angiopoietin-1, a ligand for the TIE2 receptor, during embryonic angiogenesis. *Cell* **87**, 1171–1180.

Swiatek, P. J., Lindsell, C. E., del Amo, F. F., Weinmaster, G., and Gridley, T. (1994). Notch1 is essential for postimplantation development in mice. *Genes Dev.* **8**, 707–719.

Szeto, D. P., Griffin, K. J., and Kimelman, D. (2002). HrT is required for cardiovascular development in zebrafish. *Development* **129**, 5093–5101.

Takashima, S., Kitakaze, M., Asakura, M., Asanuma, H., Sanada, S., Tashiro, F., Niwa, H., Miyazaki Ji, J., Hirota, S., Kitamura, Y., Kitsukawa, T., Fujisawa, H., Klagsbrun, M., and Hori, M. (2002). Targeting of both mouse neuropilin-1 and neuropilin-2 genes severely impairs developmental yolk sac and embryonic angiogenesis. *Proc. Natl. Acad. Sci. USA* **99**, 3657–3662.

Talbot, W. S., Trevarrow, B., Halpern, M. E., Melby, A. E., Farr, G., Postlethwait, J. H., Jowett, T., Kimmel, C. B., and Kimelman, D. (1995). A homeobox gene essential for zebrafish notochord development. *Nature* **378**, 150–157.

Tessier-Lavigne, M., and Goodman, C. S. (1996). The molecular biology of axon guidance. *Science* **274**, 1123–1133.

Trousse, F., Giess, M. C., Soula, C., Ghandour, S., Duprat, A. M., and Cochard, P. (1995). Notochord and floor plate stimulate oligodendrocyte differentiation in cultures of the chick dorsal neural tube. *J. Neurosci. Res.* **41**, 552–560.

Uyttendaele, H., Ho, J., Rossant, J., and Kitajewski, J. (2001). Vascular patterning defects associated with expression of activated Notch4 in embryonic endothelium. *Proc. Natl. Acad. Sci. USA* **98**, 5643–5648.

Vokes, S. A., and Krieg, P. A. (2002). Endoderm is required for vascular endothelial tube formation, but not for angioblast specification. *Development* **129**, 775–785.

Wang, H. U., Chen, Z. F., and Anderson, D. J. (1998). Molecular distinction and angiogenic interaction between embryonic arteries and veins revealed by ephrin-B2 and its receptor Eph-B4. *Cell* **93**, 741–753.

Ward, N. L., and Dumont, D. J. (2002). The angiopoietins and Tie2/Tek: adding to the complexity of cardiovascular development. *Semin. Cell Dev. Biol.* **13**, 19–27.

Weinstein, B. M. (2002). Plumbing the mysteries of vascular development using the zebrafish. *Semin. Cell Dev. Biol.* **13**, 515–522.

Weinstein, B. M., Stemple, D. L., Driever, W., and Fishman, M. C. (1995). Gridlock, a localized heritable vascular patterning defect in the zebrafish. *Nat. Med.* **1**, 1143–1147.

Wilkinson, D. G. (2000). Eph receptors and ephrins: Regulators of guidance and assembly. *Int. Rev. Cytol.* **196**, 177–244.

Wilting, J., Brand-Saberi, B., Huang, R., Zhi, Q., Kontges, G., Ordahl, C. P., and Christ, B. (1995). Angiogenic potential of the avian somite. *Dev. Dyn.* **202**, 165–171.

Xue, Y., Gao, X., Lindsell, C. E., Norton, C. R., Chang, B., Hicks, C., Gendron-Maguire, M., Rand, E. B., Weinmaster, G., and Gridley, T. (1999). Embryonic lethality and vascular defects in mice lacking the Notch ligand Jagged1. *Hum. Mol. Genet.* **8**, 723–730.

Yamada, T., Pfaff, S. L., Edlund, T., and Jessell, T. M. (1993). Control of cell pattern in the neural tube: motor neuron induction by diffusible factors from notochord and floor plate. *Cell* **73**, 673–686.

Yancopoulos, G. D., Davis, S., Gale, N. W., Rudge, J. S., Wiegand, S. J., and Holash, J. (2000). Vascular-specific growth factors and blood vessel formation. *Nature* **407**, 242–248.

Zhang, Y., Porat, R. M., Alon, T., Keshet, E., and Stone, J. (1999). Tissue oxygen levels control astrocyte movement and differentiation in developing retina. *Dev. Brain Res.* **118**, 135–145.

Zhong, T. P., Childs, S., Leu, J. P., and Fishman, M. C. (2001). Gridlock signalling pathway fashions the first embryonic artery. *Nature* **414**, 216–220.

Zhong, T. P., Rosenberg, M., Mohideen, M. A. P. K., Weinstein, B. M., and Fishman, M. C. (2000). Gridlock, and HLH gene required for assembly of the aorta in zebrafish. *Science* **287**, 1820–1824.

4

Wiring the Vascular Circuitry: From Growth Factors to Guidance Cues

Lisa D. Urness* and Dean Y. Li[†]
*Division of Cardiology
University of Utah School of Medicine
Salt Lake City, Utah 84112
[†]Program in Human Molecular Biology and Genetics
Departments of Medicine and Oncological Science
University of Utah School of Medicine
Salt Lake City, Utah 84112

  I. Introduction
 II. Basic Signaling Pathways Required for Forming Blood Vessels
III. The Formation of Arterial and Venous Networks
 IV. Effects of Local Cues on Arterial–Venous Identity
  V. Genetic Evidence for Vascular Guidance
 VI. Vascular Endothelial Growth Factors Are Attractive Guidance Cues
VII. Guidance as a Balance of Attractive and Repulsive Cues
VIII. Neural Guidance Pathways
 IX. Neural and Vascular Guidance Mechanisms Share Common Signaling Pathways
  X. Neuro-Vascular Development as Interdependent Processes
 XI. Independent Versus Interdependent Guidance of Vessels and Nerves
XII. Future Directions in Vascular Guidance
XIII. Summary
      Acknowledgments
      References

I. Introduction

The cardiovascular system is the first organ system to form and function during embryogenesis. The development of an efficient system to transport nutrients and oxygen is paramount for supplying the metabolic needs of a rapidly developing embryo. The organization of the vascular network and the timing of its development are highly conserved between and within species. The importance of this conservation is highlighted by the fact that the vascular system does not simply supply metabolic sustenance, but also provides critical inductive signals required for the differentiation and development of organs such as the pancreas and liver (Cleaver and Melton, 2003; Lammert *et al.*, 2001; Matsumoto *et al.*, 2001). Thus, the developing vascular network is critical for orchestrating and supporting organogenesis.

Our understanding of the formation of the vascular network has evolved from identifying the mitogens that stimulate the proliferation of the primary vascular cells, and endothelial and vascular smooth muscle cells, to investigating how the circulatory network is programmed and patterned. With the ability to culture and assay vascular cells, the initial focus revolved around determining the effects of known mitogens and cytokines on these cells and purifying vascular-specific factors. These studies identified some of the basic molecular signals required for the development of the vascular system. Second, gene-targeting experiments provided functional insight into which signaling pathways were essential in mammalian vascular development and began to order the sequence of molecular events required to make a stable vessel. This led to a unified, though elemental, view of how vessels form and sprout. Third, the molecular programs that instruct the differentiation of different types of blood vessels were identified. Blood vessels can be catalogued by their luminal caliber, the organs that they supply, and by whether they are arteries or veins. Targeted ablation of these vessel-specific programs has allowed us to identify genes that are essential for vascular development. Most recently, the field has graduated to understanding the vascular system as an integrated vascular network that is programmed and patterned in a manner similar to the neural network.

In essence, both the vascular and neural networks require path-finding mechanisms that guide sprouts from central axial structures along predetermined trajectories to their distal destinations. It is interesting that the neural and vascular networks often form parallel and super-imposable circuits, and recent evidence indicates that molecular pathways important for regulating neural patterning and guidance may play similar roles in the vasculature. As we seek to understand how the vascular circuit is formed, we must not only determine the specific role of various neural path-finding mechanisms in vascular development, but also incorporate the approaches and perspectives employed by neurobiologists to uncover them. Thus, our understanding of vascular development has evolved from characterization of the constituent endothelial cells, to the organization of mature vessels, and now to the patterning of a coordinated network of vessels.

II. Basic Signaling Pathways Required for Forming Blood Vessels

Simplistically, to form any mature vessel, endothelial cells must differentiate, proliferate, form tubes, and induce the differentiation and recruitment of the vascular smooth muscle cells that surround the endothelial tube. The ability to culture and assay primary endothelial and vascular smooth muscle cells, developed in the 1970s and 1980s, laid the groundwork for identifying factors that modulate their growth and migration. These assays ranged from generic measurements of cellular adhesion, proliferation, and migration, to vascular-specific assays such as aortic ring, chorioallantoic membrane (CAM), and corneal implant assays that evaluated endothelial tube formation and sprouting. Known factors that were purified based on non-endothelial cell assays, such as fibroblast growth factors (FGF), transforming growth factors beta (TGF-β), and platelet-derived growth factors (PDGF) were found to be active on primary vascular cell lines (Gospodarowicz et al., 1977; Pepper, 1997; Williams et al., 1982). Endothelial-specific growth factors, such as vascular endothelial growth factor (VEGF), were identified on the basis of primary vascular cell biology (Ferrara and Henzel, 1989). As observed with the discovery of all cytokines and growth factors, complexity increases with further investigation. Homologs and isoforms with different functions, multiple cognate receptors, and factors that inhibited the action of mitogens were subsequently identified. Thus, the in vitro studies that employed primary vascular cell biology were the basis for identifying the first signaling pathways required for vascular development.

Vascular development is divided into two processes. During vasculogenesis, endothelial cells differentiate, migrate, and coalesce into the central axial vessels, the dorsal aortae, and cardinal veins. The second process, termed angiogenesis, involves vessels sprouting from the primary plexus to form the mature circulatory system. Murine gene-targeting has enabled the elucidation of many of the molecular events that are essential for mammalian vascular development and helped to define whether these processes were critical for either vasculogenesis or angiogenesis (Carmeliet, 2003a; Jain, 2003). The combination of gene-targeting and cell biologic data led to a simplified model. Vascular endothelial growth factors and their respective signaling receptors play an essential role in endothelial differentiation and proliferation (Dumont et al., 1998; Ferrara et al., 1996; Fong et al., 1995; Shalaby et al., 1995). In the absence of these factors or their receptors, endothelial cells fail to develop. Next, the endothelium secretes platelet-derived growth factor that induces the recruitment and differentiation of vascular smooth muscle cells. Disruption of PDGFs or their receptors prevent endothelial beds from forming a smooth muscle medial layer (Betsholtz et al., 2001; Bostrom et al., 1996; Hellstrom et al., 1999; Soriano, 1994). Subsequently, the vascular smooth muscle cells secrete angiopoietins that act on endothelial cells to ensure proper interactions between endothelial and vascular smooth muscle cells (Maisonpierre et al., 1997; Suri et al., 1996). Targeted ablation of the angiopoietins or their cognate receptors leads to leaky vessels. Finally, partially in response to stimulation by transforming growth factors, the vascular smooth muscle cells secrete matrix proteins, such as elastin, that inhibit vascular smooth muscle cell proliferation and differentiation, thus stabilizing the mature vessel (Karnik *et al.*, 2003; Li *et al.*, 1998). Mice lacking elastin fail to maintain a patent lumen due to uncontrolled proliferation of the vascular media. Thus, to establish and maintain a mature structure, the endothelium and the vascular media interact through a variety of autocrine and paracrine signaling mechanisms.

III. The Formation of Arterial and Venous Networks

The initial molecular models of vasculogenesis and angiogenesis treated blood vessels as homogeneous populations of vascular tubes. However, a functional vascular circuit requires the partitioning of distinct arterial and venous networks that only interconnect within the capillary beds of distal target organs. Following cardiac contraction, blood is ejected under high pressure and flows through arteries to distal organ systems, while blood under low pressure returns to the heart via veins. Pathologists routinely distinguished arteries from veins well before any molecular model of vascular development was proposed. Mature arteries have a far more prominent arrangement of extracellular matrix, smooth muscle cells, and other supporting cells than do veins. It was previously thought that the primitive vasculature was shaped or remodeled by local environmental factors, for example, that hemodynamic forces such as blood pressure and blood flow dictated the differentiation of a vessel toward arterial or venous character (Glagov *et al.*, 1988; Stehbens, 1996). Evidence for genetic programs that direct the development of arterial-venous identity has emerged through the characterization of gene-targeted mice and mutational analysis in zebrafish (Adams, 2003; Lawson and Weinstein, 2002; Rossant and Hirashima, 2003).

Arteries and veins adopt distinct developmental and molecular paths at the earliest stages of vascular development. During vasculogenesis, the vessels that will ultimately give rise to the dorsal aortae and cardinal veins are patterned *de novo* and assembled from angioblasts in the lateral mesoderm (Flamme *et al.*, 1997; Pardanaud and Dieterlen-Lievre, 2000; Poole and Coffin, 1989; Risau and Flamme, 1995). Mutations in "gridlock," the zebrafish homolog of the human *HRT2* gene, result in local patterning defects wherein portions of the dorsal aorta are missing or blocked (Weinstein *et al.*, 1995; Zhong *et al.*, 2000). In zebrafish, this gene is expressed in the dorsal aorta but not the cardinal vein. Studies utilizing morpholino knock-down technology further suggest that gridlock plays a role in arteriogenesis, consistent with its early expression in pre-vasculogenic mesoderm (Zhong *et al.*, 2001). Whether angioblasts are specified to be arterial or venous before the assembly of the vessel structure occurs, or whether arterial or venous fates are assigned after the initial vascular network is laid down, is not clear. However, lineage tracing studies in zebrafish suggest that some angioblasts may be specified as arterial or venous prior to the vasculogenic consolidation of discrete vessels, given the observation that labeled angioblasts contribute to either arteries or veins, but not both (Zhong *et al.*, 2001). The fact that the major embryonic vessels in zebrafish appear to have distinct molecular identities prior to angiogenic remodeling and circulation suggests that there are additional classes of genes that are specifically required for distinguishing arteries and veins during the *de novo* formation of the central axial vessels. There are at least three gridlock family members in mice (Nakagawa *et al.*, 1999), and *in vitro* experiments have implicated them in tube formation (Henderson *et al.*, 2001) and the regulation of arterial gene programs in cultured endothelial cells (Chi *et al.*, 2003). However, though *Hey2*, the murine ortholog of gridlock, has a role in cardiogenesis, no role in murine peripheral vascular development *in vivo* has been demonstrated (Donovan *et al.*, 2002; Gessler *et al.*, 2002; Ishibashi *et al.*, 1995; Sakata *et al.*, 2002).

Another pathway clearly implicated in arterial-venous identity during zebrafish vascular development is the Notch pathway. The Notch proteins are transmembrane receptors that are activated by short-range cell-bound ligands of the DSL (Delta-Serrate-Lag2) class (Artavanis-Tsakonas *et al.*, 1999). Notch signaling was first described in the process of Drosophila neurogenesis and has since been implicated in vertebrate neurogenesis and arterial identity. Notch receptors 1 (Del Amo *et al.*, 1992; Reaume *et al.*, 1992), 3 (Lardelli *et al.*, 1994), 4 (Uyttendaele *et al.*, 1996), Notch ligands Jagged-1 (Zimrin *et al.*, 1996), and Delta-like 4 (Krebs *et al.*, 2000; Shutter *et al.*, 2000) are expressed in the vascular endothelium (Mailhos *et al.*, 2001; Villa *et al.*, 2001). Lawson *et al.* (2001, 2002) have conducted epistasis experiments in zebrafish that have demonstrated a requirement for Notch5 in arterial differentiation. They show that during zebrafish vascular development, Sonic Hedgehog (SHH) is expressed in the notochord and induces somatic tissues to express VEGF. The expression of VEGF, in turn, induces the Notch pathway in endothelial cells, thereby activating downstream arterial markers such as ephrin-B2. Overexpression of an activated Notch5 receptor leads to ectopic expansion of arterial markers on the cardinal vein; conversely, loss of Notch5 signaling results in the appearance of a large vessel that fails to express arterial markers. These data suggest that Notch functions to suppress venous fate during the arteriovenous differentiation phase of vessel development, rather than during the initial formation of the vessels themselves. Whereas Zhong *et al.* (2001) posit that gridlock functions as a fate switch downstream of Notch signaling, Lawson *et al.* (2001, 2002) have observed no change in gridlock expression as a result of Notch5 inactivation. This inconsistency may be due to incomplete inactivation of Notch signaling or to different Notch pathways utilized at the different positions analyzed in these studies.

The role of the Notch pathway in mammalian vascular development is not as well understood (for review, see Iso *et al.*, 2003). Several Notch pathway-related proteins have been shown to exhibit arterially-restricted expression in murine embryos (Mailhos *et al.*, 2001; Shutter *et al.*, 2000; Villa *et al.*, 2001), and some of these genes have been implicated in murine vascular development (D'Amore and Ng, 2002; Gridley, 2001; Krebs *et al.*, 2000; Leong *et al.*, 2002). Mice lacking Notch4 undergo vasculogenesis normally but later develop atretic cardinal veins and dorsal aortae, and other vascular remodeling defects (Krebs *et al.*, 2000). The vascular phenotypes observed in *Notch1/Notch4* double mutant embryos are more severe than those observed in mice lacking *Notch1* alone, whereas mice lacking *Notch4* show no vascular anomalies. These studies have led others to hypothesize that genetic interactions between *Notch4* and *Notch1* are important for angiogenic vascular remodeling (Gridley, 2001). In humans, defects in the Notch pathway are linked to human vascular diseases such as Alagille syndrome and cerebral autosomal dominant arteriopathy with subcortical infarcts and leukenocephalopathy (CADASIL) (Iso *et al.*, 2003). In Alagille syndrome, there are multiple congenital heart defects that involve the great vessels. Individuals suffering from CADASIL have multi-infarct dementia caused by thickened arterial walls and subsequent luminal obliteration. Thus, Notch signaling may have an important role in mammalian vascular development and disease, but a specific role in arterial differentiation has not been established.

During angiogenesis, arterial and venous sprouts come in close apposition and require stringent mechanisms to inhibit aberrant vessel fusions leading to circulatory shunts. It is likely that the molecular pathways required for maintaining boundaries between arteries and veins during angiogenesis may be different from those required for the *de novo* formation of the central axial vessels and their initial differentiation into arteries and veins during vasculogenesis (Fig. 1). Wang *et al.* (1998) and Adams *et al.* (1999) have demonstrated that a subclass of ephrin ligands and Eph receptors are expressed in the vasculature and play an important role in angiogenesis. The ephrins and Ephs were first identified on the basis of their roles in contact-mediated axonal guidance (Flanagan and Vanderhaeghen, 1998; Holder and Klein, 1999; Krull *et al.*, 1997; Wang and Anderson, 1997). These cell surface receptors allow neurons to distinguish boundaries via juxtacrine interactions (Mellitzer *et al.*, 1999). Ephrin-B2 is specifically expressed in arteries, whereas its cognate receptor, EphB4, appears to be predominantly expressed in veins and, to a lesser extent, in arteries (Adams *et al.*, 2001; Gerety *et al.*, 1999; Wang *et al.*, 1998), suggesting that these proteins may play an analogous role in maintaining distinct arterial and venous boundaries in the vasculature. The *de novo* formation of the dorsal aortae and cardinal veins is unperturbed in mice lacking these ephrin/Eph family members, indicating that these signaling molecules are not required for establishing vessel identity. The prominent phenotypic feature in *ephrin/Eph* mutant mice is the failure to remodel the primordial vascular plexus in both arterial and venous beds. Given the nearly identical and widespread angiogenic defects in *ephrin-B2−/−*, *EphB4−/−*, and *EphB2−/− /EphB3−/−* mice, it is likely that reciprocal signaling between ephrin:Eph ligand-receptor pairs is required for sorting and organizing vessels during angiogenesis. In the absence of boundaries defined by ephrins/Ephs expression, remodeling is arrested at the primitive plexus stage. Thus, these cell surface proteins may help to maintain segregation of the arterial and venous boundaries during angiogenesis via cell–cell interactions and repulsive cues (Gale and Yancopoulos, 1997, 1999; Wang *et al.*, 1998; Yancopoulos *et al.*, 1998). The molecular distinction between arteries and veins is not limited to the endothelium. The outer layers of arteries and veins, the vascular smooth muscle cells, adopt the ephrin/Eph expression pattern of its corresponding endothelial layer (Gale *et al.*, 2001; Gerety and Anderson, 2002; Shin *et al.*, 2001). Though endothelial expression of Ephrin-B2 is absolutely required for early angiogenesis, it is not known whether this medial layer expression is critical for maintaining arteries and veins in a more mature vessel network. Together, these studies suggest that disruption of vessel-specific genetic programs during angiogenesis can lead to inappropriate interactions between nascent arterial and venous sprouts.

The importance of arterial-specific programs in the *de novo* formation of the dorsal aortae and during angiogenesis is well established. Less clear is whether the venous fate is a default pathway or whether there are vein-specific programs that are essential for arterial-venous identity during either vasculogenesis or angiogenesis. Some have speculated that the expression of EphB4 in veins is required for angiogenesis in a manner analogous to the role of its cognate arterial-specific ligand, ephrin B2. However, because EphB4 is also expressed in arteries, it is difficult to conclude that venous expression of EphB4 creates a boundary between sprouting arteries and veins. There are other genes that are either preferentially expressed or specifically expressed in veins of either zebrafish or mice, such as *neuropilin-2*, *flt-4*, and *rtk-5* (Chi *et al.*, 2003; Herzog *et al.*, 2001; Lawson *et al.*, 2001; Lawson and Weinstein, 2002). The functional roles of these venous markers in development have not yet been fully elucidated. Given the outgrowth of lymphatic vessels from venous endothelium, there are likely also specific programs that distinguish lymphatics vessels from arteries and veins. Indeed, Petrova *et al.* (2002) have shown that overexpression of the homeobox transcription factor Prox-1 in blood endothelial cells drives lymphatic gene transcription. This effect was not observed in non-endothelial cells, suggesting that Prox-1 may be a fate determination factor

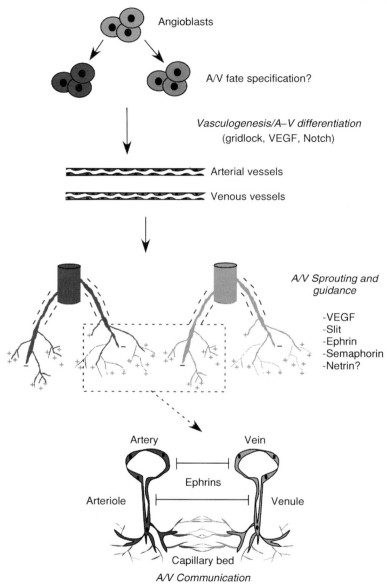

Figure 1 Differentiation and patterning of the embryonic vascular circuit. Three steps are required for the formation of a mature vascular network. First, the central axial vessels coalesce from angioblasts and differentiate into arteries and veins. Although little is known about the initial segregation of angioblasts into arterial and venous subpopulations in mammals, key insights from zebrafish studies suggest that the formation of separate arterial and venous vessels relies on Notch

that functions to reprogram differentiation of venous endothelial cells toward a lymphatic fate during development.

The importance of creating boundaries between arteries and veins during angiogenesis is illustrated by human vascular diseases. Patients afflicted with hereditary hemorrhagic telangiectasia (HHT) suffer from multiple arteriovenous malformations, and direct and abnormal connections between arterial and venous vascular beds. HHT is caused by haploinsufficiency in either activin receptor-like kinase 1 (Alk1) or endoglin; both are members of the transforming growth factor beta superfamily of receptors and ligands (Johnson et al., 1996; McAllister et al., 1994). Mice lacking *Alk1* or *endoglin* lose the morphological, molecular, and functional distinctions between arteries and veins, even though the endothelial tubes form normally during vasculogenesis at the presumptive position of the dorsal aortae and cardinal veins (Li et al., 1999b; Sorensen et al., 2003; Urness et al., 2000). These mutant mice die during mid-gestation due to angiogenic defects characterized by aberrant fusion of sprouting arterial and venous endothelial tubes. Similarly, mutations in *CCM1* are responsible for a human vascular dysplasia (Laberge-le Couteulx et al., 1999; Sahoo et al., 1999). Recent data from mice lacking *CCM1* suggest that similar mechanisms involving failure of sprouting endothelial tubes to distinguish themselves as arteries and veins are responsible for the formation of large dilated vascular malformations (Whitehead et al., 2004). Thus, molecular programs that are critical for establishing and maintaining distinct arterial and venous networks during angiogenesis may encode cues that guide arterial and venous endothelial tubes along parallel but distinct trajectories.

IV. Effects of Local Cues on Arterial–Venous Identity

It is important to emphasize that vessel-specific programs, though necessary, may not be sufficient to establish or maintain arterial-venous identity. First, after transplanting the arterial endothelium into veins and vice versa, the transplanted endothelium adopts the specific molecular programs of the host

---

signaling proteins and effectors. Evidence is accumulating for a role of VEGF in arterialization as well, perhaps as a result of Notch activation. Second, arteries and veins sprout from the central axial vessels during angiogenesis and are guided to common distal targets along parallel, but distinct, paths. It is this step in vascular network formation that is most analogous to neural guidance and appears to share many of the same attractive (+) and repellent (−) regulatory molecules and mechanisms. Finally, molecular mechanisms, such as Eph:ephrin signaling, are required to maintain the separate domains of the larger arterial and venous vessels during angiogenesis. It is not known how these mechanisms are overcome or modified at the level of the capillary bed, the only border at which the arterial and venous networks intersect. (See Color Insert.)

vessel, suggesting that local cues modulate vessel specificity (Le Noble *et al.*, 2004; Othman-Hassan *et al.*, 2001). These local cues could be either molecular or physiologic. Simple descriptive experiments that examine the expression pattern of arterial markers in saphenous veins implanted during coronary artery bypass graft may provide insight into the extent to which hemodynamic stress contributes to the assignment of arterial and venous identity. Second, though arterial and venous endothelial cells have distinct transcriptional profiles, many vessel-specific genes that are known to be functionally important *in vivo*, such as *Alk1* and *ephrin-B2*, are not expressed in a vessel-specific fashion in culture (Chi *et al.*, 2003). Whether these critical vessel-specific genes are no longer expressed in cultured cells due to absence of local physiologic conditions or cues is not known. Third, the overexpression of pan-endothelial factors such as VEGF result in vascular dysplasia and fusion (Carmeliet, 2000; Dor *et al.*, 2003; Drake and Little, 1995, 1999; Schwarz *et al.*, 2000). Thus, excessive endothelial proliferation may either inhibit or overcome vessel-specific programs that establish and maintain distinct arterial and venous beds. If vessel-specific programs are intact in these models of VEGF overexpression, it would suggest that these programs are not sufficient for maintaining arterial-venous identity during angiogenesis. More recently, by comparing the vascular patterning in the developing trunk vessels between wild-type zebrafish and transgenic *silent heart* mutants, Isogai *et al.* (2003) suggest a potential role for flow and pressure during angiogenic remodeling. The gross anatomical organization of the primary network is not affected by the loss of circulation; however, sprouting from the primary network and optimal functional segregation of the arterial and venous connections during angiogenesis is, to some extent, dependent on circulation. Thus, our relatively recent ability to manipulate arterial-venous specific genetic programs should not lead us to underestimate the potential role of local physiology in profoundly affecting vessel identity and patterning.

V. Genetic Evidence for Vascular Guidance

What was once considered to be a homogeneous array of tubes is now known in reality to be a highly complex and patterned network of distinct vessel types, the cellular interactions and morphogenic movements of which render an arborized vascular tree. The pattern with which the vasculature is laid down is highly reproducible and closely parallels the neural pattern. The prevailing view prior to the mid-1980s held that neural connections were a result of passive and random searching of axons for functional peripheral target connections. However, the identification of neural guidance cues via genetic screens, and demonstration of guidance mechanisms via *in vitro*

4. Wiring the Vascular Circuitry

assays, suggests that specific programs are hard-wired to pattern the nervous system. The evolution of our understanding of vascular development has recently undergone a similar renaissance in the last 5 years, though lagging considerably behind the molecular understanding of guidance mechanisms elucidated from neurobiology studies. Just as a functional nervous system depends on precision wiring in the assembly of neural networks, so, too, does the vascular system. A substantial body of evidence has accumulated that implies the presence of a guidance system for the vasculature as well. Undoubtedly, due to comparable developmental tasks performed throughout the body, vessel growth often parallels that of the nerve bundles and suggests that shared regulatory mechanisms appear to be utilized in disparate cellular contexts to achieve similar objectives.

The existence of vascular guidance mechanisms is directly supported by the identification or generation of mutations that manifest in the disruption of normal vessel trajectories. As described previously, the failure to maintain appropriate boundaries between arteries and veins in the ephrin/Eph mutants supports the existence of molecular guidance programs for the vasculature. Evidence suggests that additional guidance factors exist. Childs *et al.* (2002) conducted a genetic screen for vascular patterning mutations in zebrafish. Normally, intersegmental vessels migrate dorsally from the aorta and traverse between the somites. However, in *out-of-bound* (*OBD*) mutants, the intersomitic vessels sprout ectopically from the dorsal aorta and migrate anomalously into the somitic mesenchyme. Based on the non-cell-autonomous nature of this mutant, the authors postulate that the OBD gene encodes an inhibitory signal in the somites that constrains VEGF-stimulated growth to the appropriate corridors. This phenotype is remarkably reminiscent of mice and *Xenopus* lacking various ephrins and Ephs (Adams *et al.*, 1999; Helbling *et al.*, 2000), though no linkage of the OBD locus to known ephrins or other guidance genes has been established. Thus, there are a number of genetic models that strongly suggest that the trajectory of a blood vessel is defined by molecular cues that constrain sprouting along specific paths.

There are at least two different mechanisms by which vascular patterns could be sculpted: pruning of a ubiquitous, randomly patterned vascular plexus and directionally guided endothelial cell sprouting. The nonrandom vascular sprouting observed at the level of intersegmental vessels in vertebrates and the mammalian retina support the concept of angiogenic guidance, similar, in theory, to that defined by the field of neurobiology. However, before we focus on the similarities between neural and vascular patterning, there are important fundamental differences between nerves and blood vessels that add a layer of complexity to concepts of guidance. Whereas guidance in the nervous system is studied at the level of axon projection from a single neuron, in the vascular system the extension of a vessel is dependent on proliferation of the constituent endothelial cells that comprise the vessel. This is in contrast to neurons where cellular proliferation is in no way involved in axonal guidance. Given the constellation of endothelial mitogens and the tethering of these to specific tissue addresses, the question arises as to whether there is a need for invoking an elaborate guidance mechanism that is distinct from known endothelial mitogens.

VI. Vascular Endothelial Growth Factors Are Attractive Guidance Cues

The VEGF-A gene is tightly regulated at the transcriptional level (Neufeld et al., 1999) and consists of several isoforms whose differential extracellular distributions and tissue expressions make this growth factor an ideal guidance molecule (Ng et al., 2001). Moreover, the capacity of some VEGF isoforms to bind heparin enables deposition of that protein at specific sites in the extracellular matrix (Esko and Lindahl, 2001). Ruhrberg and colleagues (Ruhrberg et al., 2002) generated mice expressing single VEGF isoforms and demonstrated that localized spatial distributions of specific isoforms can direct capillary branching in mouse embryos. Mice expressing only a VEGF isoform that is missing the heparin-binding domains, $VEGF^{120}$, lack discrete deposits of VEGF and show reduced capillary branching and severe angiogenic defects. However, overall endothelial proliferation is not perturbed, and the resultant vessels exhibit larger lumenal diameters due to endothelial cell proliferation within existing vessels rather than incorporation into new vessels. In the converse experiment, mice expressing a heparin-binding VEGF isoform exhibit the reciprocal phenotype of ectopic capillary branches consisting of elongated, thin microvessels. Stalmans et al. (2002) have observed similar phenotypes in the retinal vascular beds of mice expressing single VEGF isoforms. Mice expressing only heparin-binding VEGF isoforms exhibit chaotic retinal vessel branching, whereas mice expressing only the diffuse $VEGF^{120}$ isoform appear to lack directional movement and exhibit random vessel loops. Carmeliet (2003b) posits that the diffusible $VEGF^{120}$ isoform can function as a long-range chemoattractant that provides gross directional information, while localized deposits of the heparin-binding isoforms provide short-range "guideposts" that can modulate specific twists and turns toward the source of the long-range attractant.

The effects of VEGF on the leading edge of a sprouting vessel are remarkably similar to the effect of guidance cues on axonal growth cones. Gerhardt et al. (2003) showed that filopodial extensions on specialized endothelial cells at the tips of sprouting vessels function as sensors to sample their environment and integrate information from VEGF gradients in the mammalian retina. The filopodia of axonal growth cones perform the same function in nerve cells (Mallavarapu and Mitchison, 1999; Mitchison and Kirschner, 1988; Suter and Forscher, 2000). Collectively, these data demonstrate how spatially defined deposits of growth factors can precisely pattern angiogenic sprouting and that potent endothelial mitogens such as VEGF are, by definition, positive guidance factors for vessel growth. More recently, a novel endothelial growth factor, EG-VEGF, that is specific for vessels of the endocrine organs was identified (LeCouter et al., 2001). This finding suggests additional complexity, in that vessels can be guided to supply different organ systems through distinct organ-specific endothelial mitogens.

Potential inhibitors of VEGF-mediated capillary branching are the mammalian *Sprouty* genes; orthologs of the Drosophila *Sprouty* gene. These secreted fibroblast growth factor (FGF) antagonists have been implicated in tracheal branching morphogenesis, a process that is a morphologic equivalent of vascular branching (Hacohen et al., 1998; Placzek and Skaer, 1999). Impagnatiello et al. (2001) have shown that *Spry-1* and *Spry-2* can act as negative regulators of both FGF- and VEGF-induced proliferation of endothelial cells *in vitro*. Similarly, Lee et al. (2001) have shown that *Spry-4* can inhibit FGF- and VEGF-mediated proliferation and migration in endothelial cell culture, as well as abrogate MAPK phosphorylation, a readout of tyrosine kinase signaling. Cultured mouse embryos injected with a *Sprouty-4* adenoviral construct exhibited reduced capillary sprouting. Other inhibitors of endothelial proliferation, such as angiostatin and endostatin, have been identified, though a role in embryogenesis has not been determined (O'Reilly et al., 1994, 1997). Factors that inhibit endothelial proliferation provide a potential means for modulating the attractive cues of VEGF. However, it remains to be seen whether these gene products specifically function in a guidance role for determining vascular patterning.

There are many roles played by VEGF in the development of the vasculature, from the earliest stages of vessel formation to the subsequent patterning of the mature network. As discussed previously, Lawson et al. (2001, 2002) reported a role for VEGF in arteriogenesis through its regulation of the Notch pathway. Stalmans et al. (2002) observed a reduction of arterioles in the retina of mice lacking specific VEGF isoforms, although venular development was unaffected. Similarly, mice harboring a cardiac-specific transgene for $VEGF^{164}$ exhibited enhanced ephrin-B2-expressing vessels at the expense of EphB4-expressing vessels (Visconti et al., 2002). Finally, Mukouyama et al. (2002) showed that low levels of VEGF preferentially induced ephrin-B2 expression in cultured angioblasts without altering cellular proliferation or selection. Collectively, these data strongly suggest that, in addition to its mitogenic role, VEGF isoforms can modulate arterial-venous decisions and vascular guidance.

VII. Guidance as a Balance of Attractive and Repulsive Cues

In the case of neurons, a balance of at least four different cues guide axonal outgrowth (Fig. 2). First, there are attractive cues that induce neurons to sprout axons toward the source of the cue. Second, there are cues that inhibit axon sprouting. Third, there are juxtacrine inhibitory cues that constrain axon sprouting along specific paths. Finally, there are repulsive cues that guide axons to sprout away from a factor. From the previous discussion, it is clear that the vascular system adopts many of the same mechanisms to guide its patterning. There are endothelial mitogens such as VEGF that are functionally equivalent to a neuronal attractive cue, and conversely, there are inhibitors of endothelial mitogens that can abrogate vascular sprouting. Mimicking their role in axonal path finding, the ephrin/Eph family may define arterial and venous boundaries through a juxtacrine mechanism. However, it is less clear whether there are diffusible factors that provide negative or repulsive guidance cues for the vasculature.

Prior to molecular evidence for the control of embryonic patterning, an elegant series of embryological chick–quail grafting studies revealed the presence of inhibitory domains for vascular sprouting. In chick embryos, endothelial cells only migrated ipsilaterally relative to the midline and never crossed the midline except at specific places, such as the most dorsal perineural plexus and dermis. This restriction in the distribution of endothelial sprouting suggested laterality and that there is a barrier to migration from one side of the body to the other. Wilting *et al.* (1995) showed that endothelial cells from paraxial mesoderm quail grafts, implanted in chicken embryos, sprout within the half of the chick embryo to which they are grafted, independent of the side of the quail embryo from which they were derived. These new vessels never traversed to the contralateral side of the embryonic midline. Therefore, laterality does not appear to depend on the origin of host grafts, but on the microenvironment to which they are implanted. Klessinger and Christ (1996) showed that, following notochordectomy, this barrier to migration is lost, suggesting that a chemical inhibitory factor is elicited by the notochord. Similar stereotypical axon guidance decisions made at the midline of the central nervous system was one of the fundamental observations that triggered the search for and discovery of neural guidance mechanisms.

VIII. Neural Guidance Pathways

Our primer for understanding attractive and repulsive cues has been the nervous system, where guidance mechanisms have been extensively characterized. This versatile and diversified axonal guidance paradigm has been employed from flies to vertebrates. Neurobiology has, and will continue to

4. Wiring the Vascular Circuitry

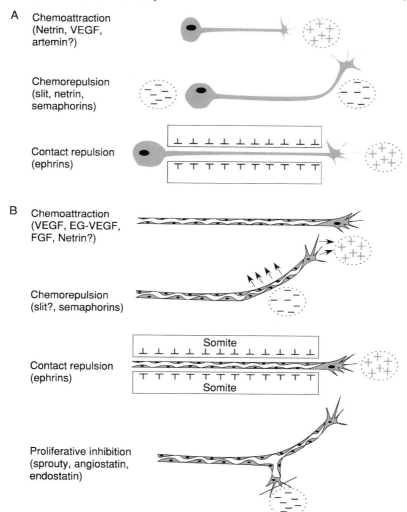

Figure 2 Axonal guidance as a paradigm for vascular patterning. Evidence for several modes of directional axon growth or neuroblast migration have been described in the literature, including chemoattraction, chemorepulsion, and contact repulsion (inhibition) (a). Similar patterning mechanisms are also apparent in the vasculature (b), particularly with regard to chemoattraction and contact repulsion. Functional studies that definitively discriminate between chemorepulsion (progressive movement of a vessel away from a point source) and proliferative inhibition have not yet been reported. (See Color Insert.)

be, a fruitful field from which to search for analogous factors and mechanisms that modulate vascular patterning. In the last decade, molecular genetics has greatly expanded our understanding of vascular patterning, and a growing number of potential vascular guidance molecules have been described. Vascular endothelial cells and neurons appear to express a strikingly similar repertoire of proteins, many of which have been shown to play comparable roles in the two systems. Before embarking on a discussion of guidance as a presumptive mechanism of vascular patterning, it is instructive to briefly review neural guidance mechanisms.

The concept of cellular guidance has been elucidated in intricate detail in the nervous system where, upon exiting the central nervous system, axons navigate specific paths through complex microenvironments in response to repellent or attractive extracellular cues. These cues are transduced by cytoskeletal dynamics at the tip of the growing axon, the growth cone. En route to their synaptic targets, axons must integrate an array of extrinsic cues, both long-range diffusible factors and short-range membrane-bound proteins, in order to make path-choice decisions (for review, see Dickson, 2002; Huber *et al.*, 2003; Tessier-Lavigne and Goodman, 1996). Without coordinated guidance cues, the individual neurons would fail to create a functional neural network. Classification of guidance cues as strictly attractive or repulsive is often difficult, as a given molecule can elicit either response under different circumstances. The action elicited is highly dependent on the complement of receptors present on the axon, the intracellular state of the growth cone, and the array of other guidance molecules in the extracellular milieu. Thus, various combinations of a relatively small number of factors can produce opposing path choices and establish complex patterns of neural connections (Culotti and Merz, 1998; Hong *et al.*, 1999).

There are at least four major classes of predominantly long-range guidance molecules or signaling pathways, as well as a number of extracellular matrix proteins that mediate short-range interactions; the latter are beyond the scope of this review. The major classes include the secreted semaphorins, netrins, slits, and the membrane-bound ephrins (for a comprehensive review, see Huber *et al.*, 2003). Each of these can signal through a number of receptor families. Semaphorins, through interaction with the neuropilin-1 receptor, primarily function as short-range inhibitory cues that deflect growth cones or restrict their growth to one path by establishing a corridor bounded by repulsive cues (Cheng *et al.*, 2001; Hedgecock *et al.*, 1990). Netrins recognize UNC-5, neogenin, and deleted in colorectal cancer (DCC) receptor families or combinations thereof, and may repel or attract axons, depending on the type of neuron and complement of receptors that are expressed on the surface of the growth cone. Signaling through UNC-5 appears to mediate repulsion exclusively, whereas DCC can mediate either repulsion or attraction (Cheng *et al.*, 2001; Hedgecock *et al.*, 1990; Keleman and Dickson, 2001).

The Slits, signaling through the Robo family of receptors (Kidd *et al.*, 1998; Seeger *et al.*, 1993), were first identified in genetic screens for axonal midline guidance defects in *Drosophila* (Battye *et al.*, 1999; Kidd *et al.*, 1998, 1999; Seeger *et al.*, 1993). Slit protein has predominantly, though not exclusively, been shown to repel axons (Battye *et al.*, 1999; Chen *et al.*, 2000; Hu, 1999; Yuan *et al.*, 1999; Zhu *et al.*, 1999) and is conserved in vertebrates (Brose *et al.*, 1999; Li *et al.*, 1999a). In addition to repulsion, secreted Slit protein has been shown to mediate axon branching and regulate neuron migration (Brose *et al.*, 1999; Kidd *et al.*, 1999; Li *et al.*, 1999a; Wu *et al.*, 1999). In the nervous system, slit protein is expressed in the floor plate of the neural tube, with some isoforms being expressed outside the nervous system as well (for review, see Wong *et al.*, 2002). The Roundabout (Robo) class of receptors was similarly identified in a *Drosophila* genetic screen for mutations affecting axon guidance at the midline (Seeger *et al.*, 1993). In the absence of Robo, axons can traverse the ventral midline heedless of the inhibitory slit activity in this domain (Kidd *et al.*, 1998; Seeger *et al.*, 1993). Three Robo genes that mediate neuronal guidance have been identified in organisms spanning invertebrates and vertebrates (Wong *et al.*, 2002) and are members of the NCAM family of single-pass transmembrane receptors. It is interesting that slit blocks the attractive cues of netrins by direct biochemical interaction between Robo and the netrin receptor, DCC (Stein and Tessier-Lavigne, 2001). Thus, combinatorial interactions between different receptor signaling systems can dramatically alter the outcome of ligand binding.

The ephrins are membrane-bound ligands for the Eph family of receptor tyrosine kinases and mediate short-range interactions requiring cell–cell contact (Cheng *et al.*, 1995; Drescher *et al.*, 1995; Wilkinson, 2001). The outcome of signaling between the receptor-bearing axon and apposing ephrin ligand-expressing cells is contact-mediated repulsion due to a collapse of the cytoskeleton of the growth cone (for review, see Frisen *et al.*, 1999; Holder and Klein, 1999; Klein, 2001; Mellitzer *et al.*, 2000; Orioli and Klein, 1997; Schmucker and Zipursky, 2001). The effect of Eph:ephrin signaling on axon navigation and patterning has been well documented via topographic mapping of retinal axons invading the tectum (Feldheim *et al.*, 2000; Knoll and Drescher, 2002; McLaughlin *et al.*, 2003). It is believed that gradients of cell membrane-bound ephrins provide a code for regulating axon migration. Wang and Anderson (1997) demonstrated that axons, expressing Eph receptors, avoid growth into the caudal half of somites where ephrins B1 and B2 are expressed. In effect, flanking domains of complementary Eph:ephrin expression can function to channel growth of a migratory population through specific permissive growth corridors. It is important to remember that the designation of ephrins as ligands is somewhat misleading, given that these membrane-bound molecules can support "reverse" signaling in the cell on which they are expressed, while simultaneously initiating a signaling cascade by activation of their cognate Eph receptor expressed on adjacent cells (Kullander and Klein, 2002). Modulation of both "forward" and "reverse" signaling, in addition to combinatorial interactions between different subclasses of ephrins and Ephs, is critical for the precise targeting of retinal axons to the correct anterior–posterior and dorsal–ventral address along the optic tectum in vertebrates (Brown et al., 2000; Feldheim et al., 2000; Hindges et al., 2002; Mann et al., 2002; Wilkinson, 2001). The ephrin/Ephs mediate both repulsion and attraction of axons in many other regions of the developing embryo as well.

IX. Neural and Vascular Guidance Mechanisms Share Common Signaling Pathways

Recently, evidence for the expression of several known neural guidance factors in the vascular system has emerged, and the functional relevance of these data is just beginning to be addressed. Neuropilin-1 was originally recognized as a neuronal cell-surface protein, the ligands for which are a family of secreted glycoproteins, the semaphorins, that function as chemorepellents for a subset of embryonic nerves (He and Tessier-Lavigne, 1997; Kolodkin et al., 1997; Raper, 2000). Klagsbrun and coworkers (Soker et al., 1998) were the first to show that Neuropilin-1 was also expressed on the surface of vascular endothelial cells and functioned as an accessory receptor to facilitate binding of a specific subset of VEGF isoforms to the canonical VEGF receptor, Flk1. This heralded our first mechanistic insight of the potential interconnectedness of the nervous and vascular systems. Evidence for the involvement of Neuropilin-1 in vascular development comes from numerous studies in the last few years. In addition to peripheral nerve defects, *neuropilin-1*-deficient embryos succumb to cardiovascular defects, including transposed axial vessels, reduced capillary branching, and neural tube vascularization (Kawasaki et al., 1999). These defects by no means phenocopy the severe vasculogenic defects observed in *flk1*−/− mice, suggesting that even though Neuropilin-1 enhances binding of VEGF to Flk1, its function appears to be specific to aspects of guidance. Accordingly, overexpression of Neuropilin-1 resulted in ectopic capillary growth and vessel dilation, in addition to neural defects (Kitsukawa et al., 1995). Additional neuropilin family members have been identified and also shown to exhibit VEGF isoform binding restrictions (Gagnon et al., 2000; Gluzman-Poltorak et al., 2000). With the identification of the neuropilins and their specificity for a restricted subset of VEGF isoforms, Shima and Mailhos (2000) suggest that the existence of multiple VEGF isoforms may now be functionally relevant in two possible ways. Combinatorial interactions between Neuropilin-1 and the canonical

4. Wiring the Vascular Circuitry

VEGF receptors, Flt1 and Flk1, may allow isoform-specific signals to be transduced. It could also be envisioned that molecules such as Neuropilin-1 could function to sequester specific VEGF isoforms to defined locales in the extracellular matrix, thereby regulating VEGF bioavailability (Houck et al., 1992) and perhaps conferring directional growth cues.

It appears that the alternative Neuropilin-1 ligands, the Semaphorins, can also modulate vascular, as well as neural, development. Miao et al. (1999) have reported that semaphorin III can compete with VEGF^{165} for binding to the neuropilin-1 receptor and can mediate a collapsin-like retraction of lamellopodia. Similarly, it is capable of disrupting the cytoskeleton of vascular endothelial cells, thereby reducing motility, reminiscent of its effect on axonal growth cones (Luo et al., 1993; Nakamura et al., 1998). Serini et al. (2003) also implicate autocrine signaling by *semaphorin-3A* to be responsible for blocking integrin-mediated endothelial cell adhesion. Furthermore, antisense knockdown of *semaphorin-3A* in zebrafish embryos severely abrogated angioblast migration, resulting in retarded development of the dorsal aorta and reduced blood flow (Shoji et al., 2003). In mice lacking semaphorin-3A, the primary defects involve neural morphogenesis; however, vascular defects that include ectopic sprouting of intersomitic vessels into the somites occur with variable penetrance (Behar et al., 1996; Serini et al., 2003; Taniguchi et al., 1997). Bates and coworkers (2003) demonstrated that semaphorin-3A-soaked beads implanted into quail embryo forelimbs cause vessels and nerves to sprout away from the beads. Collectively, these data would suggest that Neuropilin-1 binds two ligands that compete with each other and mediate opposing outcomes. However, a recent report calls into question this conclusion. Gu et al. (2003) designed an *in vivo* experiment to determine if the binding of Semaphorin-3A to Neuropilin-1 is essential for vascular development. They selectively ablated the Semaphorin-3A-binding domain in Neuropilin-1, leaving only the VEGF^{165}-binding domain intact. These knock-in mice exhibited no peripheral vascular defects but numerous axonal path-finding defects. Thus, even though semaphorin-3A and VEGF have overlapping expression patterns, semaphorin-3A does not appear to modulate vascular development via its interaction with neuropilin-1. How do these data reconcile with the vascular defects described in semaphorin mutants elsewhere? It is likely that semaphorins do indeed modulate endothelial cell activity but through an alternate transducing receptor. Neuropilin-2 is an unlikely candidate as *neuropilin-2* targeted mice are viable; however, in addition to their interaction with the neuropilin receptors, semaphorins can bind a second class of axonal guidance receptors, the plexin family (Ohta et al., 1995; Takahashi et al., 1999; Tamagnone et al., 1999; Winberg et al., 1998). It is interesting that plexin-D1 was recently shown to be expressed in both the nervous and vascular systems (van der Zwaag et al., 2002). Whether semaphorins mediate endothelial cell function via plexinD1 awaits functional studies.

Although Notch signaling has been predominantly associated with cell-fate determination, there is evidence suggesting that this pathway may play a role in the coordination of cellular movements during tissue morphogenesis. Indeed, Franklin et al. (1999) have shown that Notch signaling regulates neurite outgrowth in cultured mammalian neuroblastoma cells. Similarly, Redmond et al. (2000) and Sestan et al. (1999) have observed inhibition of neurite outgrowth by activated Notch signaling in mammalian cortical neurons. Finally, expression of Notch receptors and associated ligands have been shown to guide neuronal pathfinding in *Drosophila* (Giniger, 1998; Giniger et al., 1993; Hassan et al., 2000). In most cases, Notch signaling has been shown to exhibit a negative impact on neurite extension in both *in vitro* and *in vivo* models (for review, see Frisen and Lendahl, 2001). As discussed previously, the Notch pathway is implicated in arterial and venous identity during vasculogenesis, though this role has not been confirmed in mammalian studies. Recently, evidence has emerged to suggest that the Notch pathway may be important for regulating endothelial sprouting during angiogenesis. Mice homozygous for null mutations in Notch family genes show embryonic lethality and exhibit defects in angiogenic remodeling. Targeted inactivation of *Jagged-1*, a Notch ligand, results in abnormal remodeling of the yolk sac and head vasculature (Xue et al., 1999), and double *Notch1*−/−, *Notch4*−/− mutants display severe angiogenic defects (Krebs et al., 2000). Overexpression of a constitutively active *Notch-4* allele in endothelial cells inhibits sprouting to VEGF in *in vitro* angiogenesis assays and results in enhanced beta-1-integrin-mediated adhesion (Leong et al., 2002). Thus, Notch signaling may play a critical role in guiding vascular endothelial cell and neurite outgrowth.

The ephrin/Eph classes of proteins appear to play a similar role in guiding vascular endothelial cells as they do in neuronal pathfinding. Juxtacrine interactions between intersomitic vessels expressing Eph receptors, and the adjacent somitic mesenchyme expressing ephrin ligands, likely represent a repulsive guidance mechanism that restricts vessel growth to an intersomitic path. Similarly, the role of these proteins in delimiting arterial-venous domains described previously may relate to their ability to provide repulsive cues that prevent arteries and veins from fusing. Adams et al. (1999) described aberrant invasion of intersomitic vessels that normally channel between, not within, somites, in *EphB2/EphB3* knockout mice. EphrinB2 and EphB2, which are strongly expressed in the somitic compartment, may mediate this apparent repulsion via signaling to the appropriate partner receptors expressed on the endothelial cell surface. Similarly, Helbling et al. (2000) reported the expression of EphB4 in Xenopus intersomitic veins and reciprocal expression of the cognate ephrin-B class ligands in somites that flank the path taken by the intersomitic vessels during angiogenesis. Disruption of EphB4 by injection of dominant negative mRNAs, or by ectopic expression of ephrin ligands, also caused intersomitic veins to grow into the somites. This, too, suggests that ephrin:Eph signaling may mediate repulsive guidance cues. It is interesting that Graef *et al.* (2001) have reported lethal vascular defects in mice lacking NFATc4 and NFATc3 that are similar to the defects observed in *EphB3/EphB4* double mutants. They observed normal vasculogenic development, but highly arborized intersomitic vessels that project fanlike into the somites, suggesting a loss of boundary inhibition. The NFAT transcription factors transduce Ca^{2+} signals generated by MAP kinase activation. Given that receptor tyrosine kinases, such as the Ephs, are known to signal through MAP kinases, the Eph class of receptors are candidates for this NFAT-mediated signaling pathway. The overlapping expression of several Eph receptors and NFATs further encourages examination of this possibility.

On the basis of these observations, it is likely that members of the ephrin/Eph family are performing similar functions in vasculature patterning as they do in the nervous system, where complementary ligand-receptor domains of expression provide guidance cues by defining spatial boundaries for permissive growth. However, in addition to constraining vessel growth to defined pathways, Eph:ephrin signaling may also provide positive directional cues. Adams *et al.* (1999) showed that ephrin-B1 and ephrin-B2 had stimulatory effects on adrenal-cortex-derived (ACE) microvascular endothelial cells in an *in vitro* sprouting angiogenesis assay, with similar potency to VEGF. Similarly, Pandey *et al.* (1995) also observed sprouting of endothelial cells in the presence of ephrin-A1. These experiments provide the first documentation for a direct migratory response of endothelial cells to ephrins.

The most recent neural guidance gene family to be implicated in vascular development is that of the Robo receptor signaling pathway. A fourth member of this family was identified in mice and zebrafish, dubbed Magic Roundabout, or Robo4 (Huminiecki *et al.*, 2002; Park *et al.*, 2003). Our laboratory identified *Robo4* on the basis of its reduced expression in *Alk1*−/− mice that showed defects in vascular sprouting. *Robo4* is quite divergent in sequence from the other members of this family and, in stark contrast to other Robo receptors, is specifically expressed in the murine vasculature. We demonstrated that Robo4 binds Slit proteins and that this interaction leads to inhibition of migration, but does not affect proliferation, in heterologous cells or endothelial cells expressing Robo4 (Park *et al.*, 2003). Therefore, analogous to the role of the other Robo receptors in the nervous system, Robo4 signaling may have an inhibitory effect on vessel sprouting. The role of Robo4 in vascular development has not been characterized. Our laboratory is examining a conditional knockout of murine *Robo4* to better assess the function of this receptor *in vivo*. It is interesting that others report that Slit is chemotactic to endothelial cells and promotes tube formation via the Robo1 receptor. Surprisingly, in contrast to previous studies, they report that Slit inhibits migration of heterologous cells expressing Robo1. Although axonal path-finding defects have been reported, no vascular defects have been described in either *Slit1/Slit2* double knockout mice or in Slit3 knockout mice (Plump et al., 2002). Given their embryonic lethality and the specific expression of Robo4 in the vasculature, these mutants should be re-examined for defective angiogenesis.

We have catalogued some of the neural guidance mechanisms that have been shown to be important for vessel growth and guidance. The weight of evidence is such that it is hard to imagine other neural guidance factors not affecting angiogenesis. Conspicuously absent from vascular characterization are the netrins. The netrins are a family of neural guidance factors that interact with a number of cognate receptors including DCC, neogenin, unc5H2, and Unc5H3 (Grunwald and Klein, 2002). The netrins are attractive cues for axons; however, depending on the specific combination of receptors that predominant in a cell, netrins can also act as a repulsive cue. Preliminary *in vitro* data from our laboratory indicate that netrin-1 stimulates endothelial and vascular smooth muscle cell proliferation and migration, and induces vessel sprouting in *ex vivo* models of angiogenesis. At present, the receptors responsible for this signaling are not defined. The netrin receptor, Unc5H2, is highly expressed in the embryonic vasculature in midgestation (Engelkamp, 2002). Since Unc5H2 acts as a repulsive signaling receptor in neural guidance, one might expect exuberant angiogenesis in mice lacking this receptor. Though our laboratory has observed expression of neogenin in vascular cells in culture, at present there have been no reports of embryonic expression of neogenin and DCC *in vivo*. Complicating our understanding of netrin signaling is the recent observation that netrin can also activate the integrin family of receptors (Yebra et al., 2003). It will be important to define the expression and activities of each netrin ligand and receptor family member in the vascular system using both *in vitro* and *in vivo* models. Thus, many of the known neural guidance cues and signaling receptor pathways are expressed in the vascular system and may play analogous roles in vessel guidance during angiogenesis.

X. Neuro-Vascular Development as Interdependent Processes

The interdependence of organ systems has become increasingly apparent as our molecular understanding of developmental processes expands. Recent evidence suggests that the vascular system is not an autonomous network; rather, it can be influenced by, and in turn influence, other operations. The observation of colinear nerve:vessel tracts gives the impression of interdependent patterning mechanisms. Given the obvious similarities between vessel and nerve networks at the anatomic and molecular levels, it is not surprising that there is a requirement for some degree of coordination between them. Both systems exhibit parallel patterning and act globally throughout the body in a highly ordered and coordinated fashion (Bates et al., 2002; Feig and Guillery, 2000; Taylor et al., 2001; Wheater, 1985). Both nerves and sprouting vessels employ specialized filopodial extensions that allow them to sample the microenvironment for road signs directing their journey to target tissues (Gerhardt et al., 2003). Analogous to the two-unit cell types of the vascular system, mural and endothelial, nerve bundles consist of neurons surrounded by supporting glial cells. Moreover, an integrated neural network is comprised of two primary divisions, the sensory and motor nerves, comparable in effect to the two subdivisions of the vascular system, the arterial and venous networks. Finally, both utilize a similar array of regulatory factors during the cell fate specification and differentiation phases of development (for a comprehensive review of this topic, see Carmeliet, 2003b; Cleaver and Melton, 2003). As discussed in the previous section, there is substantial evidence to indicate that the genetic wiring for axon outgrowth and vessel navigation makes use of overlapping signaling factors. In addition to shared intrinsic regulatory mechanisms, there is also evidence to suggest that extrinsic cross talk occurs between nerves and vessels.

In addition to the anatomical similarities between the nerve and vascular networks, there is well-established evidence for an interdependent physical relationship. Autonomic nerves control vascular tone through vasodilation and constriction (Kummer and Haberberger, 1999; Zukowska-Grojec et al., 1993) and can regulate injury-induced inflammation by the release of neuropeptides into adjacent vessels (Baluk et al., 1997). Much evidence has accumulated to suggest that nerves can modulate angiogenesis. Denervation is known to result in reduced vascularization (Borisov et al., 2000), and angiogenesis occurs around nerve cell grafts (Krum and Rosenstein, 1998), spinal lesions (Bartholdi et al., 1997), and in zones of regenerating peripheral nerves (Hobson et al., 1997). Many, if not all, of these activities are likely regulated by the release of the endothelial chemoattractant, VEGF. Calza et al. (2001) have shown that VEGF released from neurons induces neoangiogenesis in the superior cervical ganglia of neonatal rats. Since this time, many additional studies have also linked VEGF expression by the nervous system to vessel growth modulation. Reciprocally, vessels provide oxygen and nutrients, and neurotrophic factors that sustain nerve viability (Donovan et al., 2000; Francis and Landis, 1999). Thus, there is a symbiotic relationship between vessels and nerves within a neurovascular bundle.

Recently, paracrine regulation of arterial vessel patterning by peripheral nerves has been established by Mukouyama and coworkers (2002). In the limb skin of embryonic mice, there is a close association of arterioles with peripheral nerves. Vascular plexus formation precedes invasion by peripheral nerves, but it is only after innervation that the vessels (1) remodel to become closely juxtaposed with nerve fibers and (2) begin to express artery-specific markers such as ephrinB2, neuropilin-1, and connexin 40. Mutations that abolish peripheral nerve or Schwann cell development cause reduced arteriogenesis, as assessed by the expression of artery-specific markers. In vitro assays of co-cultured nerve or Schwann cells with endothelial cells further corroborated these findings and implicated $VEGF^{120}$ and $VEGF^{165}$ expression from nervous cells as the mediators of this induced differentiation activity. Application of a soluble Flk-1 receptor inhibited this induction. In order to determine if the peripheral nerves can direct patterning of the arterial vessels, in addition to differentiation, *semaphorin-3A*−/− mice were examined. These mice exhibited aberrant nerve fiber patterning due to loss of semaphorin and vessel trajectories changed in kind to follow the aberrantly routed nerves. In mice lacking Schwann cells as a result of *erbB3* ablation (Riethmacher et al., 1997), the nerve fibers are still present, yet Mukouyama et al. (2002) observed greatly reduced colinearity with vessels. Given that ErbB3 is not expressed in endothelial cells, this effect was not likely due to cell-autonomous effects. The results of this study suggest that at least a subset of nerves can play dual roles in the differentiation of arterial identity as well as the guidance of arterial vessels. It is worth noting that colinearity was not apparent between venous vessels and nerves. Given that veins also express VEGF receptors, if VEGF is the mediator of this activity, why is it that veins are not similarly patterned? Potentially, the expression level of receptors may differ between the two endothelial populations, or a different combination of co-receptors may be expressed.

Gerhardt et al. (2003) have elegantly demonstrated the role of astrocytes in instructing vascular patterning in the retina. The inner retinal vascular plexus develops in close proximity to an underlying layer of astrocytes (Fruttiger, 2002; Stone and Dreher, 1987). Gerhardt and coworkers posit that the astrocytic layer serves as a scaffold for vascular outgrowth based on the observation that the polarized endothelial cells elaborate long filopodial extensions that closely associate with astrocyte cell bodies. Transgenic manipulation causes ectopic astrocyte plexi to coordinately induce reorientation of endothelial cells to these foci. Astrocytes produce heparin-binding $VEGF^{164}$ that is deposited in the astrocyte tracks and stimulates filopodial extensions and subsequent endothelial migration. Mice that only express soluble $VEGF^{120}$ show no association between the astrocyte plexus and the vascular plexus, though filopodia were still present and the endothelial cells of the vascular plexus continued to proliferate. However, endothelial filopodia extended radially rather than directionally toward the astrocyte layer. The observation of uniform vessel growth, in the absence of polarity, suggests that the $VEGF^{164}$ emanating from astrocytes normally serves as a directional guidance cue.

Just as there is evidence for the patterning of vessels by nerves, vessels can, in turn, dictate neural cell fates and guide neural outgrowth. There are

4. Wiring the Vascular Circuitry

numerous reports suggesting that vessels lay down tracks of extracellular proteins such as artemin, fibronectin, and β-netrin, on which nerves migrate, but these studies are largely correlative in nature (Koch *et al.*, 2000; Spence and Poole, 1994). Louissaint *et al.* (2002) have shown that testosterone causes up-regulation of VEGF and Flk-1 receptor in the higher vocal center (HVC) of adult songbirds throughout life. This pro-angiogenic state incites angiogenesis, followed by neurogenesis. In order to test whether these two pathways were interconnected, songbirds were treated with small molecule inhibitors of Flk-1, which resulted in significant reductions in both angiogenesis and neurogenesis. Furthermore, it was shown that FACS-isolated endothelial cells from the HVC produced a brain-derived neurotrophic factor (BDNF) that was active on HVC neurons. The authors postulate the endothelial cell-derived BDNF functions both as a neurotrophic factor as well as a chemoattractant to spatio-temporally attract nerves to microdomains of the HVC where angiogenesis is most active.

Evidence suggests that and axonal outgrowth of sympathetic nerves relies upon closely juxtapositioned vessels. Two studies have linked the activation of neuronal RET tyrosine kinase receptor by the vessel-derived artemin (ARTN) ligand to this relationship. ARTN was first identified by Baloh *et al.* (1998) as the ligand for the neural RET tyrosine kinase receptor. Durbec *et al.* (1996) reported that mice lacking the RET receptor show complete loss of all superior cervical ganglion (SCG) neurons by birth. Enomoto and colleagues (2001) have recently generated a GFP-marked allele of RET in order to precisely follow the fate of neurons in this knockout mouse. Contrary to the previous report, they find that the sympathetic ganglia are not absent but rather reduced in size, and their location is altered. Given that neurotrophic factors, such as nerve growth factor, are not expressed at the early stages of sympathetic ganglia development (Fagan *et al.*, 1996), Enomoto and colleagues suggested that ARTN may be critical for RET activation. ARTN is expressed at the right time and place at high levels in the wall of the dorsal aorta as well as in other vessels of the E12 embryo. They further show that ARTN can induce neurite outgrowth in explants of sympathetic ganglia in a dose-dependent fashion and that neurites will move directionally toward ARTN-soaked beads. These data strongly suggest that the vessel-derived ARTN functions as a guidance factor for neurons. Extending this study, Honma *et al.* (2002) have ablated *artemin* in mice by homologous recombination and observed disrupted migration and axonal projections of sympathetic neuroblasts to their target tissues. Little alteration of cell survival or proliferation was observed, and the size of the SCG was comparable to that of wild-type littermates, suggesting that ARTN functions to specifically target axons rather than as a neurotrophic factor that permits survival. Sympathetic neuroblasts were observed to extend axons in *ARTN−/−* cultured embryos that had been implanted with

ARTN-soaked beads. It would be informative to conduct the same experiment in *RET−/−* mice to determine if, as expected, this "rescue" is lost in the absence of the putative cognate receptor for ARTN. It is interesting that these authors demonstrate that ARTN is expressed predominantly in the mural cell layer of vessels. Vascular smooth muscle cells become consolidated along maturing vessels in a progressive and directional fashion; thus, it has been suggested that VSMC-derived ARTN guides sympathetic neurons to their innervation targets.

Lending further credence to the concept of neuro-vascular interdependence, Chi *et al.* (2003) have exhaustively catalogued gene expression in 50 different endothelial cell sources, from arteries, veins, and large vessels to microvessels. The microvasculature, in particular, express a wide array of secreted neural differentiation and survival factors, as well as several receptors for paracrine factors produced by neuroglial cells. These data suggest an intimate functional association between the capillary vasculature and peripheral nerves, as previously demonstrated in *in vivo* studies by Mukouyama *et al.* (2002). Larger vessels were also found to express a number of neural genes, including *Robo1, neuroregulin*, and its receptor *ErbB*. Future functional studies in the intact vasculature will shed light on whether these factors are important for neurovascular patterning.

XI. Independent Versus Interdependent Guidance of Vessels and Nerves

We have described at length how the vascular system uses signaling mechanisms originally described for neural guidance. Similarly, signaling mechanisms initially described in the vascular system can play important roles in neural guidance. As an example, numerous expression studies indicate that Schwann cells and neurons, or their precursors, neuroblasts, express receptors for VEGF (flk1 and neuropilin-1) and themselves produce VEGF. VEGF can serve as a trophic factor both during the process of neurogenesis as well as axonal outgrowth and may block the inhibitory effects of semaphorins by competing for the neuropilin-1 receptor (Bagnard *et al.*, 2001; Gu *et al.*, 2003; Honma *et al.*, 2002; Jin *et al.*, 2002; Miao *et al.*, 1999; Oosthuyse *et al.*, 2001; Schratzberger *et al.*, 2000; Sondell *et al.*, 1999, 2000; Zhang *et al.*, 2003; Zhu *et al.*, 2003). Thus, in some circumstances, the neural and vascular systems appear to utilize common factors to coordinately regulate development and guidance, without necessarily invoking direct cross talk between nerves and vessels. Bates *et al.* (2003) have documented shared, but non-obligate patterning mechanisms. Peripheral nerves and blood vessels have highly congruent and stereotypical patterns in the developing quail forelimb; however, via nerve ablations and forced aberrant vascularization, they show that there is no apparent interdependence between nerve and vessel outgrowth. Martin and Lewis (1989) reported a similar finding in embryonic chick skin. Thus, there is ample evidence for independent as well as interdependent guidance signals between vessels and nerves.

XII. Future Directions in Vascular Guidance

It is increasingly clear that in the neural system there are complex interactions between multiple ligands and receptors that provide directional information to an axon. It will be important to determine whether such interactions between analogous ligands and receptors in the vascular system are critical for vascular guidance. Moreover, as the neural field moves in the direction of precisely defining the terms *inhibition* vs. *repulsion*, and *attraction* vs. *mitogen stimulation*, so too is the vascular field. To keep pace with this growing complexity, vascular biologists need to develop tools similar to those in the neural field to distinguish between the spectrum of attractive and repulsive cues. Most vascular assays such as the aortic explant, chorioallantoic membrane, and corneal implant assays define whether a factor has pro- or antiangiogenic effects. These *in vitro* assays, in combination with real-time microscopy, could be modified to determine whether vessels can be guided toward an attractive cue along specific paths delineated by a corridor of repulsive cues. Similar assays, ganglia explant and strip assays, have been employed to distinguish the ability of factors to repulse and redirect axon sprouting, enhance branching, or inhibit sprouting. Thus, though there are no repulsive guidance factors that have yet been reported for the vasculature, this may be largely attributed to the fact that we have not assayed for this phenomenon. Nonetheless, the fact that known repulsive cues employed by the nervous system are also important to angiogenesis suggests that repulsive vascular cues do exist.

XIII. Summary

The conceptual framework for understanding the development of the vascular system has evolved dramatically. The field was initially focused on understanding the signaling pathways that regulate the activity of the individual cellular components of blood vessels. The focus shifted to understanding the interaction of these cells to form a vessel and to appreciating the heterogeneity of vessels. Most recently, it has become clear that guidance mechanisms are critical for forming an organized vascular network. We are just beginning to appreciate the complexity of function for such long-studied vascular factors as VEGF. Not only does VEGF serve as a global endothelial cell mitogen, recent data have also defined a role for VEGF isoforms in directional guidance, as well as in the preferential induction of arteriogenesis. Clearly, the stage is set for rapid advances in our understanding of the complex morphogenetic events that give rise to the vascular circuit, thanks in part to studies of the nervous system from which we may extrapolate analogous mechanisms. The architectural, as well as functional, parallels between the vascular and nervous systems foreshadowed the identification of regulatory parallels. Since the late 1990s, similar signaling molecules for the regulation of differentiation, growth, and navigation have been defined for both nerves and vessels. It appears that during the course of evolution, successful regulatory mechanisms were co-opted and recycled for patterning of these functionally unique, but structurally analogous, systems. The search for, and characterization of, vascular-specific "neural guidance analogs" promises to be an exciting venue for research in vascular development in the years to come. Though it is clear that several patterning factors are shared between the nervous and vascular systems, it is also anticipated that unique molecules may exist to specifically pattern subpopulations of endothelial cells or their derivatives, such as arterial or venous vessels, organ-specific vascular networks, and the lymphatic system.

In July 2004, two independent laboratories reported the characterization *in vivo* models for null or hypomorphic alleles of the endothelial cell-specific semaphorin receptor, Plexin-D1. Torres-Vazquez *et al.* (2004) determined that *out of bounds* mutant zebrafish harbor mutations in the Plexin-D1 gene, and that injection of knock-down morpholinos targeting this transcript causes similar intersegmental vessel patterning defects. Gitler *et al.* (2004) report that *Plexin-D1* targeted mice die shortly after birth and display coronary and aortic arch defects, as well as inappropriate invasion of intersomitic vessels into semaphorin-expressing somitic mesenchyme. Together, these data suggest that interaction of semaphorins with a Plexin-D1:neuropilin complex can orchestrate vessel path-finding and that the semaphorins may function as repellent cues in both vessel and axonal guidance.

Acknowledgments

This work was funded by the National Institutes of Health and the American Cancer Society.

References

Adams, R. H. (2003). Molecular control of arterial-venous blood vessel identity. *J. Anat.* **202**, 105–112.

Adams, R. H., Diella, F., Hennig, S., Helmbacher, F., Deutsch, U., and Klein, R. (2001). The cytoplasmic domain of the ligand ephrinB2 is required for vascular morphogenesis but not cranial neural crest migration. *Cell* **104**, 57–69.

Adams, R. H., Wilkinson, G. A., Weiss, C., Diella, F., Gale, N. W., Deutsch, U., Risau, W., and Klein, R. (1999). Roles of ephrinB ligands and EphB receptors in cardiovascular development: Demarcation of arterial/venous domains, vascular morphogenesis, and sprouting angiogenesis. *Genes Dev.* **13,** 295–306.

Artavanis-Tsakonas, S., Rand, M. D., and Lake, R. J. (1999). Notch signaling: Cell fate control and signal integration in development. *Science* **284,** 770–776.

Bagnard, D., Vaillant, C., Khuth, S. T., Dufay, N., Lohrum, M., Puschel, A. W., Belin, M. F., Bolz, J., and Thomasset, N. (2001). Semaphorin 3A-vascular endothelial growth factor-165 balance mediates migration and apoptosis of neural progenitor cells by the recruitment of shared receptor. *J. Neurosci.* **21,** 3332–3341.

Baloh, R. H., Tansey, M. G., Lampe, P. A., Fahrner, T. J., Enomoto, H., Simburger, K. S., Leitner, M. L., Araki, T., Johnson, E. M., Jr., and Milbrandt, J. (1998). Artemin, a novel member of the GDNF ligand family, supports peripheral and central neurons and signals through the GFRalpha3-RET receptor complex. *Neuron* **21,** 1291–1302.

Baluk, P., Bowden, J. J., Lefevre, P. M., and McDonald, D. M. (1997). Upregulation of substance P receptors in angiogenesis associated with chronic airway inflammation in rats. *Am. J. Physiol.* **273,** L565–L571.

Bartholdi, D., Rubin, B. P., and Schwab, M. E. (1997). VEGF mRNA induction correlates with changes in the vascular architecture upon spinal cord damage in the rat. *Eur. J. Neurosci.* **9,** 2549–2560.

Bates, D., Taylor, G. I., Minichiello, J., Farlie, P., Cichowitz, A., Watson, N., Klagsbrun, M., Mamluk, R., and Newgreen, D. F. (2003). Neurovascular congruence results from a shared patterning mechanism that utilizes Semaphorin3A and Neuropilin-1. *Dev. Biol.* **255,** 77–98.

Bates, D., Taylor, G. I., and Newgreen, D. F. (2002). The pattern of neurovascular development in the forelimb of the quail embryo. *Dev. Biol.* **249,** 300–320.

Battye, R., Stevens, A., and Jacobs, J. R. (1999). Axon repulsion from the midline of the Drosophila CNS requires slit function. *Development* **126,** 2475–2481.

Behar, O., Golden, J. A., Mashimo, H., Schoen, F. J., and Fishman, M. C. (1996). Semaphorin III is needed for normal patterning and growth of nerves, bones and heart. *Nature* **383,** 525–528.

Betsholtz, C., Karlsson, L., and Lindahl, P. (2001). Developmental roles of platelet-derived growth factors. *Bioessays* **23,** 494–507.

Borisov, A. B., Huang, S. K., and Carlson, B. M. (2000). Remodeling of the vascular bed and progressive loss of capillaries in denervated skeletal muscle. *Anat. Rec.* **258,** 292–304.

Bostrom, H., Willetts, K., Pekny, M., Leveen, P., Lindahl, P., Hedstrand, H., Pekna, M., Hellstrom, M., Gebre-Medhin, S., Schalling, M., Nilsson, M., Kurland, S., Tornell, J., Heath, J. K., and Betsholtz, C. (1996). PDGF-A signaling is a critical event in lung alveolar myofibroblast development and alveogenesis. *Cell* **85,** 863–873.

Brose, K., Bland, K. S., Wang, K. H., Arnott, D., Henzel, W., Goodman, C. S., Tessier-Lavigne, M., and Kidd, T. (1999). Slit proteins bind Robo receptors and have an evolutionarily conserved role in repulsive axon guidance. *Cell* **96,** 795–806.

Brown, A., Yates, P. A., Burrola, P., Ortuno, D., Vaidya, A., Jessell, T. M., Pfaff, S. L., O'Leary, D. D., and Lemke, G. (2000). Topographic mapping from the retina to the midbrain is controlled by relative but not absolute levels of EphA receptor signaling. *Cell* **102,** 77–88.

Calza, L., Giardino, L., Giuliani, A., Aloe, L., and Levi-Montalcini, R. (2001). Nerve growth factor control of neuronal expression of angiogenetic and vasoactive factors. *Proc. Natl. Acad. Sci. USA* **98,** 4160–4165.

Carmeliet, P. (2000). VEGF gene therapy: Stimulating angiogenesis or angioma-genesis? *Nat. Med.* **6,** 1102–1103.

Carmeliet, P. (2003a). Angiogenesis in health and disease. *Nat. Med.* **9,** 653–660.

Carmeliet, P. (2003b). Blood vessels and nerves: Common signals, pathways and diseases. *Nat. Rev. Genet.* **4**, 710–720.

Chen, J. H., Wu, W., Li, H. S., Fagaly, T., Zhou, L., Wu, J. Y., and Rao, Y. (2000). Embryonic expression and extracellular secretion of *Xenopus* slit. *Neuroscience* **96**, 231–236.

Cheng, H. J., Bagri, A., Yaron, A., Stein, E., Pleasure, S. J., and Tessier-Lavigne, M. (2001). Plexin-A3 mediates semaphorin signaling and regulates the development of hippocampal axonal projections. *Neuron* **32**, 249–263.

Cheng, H. J., Nakamoto, M., Bergemann, A. D., and Flanagan, J. G. (1995). Complementary gradients in expression and binding of ELF-1 and Mek4 in development of the topographic retinotectal projection map. *Cell* **82**, 371–381.

Chi, J. T., Chang, H. Y., Haraldsen, G., Jahnsen, F. L., Troyanskaya, O. G., Chang, D. S., Wang, Z., Rockson, S. G., van de Rijn, M., Botstein, D., *et al.* (2003). Endothelial cell diversity revealed by global expression profiling. *Proc. Natl. Acad. Sci. USA* **100**, 10623–10628.

Childs, S., Chen, J. N., Garrity, D. M., and Fishman, M. C. (2002). Patterning of angiogenesis in the zebrafish embryo. *Development* **129**, 973–982.

Cleaver, O., and Melton, D. A. (2003). Endothelial signaling during development. *Nat. Med.* **9**, 661–668.

Culotti, J. G., and Merz, D. C. (1998). DCC and netrins. *Curr. Opin. Cell Biol.* **10**, 609–613.

D'Amore, P. A., and Ng, Y. S. (2002). Won't you be my neighbor? Local induction of arteriogenesis *Cell* **110**, 289–292.

Del Amo, F. F., Smith, D. E., Swiatek, P. J., Gendron-Maguire, M., Greenspan, R. J., McMahon, A. P., and Gridley, T. (1992). Expression pattern of Motch, a mouse homolog of *Drosophila* Notch, suggests an important role in early postimplantation mouse development. *Development* **115**, 737–744.

Dickson, B. J. (2002). Molecular mechanisms of axon guidance. *Science* **298**, 1959–1964.

Donovan, J., Kordylewska, A., Jan, Y. N., and Utset, M. F. (2002). Tetralogy of fallot and other congenital heart defects in Hey2 mutant mice. *Curr. Biol.* **12**, 1605–1610.

Donovan, M. J., Lin, M. I., Wiegn, P., Ringstedt, T., Kraemer, R., Hahn, R., Wang, S., Ibanez, C. F., Rafii, S., and Hempstead, B. L. (2000). Brain derived neurotrophic factor is an endothelial cell survival factor required for intramyocardial vessel stabilization. *Development* **127**, 4531–4540.

Dor, Y., Djonov, V., and Keshet, E. (2003). Induction of vascular networks in adult organs: implications to proangiogenic therapy. *Ann. NY Acad. Sci.* **995**, 208–216.

Drake, C. J., and Little, C. D. (1995). Exogenous vascular endothelial growth factor induces malformed and hyperfused vessels during embryonic neovascularization. *Proc. Natl. Acad. Sci. USA* **92**, 7657–7661.

Drake, C. J., and Little, C. D. (1999). VEGF and vascular fusion: Implications for normal and pathological vessels. *J. Histochem. Cytochem.* **47**, 1351–1356.

Drescher, U., Kremoser, C., Handwerker, C., Loschinger, J., Noda, M., and Bonhoeffer, F. (1995). *In vitro* guidance of retinal ganglion cell axons by RAGS, a 25 kDa tectal protein related to ligands for Eph receptor tyrosine kinases. *Cell* **82**, 359–370.

Dumont, D. J., Jussila, L., Taipale, J., Lymboussaki, A., Mustonen, T., Pajusola, K., Breitman, M., and Alitalo, K. (1998). Cardiovascular failure in mouse embryos deficient in VEGF receptor-3. *Science* **282**, 946–949.

Durbec, P. L., Larsson-Blomberg, L. B., Schuchardt, A., Costantini, F., and Pachnis, V. (1996). Common origin and developmental dependence on c-ret of subsets of enteric and sympathetic neuroblasts. *Development* **122**, 349–358.

Engelkamp, D. (2002). Cloning of three mouse Unc5 genes and their expression patterns at mid-gestation. *Mech. Dev.* **118**, 191–197.

4. Wiring the Vascular Circuitry

Enomoto, H., Crawford, P. A., Gorodinsky, A., Heuckeroth, R. O., Johnson, E. M., Jr., and Milbrandt, J. (2001). RET signaling is essential for migration, axonal growth and axon guidance of developing sympathetic neurons. *Development* **128,** 3963–3974.

Esko, J. D., and Lindahl, U. (2001). Molecular diversity of heparan sulfate. *J. Clin. Invest.* **108,** 169–173.

Fagan, A. M., Zhang, H., Landis, S., Smeyne, R. J., Silos-Santiago, I., and Barbacid, M. (1996). TrkA, but not TrkC, receptors are essential for survival of sympathetic neurons in vivo. *J. Neurosci.* **16,** 6208–6218.

Feig, S. L., and Guillery, R. W. (2000). Corticothalamic axons contact blood vessels as well as nerve cells in the thalamus. *Eur. J. Neurosci.* **12,** 2195–2198.

Feldheim, D. A., Kim, Y. I., Bergemann, A. D., Frisen, J., Barbacid, M., and Flanagan, J. G. (2000). Genetic analysis of ephrin-A2 and ephrin-A5 shows their requirement in multiple aspects of retinocollicular mapping. *Neuron* **25,** 563–574.

Ferrara, N., Carver-Moore, K., Chen, H., Dowd, M., Lu, L., O'Shea, K. S., Powell-Braxton, L., Hillan, K. J., and Moore, M. W. (1996). Heterozygous embryonic lethality induced by targeted inactivation of the VEGF gene. *Nature* **380,** 439–442.

Ferrara, N., and Henzel, W. J. (1989). Pituitary follicular cells secrete a novel heparin-binding growth factor specific for vascular endothelial cells. *Biochem. Biophys. Res. Commun.* **161,** 851–858.

Flamme, I., Frolich, T., and Risau, W. (1997). Molecular mechanisms of vasculogenesis and embryonic angiogenesis. *J. Cell Physiol.* **173,** 206–210.

Flanagan, J. G., and Vanderhaeghen, P. (1998). The ephrins and Eph receptors in neural development. *Annu. Rev. Neurosci.* **21,** 309–345.

Fong, G. H., Rossant, J., Gertsenstein, M., and Breitman, M. L. (1995). Role of the Flt-1 receptor tyrosine kinase in regulating the assembly of vascular endothelium. *Nature* **376,** 66–70.

Francis, N. J., and Landis, S. C. (1999). Cellular and molecular determinants of sympathetic neuron development. *Annu. Rev. Neurosci.* **22,** 541–566.

Franklin, J. L., Berechid, B. E., Cutting, F. B., Presente, A., Chambers, C. B., Foltz, D. R., Ferreira, A., and Nye, J. S. (1999). Autonomous and non-autonomous regulation of mammalian neurite development by Notch1 and Delta1. *Curr. Biol.* **9,** 1448–1457.

Frisen, J., Holmberg, J., and Barbacid, M. (1999). Ephrins and their Eph receptors: Multitalented directors of embryonic development. *EMBO J.* **18,** 5159–5165.

Frisen, J., and Lendahl, U. (2001). Oh no, Notch again! *Bioessays* **23,** 3–7.

Fruttiger, M. (2002). Development of the mouse retinal vasculature: angiogenesis versus vasculogenesis. *Invest. Ophthalmol. Vis. Sci.* **43,** 522–527.

Gagnon, M. L., Bielenberg, D. R., Gechtman, Z., Miao, H. Q., Takashima, S., Soker, S., and Klagsbrun, M. (2000). Identification of a natural soluble neuropilin-1 that binds vascular endothelial growth factor: In vivo expression and antitumor activity. *Proc. Natl. Acad. Sci. USA* **97,** 2573–2578.

Gale, N. W., Baluk, P., Pan, L., Kwan, M., Holash, J., DeChiara, T. M., McDonald, D. M., and Yancopoulos, G. D. (2001). Ephrin-B2 selectively marks arterial vessels and neovascularization sites in the adult, with expression in both endothelial and smooth-muscle cells. *Dev. Biol.* **230,** 151–160.

Gale, N. W., and Yancopoulos, G. D. (1997). Ephrins and their receptors: A repulsive topic? *Cell Tissue Res.* **290,** 227–241.

Gale, N. W., and Yancopoulos, G. D. (1999). Growth factors acting via endothelial cell-specific receptor tyrosine kinases: VEGFs, angiopoietins, and ephrins in vascular development. *Genes Dev.* **13,** 1055–1066.

Gerety, S. S., and Anderson, D. J. (2002). Cardiovascular ephrinB2 function is essential for embryonic angiogenesis. *Development* **129,** 1397–1410.

Gerety, S. S., Wang, H. U., Chen, Z. F., and Anderson, D. J. (1999). Symmetrical mutant phenotypes of the receptor EphB4 and its specific transmembrane ligand ephrin-B2 in cardiovascular development. *Mol. Cell* **4**, 403–414.

Gerhardt, H., Golding, M., Fruttiger, M., Ruhrberg, C., Lundkvist, A., Abramsson, A., Jeltsch, M., Mitchell, C., Alitalo, K., Shima, D., *et al.* (2003). VEGF guides angiogenic sprouting utilizing endothelial tip cell filopodia. *J. Cell Biol.* **161**, 1163–1177.

Gessler, M., Knobeloch, K. P., Helisch, A., Amann, K., Schumacher, N., Rohde, E., Fischer, A., and Leimeister, C. (2002). Mouse gridlock: No aortic coarctation or deficiency, but fatal cardiac defects in Hey2 −/− mice. *Curr. Biol.* **12**, 1601–1604.

Giniger, E. (1998). A role for Abl in Notch signaling. *Neuron* **20**, 667–681.

Giniger, E., Jan, L. Y., and Jan, Y. N. (1993). Specifying the path of the intersegmental nerve of the Drosophila embryo: A role for Delta and Notch. *Development* **117**, 431–440.

Gitler, A. D., Lu, M. M., and Epstein, J. A. (2004). PlexinD1 and Semaphorin signaling are required in endothelial cells for cardiovascular development. *Dev. Cell* **7**, 107–116.

Glagov, S., Zarins, C., Giddens, D. P., and Ku, D. N. (1988). Hemodynamics and atherosclerosis. Insights and perspectives gained from studies of human arteries. *Arch. Pathol. Lab Med.* **112**, 1018–1031.

Gluzman-Poltorak, Z., Cohen, T., Herzog, Y., and Neufeld, G. (2000). Neuropilin-2 is a receptor for the vascular endothelial growth factor (VEGF) forms VEGF-145 and VEGF-165. *J. Biol. Chem.* **275**, 29922.

Gospodarowicz, D., Moran, J. S., and Braun, D. L. (1977). Control of proliferation of bovine vascular endothelial cells. *J. Cell Physiol.* **91**, 377–385.

Graef, I. A., Chen, F., Chen, L., Kuo, A., and Crabtree, G. R. (2001). Signals transduced by Ca(2+)/calcineurin and NFATc3/c4 pattern the developing vasculature. *Cell* **105**, 863–875.

Gridley, T. (2001). Notch signaling during vascular development. *Proc. Natl. Acad. Sci. USA* **98**, 5377–5378.

Grunwald, I. C., and Klein, R. (2002). Axon guidance: Receptor complexes and signaling mechanisms. *Curr. Opin. Neurobiol.* **12**, 250–259.

Gu, C., Rodriguez, E. R., Reimert, D. V., Shu, T., Fritzsch, B., Richards, L. J., Kolodkin, A. L., and Ginty, D. D. (2003). Neuropilin-1 conveys semaphorin and VEGF signaling during neural and cardiovascular development. *Dev. Cell* **5**, 45–57.

Hacohen, N., Kramer, S., Sutherland, D., Hiromi, Y., and Krasnow, M. A. (1998). sprouty encodes a novel antagonist of FGF signaling that patterns apical branching of the Drosophila airways. *Cell* **92**, 253–263.

Hassan, B. A., Bermingham, N. A., He, Y., Sun, Y., Jan, Y. N., Zoghbi, H. Y., and Bellen, H. J. (2000). Atonal regulates neurite arborization but does not act as a proneural gene in the Drosophila brain. *Neuron* **25**, 549–561.

He, Z., and Tessier-Lavigne, M. (1997). Neuropilin is a receptor for the axonal chemorepellent Semaphorin III. *Cell* **90**, 739–751.

Hedgecock, E. M., Culotti, J. G., and Hall, D. H. (1990). The unc-5, unc-6, and unc-40 genes guide circumferential migrations of pioneer axons and mesodermal cells on the epidermis in *C. elegans*. *Neuron* **4**, 61–85.

Helbling, P. M., Saulnier, D. M., and Brandli, A. W. (2000). The receptor tyrosine kinase EphB4 and ephrin-B ligands restrict angiogenic growth of embryonic veins in *Xenopus laevis*. *Development* **127**, 269–278.

Hellstrom, M., Kalen, M., Lindahl, P., Abramsson, A., and Betsholtz, C. (1999). Role of PDGF-B and PDGFR-beta in recruitment of vascular smooth muscle cells and pericytes during embryonic blood vessel formation in the mouse. *Development* **126**, 3047–3055.

Henderson, A. M., Wang, S. J., Taylor, A. C., Aitkenhead, M., and Hughes, C. C. (2001). The basic helix-loop-helix transcription factor HESR1 regulates endothelial cell tube formation. *J. Biol. Chem.* **276**, 6169–6176.

Herzog, Y., Kalcheim, C., Kahane, N., Reshef, R., and Neufeld, G. (2001). Differential expression of neuropilin-1 and neuropilin-2 in arteries and veins. *Mech. Dev.* **109**, 115–119.

Hindges, R., McLaughlin, T., Genoud, N., Henkemeyer, M., and O'Leary, D. D. (2002). EphB forward signaling controls directional branch extension and arborization required for dorsal-ventral retinotopic mapping. *Neuron* **35**, 475–487.

Hobson, M. I., Brown, R., Green, C. J., and Terenghi, G. (1997). Inter-relationships between angiogenesis and nerve regeneration: A histochemical study. *Br. J. Plast. Surg.* **50**, 125–131.

Holder, N., and Klein, R. (1999). Eph receptors and ephrins: Effectors of morphogenesis. *Development* **126**, 2033–2044.

Hong, K., Hinck, L., Nishiyama, M., Poo, M. M., Tessier-Lavigne, M., and Stein, E. (1999). A ligand-gated association between cytoplasmic domains of UNC5 and DCC family receptors converts netrin-induced growth cone attraction to repulsion. *Cell* **97**, 927–941.

Honma, Y., Araki, T., Gianino, S., Bruce, A., Heuckeroth, R., Johnson, E., and Milbrandt, J. (2002). Artemin is a vascular-derived neurotropic factor for developing sympathetic neurons. *Neuron* **35**, 267–282.

Houck, K. A., Leung, D. W., Rowland, A. M., Winer, J., and Ferrara, N. (1992). Dual regulation of vascular endothelial growth factor bioavailability by genetic and proteolytic mechanisms. *J. Biol. Chem.* **267**, 26031–26037.

Hu, H. (1999). Chemorepulsion of neuronal migration by Slit2 in the developing mammalian forebrain. *Neuron* **23**, 703–711.

Huber, A. B., Kolodkin, A. L., Ginty, D. D., and Cloutier, J. F. (2003). Signaling at the growth cone: Ligand-receptor complexes and the control of axon growth and guidance. *Annu. Rev. Neurosci.* **26**, 509–563.

Huminiecki, L., Gorn, M., Suchting, S., Poulsom, R., and Bicknell, R. (2002). Magic roundabout is a new member of the roundabout receptor family that is endothelial specific and expressed at sites of active angiogenesis. *Genomics* **79**, 547–552.

Impagnatiello, M. A., Weitzer, S., Gannon, G., Compagni, A., Cotten, M., and Christofori, G. (2001). Mammalian sprouty-1 and -2 are membrane-anchored phosphoprotein inhibitors of growth factor signaling in endothelial cells. *J. Cell Biol.* **152**, 1087–1098.

Ishibashi, M., Ang, S. L., Shiota, K., Nakanishi, S., Kageyama, R., and Guillemot, F. (1995). Targeted disruption of mammalian hairy and Enhancer of split homolog-1 (HES-1) leads to up-regulation of neural helix-loop-helix factors, premature neurogenesis, and severe neural tube defects. *Genes Dev.* **9**, 3136–3148.

Iso, T., Hamamori, Y., and Kedes, L. (2003). Notch signaling in vascular development. *Arterioscler. Thromb. Vasc. Biol.* **23**, 543–553.

Isogai, S., Lawson, N. D., Torrealday, S., Horiguchi, M., and Weinstein, B. M. (2003). Angiogenic network formation in the developing vertebrate trunk. *Development* **130**, 5281–5290.

Jain, R. K. (2003). Molecular regulation of vessel maturation. *Nat. Med.* **9**, 685–693.

Jin, K., Zhu, Y., Sun, Y., Mao, X. O., Xie, L., and Greenberg, D. A. (2002). Vascular endothelial growth factor (VEGF) stimulates neurogenesis *in vitro* and *in vivo*. *Proc. Natl. Acad. Sci. USA* **99**, 11946–11950.

Johnson, D. W., Berg, J. N., Baldwin, M. A., Gallione, C. J., Marondel, I., Yoon, S. J., Stenzel, T. T., Speer, M., Pericak-Vance, M. A., Diamond, A., Guttmacher, A. E., Jackson, C. E., Attisano, L., Kucherlapati, R., Porteous, M. E., and Marchuk, D. A. (1996). Mutations in the activin receptor-like kinase 1 gene in hereditary haemorrhagic telangiectasia type 2. *Nat. Genet.* **13**, 189–195.

Karnik, S. K., Brooke, B. S., Bayes-Genis, A., Sorensen, L., Wythe, J. D., Schwartz, R. S., Keating, M. T., and Li, D. Y. (2003). A critical role for elastin signaling in vascular morphogenesis and disease. *Development* **130**, 411–423.

Kawasaki, T., Kitsukawa, T., Bekku, Y., Matsuda, Y., Sanbo, M., Yagi, T., and Fujisawa, H. (1999). A requirement for neuropilin-1 in embryonic vessel formation. *Development* **126,** 4895–4902.

Keleman, K., and Dickson, B. J. (2001). Short- and long-range repulsion by the *Drosophila* Unc5 netrin receptor. *Neuron* **32,** 605–617.

Kidd, T., Bland, K. S., and Goodman, C. S. (1999). Slit is the midline repellent for the Robo receptor in *Drosophila*. *Cell* **96,** 785–794.

Kidd, T., Brose, K., Mitchell, K. J., Fetter, R. D., Tessier-Lavigne, M., Goodman, C. S., and Tear, G. (1998). Roundabout controls axon crossing of the CNS midline and defines a novel subfamily of evolutionarily conserved guidance receptors. *Cell* **92,** 205–215.

Kitsukawa, T., Shimono, A., Kawakami, A., Kondoh, H., and Fujisawa, H. (1995). Overexpression of a membrane protein, neuropilin, in chimeric mice causes anomalies in the cardiovascular system, nervous system and limbs. *Development* **121,** 4309–4318.

Klein, R. (2001). Excitatory Eph receptors and adhesive ephrin ligands. *Curr. Opin. Cell Biol.* **13,** 196–203.

Klessinger, S., and Christ, B. (1996). Axial structures control laterality in the distribution pattern of endothelial cells. *Anat. Embryol. (Berl.)* **193,** 319–330.

Knoll, B., and Drescher, U. (2002). Ephrin-As as receptors in topographic projections. *Trends Neurosci.* **25,** 145–149.

Koch, M., Murrell, J. R., Hunter, D. D., Olson, P. F., Jin, W., Keene, D. R., Brunken, W. J., and Burgeson, R. E. (2000). A novel member of the netrin family, beta-netrin, shares homology with the beta chain of laminin: Identification, expression, and functional characterization. *J. Cell Biol.* **151,** 221–234.

Kolodkin, A. L., Levengood, D. V., Rowe, E. G., Tai, Y. T., Giger, R. J., and Ginty, D. D. (1997). Neuropilin is a Semaphorin III receptor. *Cell* **90,** 753–762.

Krebs, L. T., Xue, Y., Norton, C. R., Shutter, J. R., Maguire, M., Sundberg, J. P., Gallahan, D., Closson, V., Kitajewski, J., Callahan, R., *et al.* (2000). Notch signaling is essential for vascular morphogenesis in mice. *Genes Dev.* **14,** 1343–1352.

Krull, C. E., Lansford, R., Gale, N. W., Collazo, A., Marcelle, C., Yancopoulos, G. D., Fraser, S. E., and Bronner-Fraser, M. (1997). Interactions of Eph-related receptors and ligands confer rostrocaudal pattern to trunk neural crest migration. *Curr. Biol.* **7,** 571–580.

Krum, J. M., and Rosenstein, J. M. (1998). VEGF mRNA and its receptor flt-1 are expressed in reactive astrocytes following neural grafting and tumor cell implantation in the adult CNS. *Exp. Neurol.* **154,** 57–65.

Kullander, K., and Klein, R. (2002). Mechanisms and functions of Eph and ephrin signalling. *Nat. Rev. Mol. Cell Biol.* **3,** 475–486.

Kummer, W., and Haberberger, R. (1999). Extrinsic and intrinsic cholinergic systems of the vascular wall. *Eur. J. Morphol.* **37,** 223–226.

Laberge-le Couteulx, S., Jung, H. H., Labauge, P., Houtteville, J. P., Lescoat, C., Cecillon, M., Marechal, E., Joutel, A., Bach, J. F., and Tournier-Lasserve, E. (1999). Truncating mutations in CCM1, encoding KRIT1, cause hereditary cavernous angiomas. *Nat. Genet.* **23,** 189–193.

Lammert, E., Cleaver, O., and Melton, D. (2001). Induction of pancreatic differentiation by signals from blood vessels. *Science* **294,** 564–567.

Lardelli, M., Dahlstrand, J., and Lendahl, U. (1994). The novel Notch homologue mouse Notch 3 lacks specific epidermal growth factor-repeats and is expressed in proliferating neuroepithelium. *Mech. Dev.* **46,** 123–136.

Lawson, N. D., Scheer, N., Pham, V. N., Kim, C. H., Chitnis, A. B., Campos-Ortega, J. A., and Weinstein, B. M. (2001). Notch signaling is required for arterial-venous differentiation during embryonic vascular development. *Development* **128,** 3675–3683.

Lawson, N. D., Vogel, A. M., and Weinstein, B. M. (2002). Sonic hedgehog and vascular endothelial growth factor act upstream of the Notch pathway during arterial endothelial differentiation. *Dev. Cell* **3**, 127–136.
Lawson, N. D., and Weinstein, B. M. (2002). Arteries and veins: Making a difference with zebrafish. *Nat. Rev. Genet.* **3**, 674–682.
Le Noble, F., Moyon, D., Pardanaud, L., Yuan, L., Djonov, V., Matthijsen, R., Breant, C., Fleury, V., and Eichmann, A. (2004). Flow regulates arterial-venous differentiation in the chick embryo yolk sac. *Development* **131**, 361–375.
LeCouter, J., Kowalski, J., Foster, J., Hass, P., Zhang, Z., Dillard-Telm, L., Frantz, G., Rangell, L., DeGuzman, L., Keller, G. A., *et al.* (2001). Identification of an angiogenic mitogen selective for endocrine gland endothelium. *Nature* **412**, 877–884.
Lee, S. H., Schloss, D. J., Jarvis, L., Krasnow, M. A., and Swain, J. L. (2001). Inhibition of angiogenesis by a mouse sprouty protein. *J. Biol. Chem.* **276**, 4128–4133.
Leong, K. G., Hu, X., Li, L., Noseda, M., Larrivee, B., Hull, C., Hood, L., Wong, F., and Karsan, A. (2002). Activated Notch4 inhibits angiogenesis: Role of beta 1-integrin activation. *Mol. Cell Biol.* **22**, 2830–2841.
Li, D. Y., Brooke, B., Davis, E. C., Mecham, R. P., Sorensen, L. K., Boak, B. B., Eichwald, E., and Keating, M. T. (1998). Elastin is an essential determinant of arterial morphogenesis. *Nature* **393**, 276–280.
Li, D. Y., Sorensen, L. K., Brooke, B. S., Urness, L. D., Davis, E. C., Taylor, D. G., Boak, B. B., and Wendel, D. P. (1999b). Defective angiogenesis in mice lacking endoglin. *Science* **284**, 1534–1537.
Li, H. S., Chen, J. H., Wu, W., Fagaly, T., Zhou, L., Yuan, W., Dupuis, S., Jiang, Z. H., Nash, W., Gick, C., Ornitz, D. M., Wu, J. Y., and Rao, Y. (1999a). Vertebrate Slit, a secreted ligand for the transmembrane protein roundabout, is a repellent for olfactory bulb axons. *Cell* **96**, 807–818.
Louissaint, A., Jr., Rao, S., Leventhal, C., and Goldman, S. A. (2002). Coordinated interaction of neurogenesis and angiogenesis in the adult songbird brain. *Neuron* **34**, 945–960.
Luo, Y., Raible, D., and Raper, J. A. (1993). Collapsin: A protein in brain that induces the collapse and paralysis of neuronal growth cones. *Cell* **75**, 217–227.
Mailhos, C., Modlich, U., Lewis, J., Harris, A., Bicknell, R., and Ish-Horowicz, D. (2001). Delta4, an endothelial specific Notch ligand expressed at sites of physiological and tumor angiogenesis. *Differentiation* **69**, 135–144.
Maisonpierre, P. C., Suri, C., Jones, P. F., Bartunkova, S., Wiegand, S. J., Radziejewski, C., Compton, D., McClain, J., Aldrich, T. H., Papadopoulos, N., Daly, T. J., Davis, S., Sato, T. N., and Yancopoulos, G. D. (1997). Angiopoietin-2, a natural antagonist for Tie2 that disrupts *in vivo* angiogenesis. *Science* **277**, 55–60.
Mallavarapu, A., and Mitchison, T. (1999). Regulated actin cytoskeleton assembly at filopodium tips controls their extension and retraction. *J. Cell Biol.* **146**, 1097–1106.
Mann, F., Ray, S., Harris, W., and Holt, C. (2002). Topographic mapping in dorsoventral axis of the Xenopus retinotectal system depends on signaling through-ephrin-B ligands. *Neuron* **35**, 461–473.
Martin, P., and Lewis, J. (1989). Origins of the neurovascular bundle: Interactions between developing nerves and blood vessels in embryonic chick skin. *Int. J. Dev. Biol.* **33**, 379–387.
Matsumoto, K., Yoshitomi, H., Rossant, J., and Zaret, K. S. (2001). Liver organogenesis promoted by endothelial cells prior to vascular function. *Science* **294**, 559–563.
McAllister, K. A., Grogg, K. M., Johnson, D. W., Gallione, C. J., Baldwin, M. A., Jackson, C. E., Helmbold, E. A., Markel, D. S., McKinnon, W. C., Murrell, J., *et al.* (1994). Endoglin,

a TGF-beta binding protein of endothelial cells, is the gene for hereditary haemorrhagic telangiectasia type 1. *Nat. Genet.* **8,** 345–351.

McLaughlin, T., Hindges, R., and O'Leary, D. D. (2003). Regulation of axial patterning of the retina and its topographic mapping in the brain. *Curr. Opin. Neurobiol.* **13,** 57–69.

Mellitzer, G., Xu, Q., and Wilkinson, D. G. (1999). Eph receptors and ephrins restrict cell intermingling and communication. *Nature* **400,** 77–81.

Mellitzer, G., Xu, Q., and Wilkinson, D. G. (2000). Control of cell behaviour by signalling through eph receptors and ephrins. *Curr. Opin. Neurobiol.* **10,** 400–408.

Miao, H. Q., Soker, S., Feiner, L., Alonso, J. L., Raper, J. A., and Klagsbrun, M. (1999). Neuropilin-1 mediates collapsin-1/semaphorin III inhibition of endothelial cell motility: Functional competition of collapsin-1 and vascular endothelial growth factor-165. *J. Cell Biol.* **146,** 233–242.

Mitchison, T., and Kirschner, M. (1988). Cytoskeletal dynamics and nerve growth. *Neuron* **1,** 761–772.

Mukouyama, Y. S., Shin, D., Britsch, S., Taniguchi, M., and Anderson, D. J. (2002). Sensory nerves determine the pattern of arterial differentiation and blood vessel branching in the skin. *Cell* **109,** 693–705.

Nakagawa, O., Nakagawa, M., Richardson, J. A., Olson, E. N., and Srivastava, D. (1999). HRT1, HRT2, and HRT3: A new subclass of bHLH transcription factors marking specific cardiac, somitic, and pharyngeal arch segments. *Dev. Biol.* **216,** 72–84.

Nakamura, F., Tanaka, M., Takahashi, T., Kalb, R. G., and Strittmatter, S. M. (1998). Neuropilin-1 extracellular domains mediate semaphorin D/III-induced growth cone collapse. *Neuron* **21,** 1093–1100.

Neufeld, G., Cohen, T., Gengrinovitch, S., and Poltorak, Z. (1999). Vascular endothelial growth factor (VEGF) and its receptors. *FASEB J.* **13,** 9–22.

Ng, Y. S., Rohan, R., Sunday, M. E., Demello, D. E., and D'Amore, P. A. (2001). Differential expression of VEGF isoforms in mouse during development and in the adult. *Dev. Dyn.* **220,** 112–121.

O'Reilly, M. S., Boehm, T., Shing, Y., Fukai, N., Vasios, G., Lane, W. S., Flynn, E., Birkhead, J. R., Olsen, B. R., and Folkman, J. (1997). Endostatin: An endogenous inhibitor of angiogenesis and tumor growth. *Cell* **88,** 277–285.

O'Reilly, M. S., Holmgren, L., Shing, Y., Chen, C., Rosenthal, R. A., Moses, M., Lane, W. S., Cao, Y., Sage, E. H., and Folkman, J. (1994). Angiostatin: A novel angiogenesis inhibitor that mediates the suppression of metastases by a Lewis lung carcinoma. *Cell* **79,** 315–328.

Ohta, K., Mizutani, A., Kawakami, A., Murakami, Y., Kasuya, Y., Takagi, S., Tanaka, H., and Fujisawa, H. (1995). Plexin: A novel neuronal cell surface molecule that mediates cell adhesion via a homophilic binding mechanism in the presence of calcium ions. *Neuron* **14,** 1189–1199.

Oosthuyse, B., Moons, L., Storkebaum, E., Beck, H., Nuyens, D., Brusselmans, K., Van Dorpe, J., Hellings, P., Gorselink, M., Heymans, S., Theilmeier, G., Dewerchin, M., Laudenbach, V., Vermylen, P., Raat, H., Acker, T., Vleminckx, V., Van Den Bosch, L., Cashman, N., Fujisawa, H., Drost, M. R., Sciot, R., Bruyninckx, F., Hicklin, D. J., Ince, C., Gressens, P., Lupu, F., Plate, K. H., Robberecht, W., Herbert, J. M., Collen, D., and Carmeliet, P. (2001). Deletion of the hypoxia-response element in the vascular endothelial growth factor promoter causes motor neuron degeneration. *Nat. Genet.* **28,** 131–138.

Orioli, D., and Klein, R. (1997). The Eph receptor family: axonal guidance by contact repulsion. *Trends Genet.* **13,** 354–359.

Othman-Hassan, K., Patel, K., Papoutsi, M., Rodriguez-Niedenfuhr, M., Christ, B., and Wilting, J. (2001). Arterial identity of endothelial cells is controlled by local cues. *Dev. Biol.* **237,** 398–409.

Pandey, A., Shao, H., Marks, R. M., Polverini, P. J., and Dixit, V. M. (1995). Role of B61, the ligand for the Eck receptor tyrosine kinase, in TNF-alpha-induced angiogenesis. *Science* **268,** 567–569.

Pardanaud, L., and Dieterlen-Lievre, F. (2000). Ontogeny of the endothelial system in the avian model. *Adv. Exp. Med. Biol.* **476**, 67–78.

Park, K. W., Morrison, C. M., Sorensen, L. K., Jones, C. A., Rao, Y., Chien, C. B., Wu, J. Y., Urness, L. D., and Li, D. Y. (2003). Robo4 is a vascular-specific receptor that inhibits endothelial migration. *Dev. Biol.* **261**, 251–267.

Pepper, M. S. (1997). Transforming growth factor-beta: Vasculogenesis, angiogenesis, and vessel wall integrity. *Cytokine Growth Factor Rev.* **8**, 21–43.

Petrova, T. V., Makinen, T., Makela, T. P., Saarela, J., Virtanen, I., Ferrell, R. E., Finegold, D. N., Kerjaschki, D., Yla-Herttuala, S., and Alitalo, K. (2002). Lymphatic endothelial reprogramming of vascular endothelial cells by the Prox-1 homeobox transcription factor. *EMBO J.* **21**, 4593–4599.

Placzek, M., and Skaer, H. (1999). Airway patterning: A paradigm for restricted signalling. *Curr. Biol.* **9**, R506–R510.

Plump, A. S., Erskine, L., Sabatier, C., Brose, K., Epstein, C. J., Goodman, C. S., Mason, C. A., and Tessier-Lavigne, M. (2002). Slit1 and Slit2 cooperate to prevent premature midline crossing of retinal axons in the mouse visual system. *Neuron* **33**, 219–232.

Poole, T. J., and Coffin, J. D. (1989). Vasculogenesis and angiogenesis: two distinct morphogenetic mechanisms establish embryonic vascular pattern. *J. Exp. Zool.* **251**, 224–231.

Raper, J. A. (2000). Semaphorins and their receptors in vertebrates and invertebrates. *Curr. Opin. Neurobiol.* **10**, 88–94.

Reaume, A. G., Conlon, R. A., Zirngibl, R., Yamaguchi, T. P., and Rossant, J. (1992). Expression analysis of a Notch homologue in the mouse embryo. *Dev. Biol.* **154**, 377–387.

Redmond, L., Oh, S. R., Hicks, C., Weinmaster, G., and Ghosh, A. (2000). Nuclear Notch1 signaling and the regulation of dendritic development. *Nat. Neurosci.* **3**, 30–40.

Riethmacher, D., Sonnenberg-Riethmacher, E., Brinkmann, V., Yamaai, T., Lewin, G. R., and Birchmeier, C. (1997). Severe neuropathies in mice with targeted mutations in the ErbB3 receptor. *Nature* **389**, 725–730.

Risau, W., and Flamme, I. (1995). Vasculogenesis. *Annu. Rev. Cell Dev. Biol.* **11**, 73–91.

Rossant, J., and Hirashima, M. (2003). Vascular development and patterning: Making the right choices. *Curr. Opin. Genet. Dev.* **13**, 408–412.

Ruhrberg, C., Gerhardt, H., Golding, M., Watson, R., Ioannidou, S., Fujisawa, H., Betsholtz, C., and Shima, D. T. (2002). Spatially restricted patterning cues provided by heparin-binding VEGF-A control blood vessel branching morphogenesis. *Genes Dev.* **16**, 2684–2698.

Sahoo, T., Johnson, E. W., Thomas, J. W., Kuehl, P. M., Jones, T. L., Dokken, C. G., Touchman, J. W., Gallione, C. J., Lee-Lin, S. Q., Kosofsky, B., *et al.* (1999). Mutations in the gene encoding KRIT1, a Krev-1/rap1a binding protein, cause cerebral cavernous malformations (CCM1). *Hum. Mol. Genet.* **8**, 2325–2333.

Sakata, Y., Kamei, C. N., Nakagami, H., Bronson, R., Liao, J. K., and Chin, M. T. (2002). Ventricular septal defect and cardiomyopathy in mice lacking the transcription factor CHF1/Hey2. *Proc. Natl. Acad. Sci. USA* **99**, 16197–16202.

Schmucker, D., and Zipursky, S. L. (2001). Signaling downstream of Eph receptors and ephrin ligands. *Cell* **105**, 701–704.

Schratzberger, P., Schratzberger, G., Silver, M., Curry, C., Kearney, M., Magner, M., Alroy, J., Adelman, L. S., Weinberg, D. H., Ropper, A. H., and Isner, J. M. (2000). Favorable effect of VEGF gene transfer on ischemic peripheral neuropathy. *Nat. Med.* **6**, 405–413.

Schwarz, E. R., Speakman, M. T., Patterson, M., Hale, S. S., Isner, J. M., Kedes, L. H., and Kloner, R. A. (2000). Evaluation of the effects of intramyocardial injection of DNA expressing vascular endothelial growth factor (VEGF) in a myocardial infarction model in the rat–angiogenesis and angioma formation. *J. Am. Coll. Cardiol.* **35**, 1323–1330.

Seeger, M., Tear, G., Ferres-Marco, D., and Goodman, C. S. (1993). Mutations affecting growth cone guidance in *Drosophila*: Genes necessary for guidance toward or away from the midline. *Neuron* **10**, 409–426.

Serini, G., Valdembri, D., Zanivan, S., Morterra, G., Burkhardt, C., Caccavari, F., Zammataro, L., Primo, L., Tamagnone, L., Logan, M., Tessier-Lavigne, M., Taniguchi, M., Puschel, A. W., and Bussolino, F. (2003). Class 3 semaphorins control vascular morphogenesis by inhibiting integrin function. *Nature* **424**, 391–397.

Sestan, N., Artavanis-Tsakonas, S., and Rakic, P. (1999). Contact-dependent inhibition of cortical neurite growth mediated by Notch signaling. *Science* **286**, 741–746.

Shalaby, F., Rossant, J., Yamaguchi, T. P., Gertsenstein, M., Wu, X. F., Breitman, M. L., and Schuh, A. C. (1995). Failure of blood-island formation and vasculogenesis in Flk-1-deficient mice. *Nature* **376**, 62–66.

Shima, D. T., and Mailhos, C. (2000). Vascular developmental biology: Getting nervous. *Curr. Opin. Genet. Dev.* **10**, 536–542.

Shin, D., Garcia-Cardena, G., Hayashi, S., Gerety, S., Asahara, T., Stavrakis, G., Isner, J., Folkman, J., Gimbrone, M. A., Jr., and Anderson, D. J. (2001). Expression of ephrinB2 identifies a stable genetic difference between arterial and venous vascular smooth muscle as well as endothelial cells, and marks subsets of microvessels at sites of adult neovascularization. *Dev. Biol.* **230**, 139–150.

Shoji, W., Isogai, S., Sato-Maeda, M., Obinata, M., and Kuwada, J. Y. (2003). Semaphorin3a1 regulates angioblast migration and vascular development in zebrafish embryos. *Development* **130**, 3227–3236.

Shutter, J. R., Scully, S., Fan, W., Richards, W. G., Kitajewski, J., Deblandre, G. A., Kintner, C. R., and Stark, K. L. (2000). Dll4, a novel Notch ligand expressed in arterial endothelium. *Genes Dev.* **14**, 1313–1318.

Soker, S., Takashima, S., Miao, H. Q., Neufeld, G., and Klagsbrun, M. (1998). Neuropilin-1 is expressed by endothelial and tumor cells as an isoform-specific receptor for vascular endothelial growth factor. *Cell* **92**, 735–745.

Sondell, M., Lundborg, G., and Kanje, M. (1999). Vascular endothelial growth factor has neurotrophic activity and stimulates axonal outgrowth, enhancing cell survival and Schwann cell proliferation in the peripheral nervous system. *J. Neurosci.* **19**, 5731–5740.

Sondell, M., Sundler, F., and Kanje, M. (2000). Vascular endothelial growth factor is a neurotrophic factor which stimulates axonal outgrowth through the flk-1 receptor. *Eur. J. Neurosci.* **12**, 4243–4254.

Sorensen, L. K., Brooke, B. S., Li, D. Y., and Urness, L. D. (2003). Loss of distinct arterial and venous boundaries in mice lacking endoglin, a vascular-specific TGFbeta coreceptor. *Dev. Biol.* **261**, 235–250.

Soriano, P. (1994). Abnormal kidney development and hematological disorders in PDGF beta-receptor mutant mice. *Genes Dev.* **8**, 1888–1896.

Spence, S. G., and Poole, T. J. (1994). Developing blood vessels and associated extracellular matrix as substrates for neural crest migration in Japanese quail, *Coturnix coturnix japonica*. *Int. J. Dev. Biol.* **38**, 85–98.

Stalmans, I., Ng, Y. S., Rohan, R., Fruttiger, M., Bouche, A., Yuce, A., Fujisawa, H., Hermans, B., Shani, M., Jansen, S., Hicklin, D., Anderson, D. J., Gardiner, T., Hammes, H. P., Moons, L., Dewerchin, M., Collen, D., Carmeliet, P., and D'Amore, P. A. (2002). Arteriolar and venular patterning in retinas of mice selectively expressing VEGF isoforms. *J. Clin. Invest.* **109**, 327–336.

Stehbens, W. E. (1996). Structural and architectural changes during arterial development and the role of hemodynamics. *Acta Anat. (Basel)* **157**, 261–274.

Stein, E., and Tessier-Lavigne, M. (2001). Hierarchical organization of guidance receptors: silencing of netrin attraction by slit through a Robo/DCC receptor complex. *Science* **291**, 1928–1938.

Stone, J., and Dreher, Z. (1987). Relationship between astrocytes, ganglion cells and vasculature of the retina. *J. Comp. Neurol.* **255**, 35–49.

Suri, C., Jones, P. F., Patan, S., Bartunkova, S., Maisonpierre, P. C., Davis, S., Sato, T. N., and Yancopoulos, G. D. (1996). Requisite role of angiopoietin-1, a ligand for the TIE2 receptor, during embryonic angiogenesis. *Cell* **87**, 1171–1180.

Suter, D. M., and Forscher, P. (2000). Substrate-cytoskeletal coupling as a mechanism for the regulation of growth cone motility and guidance. *J. Neurobiol.* **44**, 97–113.

Takahashi, T., Fournier, A., Nakamura, F., Wang, L. H., Murakami, Y., Kalb, R. G., Fujisawa, H., and Strittmatter, S. M. (1999). Plexin-neuropilin-1 complexes form functional semaphorin-3A receptors. *Cell* **99**, 59–69.

Tamagnone, L., Artigiani, S., Chen, H., He, Z., Ming, G. I., Song, H., Chedotal, A., Winberg, M. L., Goodman, C. S., Poo, M., Tessier-Lavigne, M., and Comoglio, P. M. (1999). Plexins are a large family of receptors for transmembrane, secreted, and GPI-anchored semaphorins in vertebrates. *Cell* **99**, 71–80.

Taniguchi, M., Yuasa, S., Fujisawa, H., Naruse, I., Saga, S., Mishina, M., and Yagi, T. (1997). Disruption of semaphorin III/D gene causes severe abnormality in peripheral nerve projection. *Neuron* **19**, 519–530.

Taylor, G. I., Bates, D., and Newgreen, D. F. (2001). The developing neurovascular anatomy of the embryo: a technique of simultaneous evaluation using fluorescent labeling, confocal microscopy, and three-dimensional reconstruction. *Plast. Reconstr. Surg.* **108**, 597–604.

Tessier-Lavigne, M., and Goodman, C. S. (1996). The molecular biology of axon guidance. *Science* **274**, 1123–1133.

Torres-Vazquez, J., Gitler, A. D., Fraser, S. D., Berk, J. D., Pham, V. N., Fishman, M. C., Childs, S., Epstein, J. A., and Weinstein, B. M. (2004). Semaphorin-Plexin signaling guides patterning of the developing vasculature. *Dev. Cell* **7**, 117–123.

Urness, L. D., Sorensen, L. K., and Li, D. Y. (2000). Arteriovenous malformations in mice lacking activin receptor-like kinase-1. *Nat. Genet.* **26**, 328–331.

Uyttendaele, H., Marazzi, G., Wu, G., Yan, Q., Sassoon, D., and Kitajewski, J. (1996). Notch4/int-3, a mammary proto-oncogene, is an endothelial cell-specific mammalian Notch gene. *Development* **122**, 2251–2259.

Van der Zwaag, B., Hellemons, A. J., Leenders, W. P., Burbach, J. P., Brunner, H. G., Padberg, G. W., and Van Bokhoven, H. (2002). PLEXIN-D1, a novel plexin family member, is expressed in vascular endothelium and the central nervous system during mouse embryogenesis. *Dev. Dyn.* **225**, 336–343.

Villa, N., Walker, L., Lindsell, C. E., Gasson, J., Iruela-Arispe, M. L., and Weinmaster, G. (2001). Vascular expression of Notch pathway receptors and ligands is restricted to arterial vessels. *Mech. Dev.* **108**, 161–164.

Visconti, R. P., Richardson, C. D., and Sato, T. N. (2002). Orchestration of angiogenesis and arteriovenous contribution by angiopoietins and vascular endothelial growth factor (VEGF). *Proc. Natl. Acad. Sci. USA* **99**, 8219–8224.

Wang, H. U., and Anderson, D. J. (1997). Eph family transmembrane ligands can mediate repulsive guidance of trunk neural crest migration and motor axon outgrowth. *Neuron* **18**, 383–396.

Wang, H. U., Chen, Z. F., and Anderson, D. J. (1998). Molecular distinction and angiogenic interaction between embryonic arteries and veins revealed by ephrin-B2 and its receptor Eph-B4. *Cell* **93**, 741–753.

Weinstein, B. M., Stemple, D. L., Driever, W., and Fishman, M. C. (1995). Gridlock, a localized heritable vascular patterning defect in the zebrafish. *Nat. Med.* **1**, 1143–1147.

Wheater, P., Stevens, A., and Lowe, J. (1985). Basic histopathology: A color atlas and text. Churchill Livingstone, New York.

Whitehead, K., Plummer, N., Adams, J., Marchuk, D., and Li, D. Y. (2004). Ccm1 is required for arterial morphogenesis: Implications for the etiology of human cavernous malformations. *Development* **131**, 1437–1448.

Wilkinson, D. G. (2001). Multiple roles of EPH receptors and ephrins in neural development. *Nat. Rev. Neurosci.* **2**, 155–164.

Williams, L. T., Tremble, P., and Antoniades, H. N. (1982). Platelet-derived growth factor binds specifically to receptors on vascular smooth muscle cells and the binding becomes nondissociable. *Proc. Natl. Acad. Sci. USA* **79**, 5867–5870.

Wilting, J., Brand-Saberi, B., Huang, R., Zhi, Q., Kontges, G., Ordahl, C. P., and Christ, B. (1995). Angiogenic potential of the avian somite. *Dev. Dyn.* **202**, 165–171.

Winberg, M. L., Noordermeer, J. N., Tamagnone, L., Comoglio, P. M., Spriggs, M. K., Tessier-Lavigne, M., and Goodman, C. S. (1998). Plexin A is a neuronal semaphorin receptor that controls axon guidance. *Cell* **95**, 903–916.

Wong, K., Park, H. T., Wu, J. Y., and Rao, Y. (2002). Slit proteins: molecular guidance cues for cells ranging from neurons to leukocytes. *Curr. Opin. Genet. Dev.* **12**, 583–591.

Wu, W., Wong, K., Chen, J., Jiang, Z., Dupuis, S., Wu, J. Y., and Rao, Y. (1999). Directional guidance of neuronal migration in the olfactory system by the protein Slit. *Nature* **400**, 331–336.

Xue, Y., Gao, X., Lindsell, C. E., Norton, C. R., Chang, B., Hicks, C., Gendron-Maguire, M., Rand, E. B., Weinmaster, G., and Gridley, T. (1999). Embryonic lethality and vascular defects in mice lacking the Notch ligand Jagged1. *Hum. Mol. Genet.* **8**, 723–730.

Yancopoulos, G. D., Klagsbrun, M., and Folkman, J. (1998). Vasculogenesis, angiogenesis, and growth factors: Ephrins enter the fray at the border. *Cell* **93**, 661–664.

Yebra, M., Montgomery, A. M., Diaferia, G. R., Kaido, T., Silletti, S., Perez, B., Just, M. L., Hildbrand, S., Hurford, R., Florkiewicz, E., *et al.* (2003). Recognition of the neural chemoattractant Netrin-1 by integrins alpha6beta4 and alpha3beta1 regulates epithelial cell adhesion and migration. *Dev. Cell* **5**, 695–707.

Yuan, W., Zhou, L., Chen, J. H., Wu, J. Y., Rao, Y., and Ornitz, D. M. (1999). The mouse SLIT family: Secreted ligands for ROBO expressed in patterns that suggest a role in morphogenesis and axon guidance. *Dev. Biol.* **212**, 290–306.

Zhang, H., Vutskits, L., Pepper, M. S., and Kiss, J. Z. (2003). VEGF is a chemoattractant for FGF-2-stimulated neural progenitors. *J. Cell Biol.* **163**, 1375–1384.

Zhong, T. P., Childs, S., Leu, J. P., and Fishman, M. C. (2001). Gridlock signalling pathway fashions the first embryonic artery. *Nature* **414**, 216–220.

Zhong, T. P., Rosenberg, M., Mohideen, M. A., Weinstein, B., and Fishman, M. C. (2000). gridlock, an HLH gene required for assembly of the aorta in zebrafish. *Science* **287**, 1820–1824.

Zhu, Y., Jin, K., Mao, X. O., and Greenberg, D. A. (2003). Vascular endothelial growth factor promotes proliferation of cortical neuron precursors by regulating E2F expression. *FASEB J.* **17**, 186–193.

Zhu, Y., Li, H., Zhou, L., Wu, J. Y., and Rao, Y. (1999). Cellular and molecular guidance of GABAergic neuronal migration from an extracortical origin to the neocortex. *Neuron* **23**, 473–485.

Zimrin, A. B., Pepper, M. S., McMahon, G. A., Nguyen, F., Montesano, R., and Maciag, T. (1996). An antisense oligonucleotide to the Notch ligand jagged enhances fibroblast growth factor-induced angiogenesis *in vitro*. *J. Biol. Chem.* **271**, 32499–32502.

Zukowska-Grojec, Z., Pruszczyk, P., Colton, C., Yao, J., Shen, G. H., Myers, A. K., and Wahlestedt, C. (1993). Mitogenic effect of neuropeptide Y in rat vascular smooth muscle cells. *Peptides* **14**, 263–268.

5
Vascular Endothelial Growth Factor and Its Receptors in Embryonic Zebrafish Blood Vessel Development

Katsutoshi Goishi[*,†] *and Michael Klagsbrun*[*,†,‡]
[*]Vascular Biology Program
[†]Department of Surgery
[‡]Department of Pathology
Children's Hospital and Harvard Medical School
Boston, Massachusetts 02115

    I. Introduction
   II. Vascular Growth Factors
  III. Blood Vessel Development
  IV. Advantages of the Zebrafish Model for Vascular Research
   V. Methods for Analyzing the Zebrafish Vasculature
      A. Microangiography
      B. Alkaline Phosphatase Staining
      C. *In Situ* Hybridization
      D. Transgenic Zebrafish
      E. Morpholino Antisense Knockdown
      F. Small Chemical Compounds
  VI. Vascular Endothelial Growth Factor
 VII. VEGF Receptors
VIII. Neuropilins
  IX. Semaphorins
   X. Future Perspectives
      Acknowledgments
      References

There is intense interest in how blood vessel development is regulated. A number of vascular growth factors and their receptors have been described. The vascular endothelial growth factor (VEGF) and its receptors are major contributors to normal mammalian vascular development. These receptors include VEGFR-1, VEGFR-2, VEGFR-3, neuropilin-1 (NRP1), and NRP2. The function of these genes have been determined to some degree in mouse gene targeting studies. These knockouts are embryonicly lethal, and early death can be attributed in part to lack of normal blood and lymphatic vessel development. More recently, it has been demonstrated that zebrafish are an excellent model for studying the genes and proteins that regulate embryonic vascular development. Zebrafish have a number of advantages compared to mice, including rapid embryonic development and the ability to examine and manipulate embryos outside of the animal. In this review, we describe some of the earlier mouse VEGF/receptor functional studies and emphasize the development of the zebrafish vasculature. We describe the zebrafish vasculature, zebrafish VEGF and VEGF receptors, advantages of the zebrafish model, resources, and methods of determining growth factor and receptor function. © 2004, Elsevier Inc.

I. Introduction

The circulation of blood cells via blood vessels is critical for maintaining vertebrate homeostasis by supplying oxygen and nutrients to the various organs and by providing a conduit for the excretion of waste products. Capillaries are made up of endothelial cells (EC) that form a lumen and become covered with pericytes. Larger blood vessels, the veins and arteries, contain an inner layer of EC and outer layers of smooth muscle cells. More than just a lining, the endothelium acts as a selective barrier for the transport of molecules between the blood and tissues, an anticoagulant, an antithrombotic, and a regulator of vascular tone. There is a good deal of heterogeneity in EC phenotype from organ to organ (Aird, 2002, 2003). Various EC types express different markers, such as *ephrin-B2* and *neuropilin-1* (*NRP1*) in arteries, and *ephB4* and *NRP2* in veins (Lawson and Weinstein, 2002a; Torres-Vazquez *et al.*, 2003). The EC phenotype is often dependent on the environment; for example, venous EC exposed to arterial blood flow undergo profound changes in gene expression (Kwei *et al.*, 2004).

It has become apparent that normal blood vessel formation is critical for embryonic development and that aberrant blood vessel formation in the adult contributes to pathologies such as tumor growth and retinopathies. Thus, there is major interest in identifying vascular growth factors and their receptors that regulate EC proliferation, migration, and survival (Conway *et al.*, 2001; Klagsbrun and Moses, 1999; Ribatti *et al.*, 2000; Yancopoulos *et al.*, 1998, 2000). In the last 10 years, gene knockout studies in mice have been successfully used to determine vascular growth factor and receptor function during mouse development. In the last few years, however, the zebrafish has also become recognized as an important model for the study of embryonic vascular development. This chapter emphasizes recent studies of zebrafish vascular growth factors and receptors.

II. Vascular Growth Factors

During embryogenesis, the spatiotemporal expression of vascular growth factors and their receptors is strictly regulated. Knockouts of vascular growth factors result in embryonic lethality (Cleaver and Melton, 2003;

Conway *et al.*, 2001; Ferrara *et al.*, 2003; Wilting *et al.*, 2003; Yancopoulos *et al.*, 2000). On the other hand, overexpression of vascular growth factors and suppression of angiogenesis inhibitors can result in enhanced tumor progression (Kerbel and Folkman, 2002). It has been suggested that there is a balance between angiogenesis stimulators and inhibitors that regulates normal angiogenesis (Hanahan and Folkman, 1996). There are a number of cognate growth factor–receptor pairs that regulate vasculogenesis and angiogenesis. These are VEGF and VEGF receptors, including the receptor tyrosine kinases and neuropilins (NRP), angiopoietins and Tie receptors, fibroblast growth factors (FGF) and FGF receptors, and ephrins and EPH receptors. Since VEGF and its receptors are so far the most studied angiogenesis factors in zebrafish, this review will concentrate on these vascular factors.

III. Blood Vessel Development

Hematopoiesis, the genesis of blood cells from pluripotent hematopoietic stem cells, and vasculogenesis are linked and share some pathways (Baron, 2003; Choi, 2002; Eichmann *et al.*, 2002; Ema and Rossant, 2003; Mikkola and Orkin, 2002; Ribatti *et al.*, 2002). The close developmental association of the hematopoietic and EC lineages within the blood island has led to the hypothesis that they arise from a common precursor, the hemangioblast (Sabin, 1917). For example, hematopoietic and endothelial precursor cells share a number of cell surface markers in common, such as vascular endothelial growth factor receptor1 (VEGFR-1) (Peters *et al.*, 1993), VEGFR-2 (Kaipainen *et al.*, 1993; Shalaby *et al.*, 1995; Yamaguchi *et al.*, 1993), Tie1, Tie2 (Dumont *et al.*, 1995), and CD34 (Fina *et al.*, 1990). However, evidence for the *in vivo* existence of hemangioblasts remains elusive.

In mammals, the yolk sac is the initial site of primitive blood vessel formation beginning at E7.5 in the mouse. It has been suggested that there is an EC precursor, the angioblast, which matures to form blood islands, the external cells of which flatten and differentiate into endothelial cells in a process known as vasculogenesis (Baron, 2003; Choi, 2002; Eichmann *et al.*, 2002; Ema and Rossant, 2003; Mikkola and Orkin, 2002; Ribatti *et al.*, 2002; Risau, 1997; Risau and Flamme, 1995). Vasculogenesis is also characterized by the differentiation of precursor cells into vascular EC, which assemble to form a vascular plexus that supports blood cell circulation and which matures into a vascular network by extensive pruning and remodeling. Angiogenesis is the other major process in blood vessel development. Angiogenesis is the sprouting of capillaries from preexisting vasculogenic blood vessels, for example, the sprouting and branching of capillaries from pre-existing arteries and veins (Risau, 1997; Risau and Flamme, 1995).

These sprouts mature into larger blood vessels by attracting pericytes and smooth muscle cells.

Originally thought to be strictly an embryonic process, it is now realized that vasculogenesis can occur in the adult by the differentiation of circulating endothelial progenitor cells into mature EC and their incorporation into new blood vessels (Asahara et al., 1997). Angiogenesis is limited in the adult, for example, in the female reproductive system and wound healing. Angiogenesis also contributes to pathology, for example, tumor angiogenesis. Invasion or co-option of normal capillaries and their proliferation in tumors is an essential step in tumor progression and metastasis (Hanahan and Folkman, 1996).

Unlike mammals, zebrafish do not form yolk sac blood islands; instead, hematopoietic cells and endothelial precursor cells are formed in the embryo proper. Angioblasts, the precursors of EC in zebrafish, are derived primarily from the zebrafish cephalic and posterior trunk ventrolateral mesoderm (Fouquet et al., 1997). The angioblasts migrate and differentiate into mature EC and form the major arterial and venous blood vessels, a vasculogenic process. At 24 h post-fertilization (hpf), blood is formed in the intermediate cell mass at a dorsal location above the yolk extension (Davidson and Zon, 2000; Traver and Zon, 2002). The heart starts beating and blood circulation begins in the zebrafish embryo at approximately 24 hpf. Blood flows through a simple circuit via the outflow tract of the heart into the dorsal aorta, continues into the trunk/tail, and then into the caudal and cardinal veins before returning to the heart. Inter-somitic angiogenesis starts at 24 hpf. Intersegmental vessels (ISV) sprout from the dorsal aorta and posterior caudal vein between each of the somites and connect to the dorsal longitudinal anastomotic vessel (DLAV). The zebrafish ISV consists of three EC, an inverted T-shaped cell based in the aorta and, branching dorsally, a connecting cell, and most dorsally, a T-shaped cell based in the DLAV and branching ventrally (Isogai et al., 2001, 2003).

There is a mutant zebrafish known as *cloche*, in which virtually no blood cells or blood vessels are produced (Liao et al., 1997, 1998; Stainier et al., 1995, 1996; Thompson et al., 1998). Overexpression of *scl/tall*, one of the earliest markers of hematopoietic cell lineage, in the *cloche* mutant partially rescues the hematopoietic and EC defects (Liao et al., 1998). Together, these results suggest that the hematopoietic and endothelial cell lineages are closely related and support the hemangioblast hypothesis.

IV. Advantages of the Zebrafish Model for Vascular Research

Zebrafish embryos are excellent models for studying the genes and proteins that regulate embryonic development (Ackermann and Paw, 2003; Dooley and Zon, 2000; Talbot and Hopkins, 2000; Thisse and Zon, 2002). There are

5. VEGF and Blood Vessel Development

Table I Advantages of the Zebrafish Model for Vascular Studies

- **Small size:** Adults are approximately 3 cm. They are easy to maintain on a large scale. They can survive loss of the circulatory system for several days by O_2 diffusion.
- **Rapid embryonic development:** Phenotype can be observed in 1–3 days.
- **Transparency:** Zebrafish embryos are transparent, allowing real-time visualization of the cardiovascular system, including the heart and the circulation of blood.
- **Embryos are extracorporeal:** It is possible to manipulate the embryos, For example, injection of mRNA, morpholinos, and small compounds.
- **Statistical analysis:** Approximately 100–200 embryos are produced per mating, enabling phenotypes to be analyzed in a statistical manner.
- **Function studies:** Administration of antisense morpholinos and low molecular inhibitors inhibit protein synthesis and activity, respectively, and loss of function can be determined.
- **Forward genetics:** It is possible to generate mutants on a large scale, choose phenotypes of interest, and positional clone the corresponding gene.
- **Extensive databases:** There are comprehensive Web-based databases that access a variety of resources, including an EST database, a genome database, a mutant database, a developmental atlas, and gene expression patterns.

a number of advantages for studying the zebrafish cardiovascular system (summarized in Table I). For example, (1) zebrafish are small, (approximately 3 cm as adults) and thus easy to care for and maintain on a large scale; (2) embryonic development is rapid, with much of it occurring within 6 days postfertilization (Langheinrich, 2003); (3) zebrafish embryos are transparent, allowing visualization of the zebrafish cardiovascular system, including the heart and circulation of the blood; (4) zebrafish embryos are extracorporeal, unlike mice, making it possible to analyze normal and abnormal cardiovascular development in living embryos by microscopy; (5) since zebrafish embryos are extracorporeal, they are readily manipulated, for example, by microinjection of mRNA or DNA; (6) approximately 100–200 embryos are produced per mating, enabling phenotypes to be analyzed in a statistical manner; (7) function can be studied readily by administration of antisense morpholinos and low molecular weight chemical compounds (discussed later); and (8) although zebrafish embryos have a circulatory system that develops by 24 hpf, their small size enables them to survive by passive oxygen diffusion for up to 4–5 days. Thus, it is possible to analyze the effects of mutants or angiogenesis inhibitors in zebrafish lacking a functional circulatory system. In contrast, mammalian mutations and knockouts are often embryonic lethal, precluding further analysis; and (9) zebrafish are suitable for forward genetics. In this approach, mutations are induced by mutagens such an ethylnitrosourea (ENU) (Grunwald and Streisinger, 1992), mutant fish are isolated, and the genes affected are identified by positional cloning. Examples of mutants include *one-eyed pinhead* (defects in the EGF-CFC gene; Zhang et al., 1998), *jekyll* (UDP-glucose dehydrogenase) (Walsh and Stainier, 2001), *gridlock* (Hey2) (Zhong et al., 2000), and *sonic-you* (sonic hedgehog) (Schauerte et al., 1998).

The power and popularity of the zebrafish is largely attributable to the ability to conduct large-scale mutagenesis screens. In Boston and Tübingen, more than 4000 embryonic lethal mutants phenotype were screened, resulting in the isolation of over 100 mutations affecting the formation and function of the cardiovascular system (Chen et al., 1996; Driever et al., 1996; Haffter et al., 1996; Ransom et al., 1996; Stainier et al., 1996; Weinstein et al., 1996). More recently, in the Tübingen 2000 screen consortium, 4500 mutagenized genomes were screened and more than 700 cardiovascular mutants were identified, including alterations in the *flk1* gene (VEGFR-2) (Habeck et al., 2002).

In addition, there are zebrafish databases that facilitate zebrafish research, for example, the zebrafish information network (ZFIN). This centralized database includes genomic data, mutant phenotype, anatomical atlases, gene expression patterns, and other useful zebrafish information. There are zebrafish stock centers, one at the University of Oregon (the zebrafish International Resource Center) and another in Tübingen (the Tübingen zebrafish stock center), which maintain wild-type and mutant strains.

For analysis of the zebrafish vasculature, a very detailed atlas of the zebrafish microvasculature using microangiography and confocal microscopy has been published by Weinstein and colleagues (Isogai et al., 2001). The resolution of this technique makes it possible to see blood vessels less than 1 micrometer in diameter. Their online interactive atlas displays the vascular anatomy of the developing zebrafish from 1–7 days postfertilization in three-dimensional detail. Wiring patterns and the nomenclature of major vessels are described in detail. The Weinstein Web site links to "The Interactive Atlas of Zebrafish Vascular Anatomy" and allows access to the latest developments in zebrafish vascular biology.

V. Methods for Analyzing the Zebrafish Vasculature

A. Microangiography

The transparency of zebrafish embryos makes it possible to analyze real-time blood flow under the microscope or by video microscopy. However, microangiography is more practical for high-resolution analysis of the circulation in multiple embryos. Microangiography is performed by injection of fluorescent dyes into the blood circulation. Zebrafish are not harmed by these injections, and no defects have been noted. Types of dyes used include dextran conjugates, which can vary in molecular size ($3 \times 10^4 - 2 \times 10^6$

5. VEGF and Blood Vessel Development

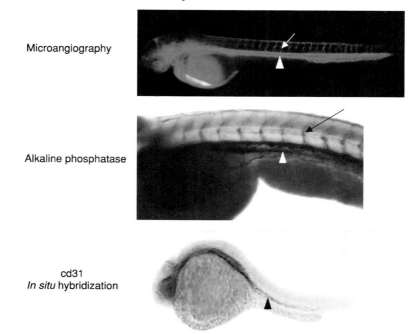

Figure 1 Detection of the zebrafish zasculature. (Upper) Microangiography showing blood circulation. FITC-dextran (green color) was injected in the cardinal vein at 56 hpf. Green fluorescence visualizes functional blood flow in axial vessels (arrowhead) and ISV (arrow). (Middle) Alkaline phosphatase staining showing blood vessel structure. Zebrafish embryos (72 hpf) were stained with substrates of alkaline phosphatase (BCIP and NBT). The dark purple stained structures are axial vessels (arrowhead) and ISV (arrow). (Lower) *In situ* hybridization (ISH) with *cd31*, a specific EC marker. Whole-mount ISH with 24 hpf embryos was performed. Arrowhead indicates axial vessels. ISV are not formed yet at this 24 hpf stage. (See Color Insert.)

daltons), charge, and color (Goishi *et al.*, 2003; Lee *et al.*, 2002; Nasevicius *et al.*, 2000). Microangiography of zebrafish blood vessels is shown in Fig. 1. Another method uses sized fluorescent microbeads, typically 0.02–0.04 microns (Isogai *et al.*, 2001; Weinstein *et al.*, 1995). In both cases, confocal microscopy imaging needs to be done rapidly (within 15 min postinjection) due to sticking of the dye to blood vessels or leakage of dye out of the blood vessels. Non-invasive methods not requiring confocal or fluorescent microscopy have also been reported (Lee *et al.*, 2002; Pelster, 2003). For example, in digital motion analysis, zebrafish are magnified 40-fold using transmitted light on an inverted microscope, and serial time-lapse images are collected using a CCD digital camera. The mean image intensity for these images is calculated on a pixel-by-pixel basis by using the IPLAB

RATIOPLUS program (Scanalytics). Images depicting the variance from the mean are produced to display areas of movement, which correspond to blood cell flow (Lee et al., 2002).

B. Alkaline Phosphatase Staining

Endogenous alkaline phosphatase staining is a convenient, although not specific, method for analyzing the zebrafish vasculature (Chen et al., 1996; Habeck et al., 2002; Stainier et al., 1996) (Fig. 1). Zebrafish embryos are stained in Bromo-4-chloro-3-indolyl phosphate (BCIP) and nitroblue tetrazolium (NBT) to detect endogenous alkaline phosphatase present in EC (Fig. 1). Alkaline phosphatase provides excellent resolution in trunk blood vessels; however, a limitation to this approach is that endogenous alkaline phosphatase is expressed strongly in the brain so that it is difficult to analyze the head vasculature.

C. *In Situ* Hybridization

In situ hybridization (ISH) is highly utilized in zebrafish gene expression studies. This method has advantages compared to immunochemistry, since available antibodies are often species-specific and do not always detect zebrafish proteins. For ISH, zebrafish gene sequences are used to design specific probes. A number of markers are used to detect blood vessels, including pan-endothelial markers such as *cd31* (also known as platelet/endothelial cell adhesion molecule; *pecam1*, Fig. 1), *tie1/tie2* (Liao et al., 1997; Lyons et al., 1998), and *fli1* (Thompson et al., 1998). There are also specific markers for arteries that include *delta c* (Smithers et al., 2000), *ephrinb2* (Lawson et al., 2001), *flk1* (Fouquet et al., 1997; Liao et al., 1997; Sumoy et al., 1997; Thompson et al., 1998), *gridlock* (Liao et al., 1997; Parker and Stainier, 1999; Stainier et al., 1995; Weinstein et al., 1995; Zhong et al., 2000, 2001), *Notch5* (Lawson et al., 2001), and *tbx20* (Ahn et al., 2000). Another group of genes are expressed specifically in veins, for example-*ephb4* and *flt4* (Lawson et al., 2001; Thompson et al., 1998).

D. Transgenic Zebrafish

The application of transgenic technology to the zebrafish has shown the possibility and feasibility of *in vivo* endothelial imaging in living animals. A murine *Tie2* promoter has been used to drive green fluorescent protein (GFP) expression in zebrafish EC, and stable germ line transgenic lines have

5. VEGF and Blood Vessel Development

been prepared (Motoike et al., 2000). More recently, GFP transgenic zebrafish driven by the zebrafish *fli1* promoter have been created (Lawson and Weinstein, 2002b). This transgenic line expresses GFP at high levels specifically in EC and enables high-resolution, long-term, time-lapse analysis of the EC in developing blood vessels (Lawson and Weinstein, 2002b).

E. Morpholino Antisense Knockdown

Morpholino phosphorodiamidate oligos (morpholino, MO) are synthetic antisense DNA analogues (Summerton, 1999) that are used to analyze loss of function (Nasevicius and Ekker, 2000). They inhibit protein translation by blocking mRNA binding to ribosomes (referred to as knockdown). MO do not inhibit gene expression, so this approach is not a knockout. Typically, an MO oligo corresponds to the first 25 bases of coding sequence (Summerton, 1999). However, MO have also been used to target exon-intron boundaries to cause mis-splicing defects (Draper et al., 2001). The effectiveness of an MO as an inhibitor of translation can be analyzed quantitatively using an *in vitro* transcription and translation assay (Goishi et al., 2003; Lee et al., 2002). Western blotting can be used to analyze protein expression both *in vitro* and *in vivo* if antibodies that recognize the zebrafish protein are available (Goishi et al., 2003; Liu et al., 2003; Nasevicius and Ekker, 2000). Specific and nonspecific effects of MO can be distinguished by the use of appropriate controls (Heasman, 2002). Controls include the use of a standard MO, or even better, use of a 4 or 5 base mismatch MO. The efficacy of control MO can be measured in the *in vitro* transcription and translation assay (Lee et al., 2002). In addition, confidence in the accuracy of MO inhibition can be obtained by rescuing the phenotype with injection of mRNA corresponding to the protein knocked down (Heasman, 2002). When possible, an MO-induced phenotype can be compared to a phenotype induced by another relevant inhibitor, for example, an EGF receptor MO compared to an EGF receptor kinase inhibitor or compared to overexpression of a dominant negative EGF receptor (Goishi et al., 2003).

MO are injected at the 1–4 cell stage. One of the first MO-mediated knockdowns shown to affect the vasculature was the VEGF MO. The resulting morphant displayed a disruption of axial, ISV, and DLAV formation. (Nasevicius and Ekker, 2000; Nasevicius et al., 2000). It should be noted that many zebrafish genes are duplicated as a result of a tetraploidization event that occurred 350 million years ago with different sequences. For example, there have been reports of two NRP1s, and two NRP2s (Bovenkamp et al., 2004). Since there may be overlapping functions for a gene pair, it may be necessary to knock down both of the duplicates to observe loss of function.

F. Small Chemical Compounds

One advantage of the zebrafish is that it is possible to introduce small soluble molecules into the embryo media and analyze for their effects on phenotype (MacRae and Peterson, 2003; Milan *et al.*, 2003; Peterson *et al.*, 2000; Tanguay *et al.*, 2000; Yeh and Crews, 2003). Thus, zebrafish may be amenable for high-throughput assays (MacRae and Peterson, 2003; Milan *et al.*, 2003; Peterson *et al.*, 2000, 2001; Tanguay *et al.*, 2000; Yeh and Crews, 2003). As an example, adding small molecular weight VEGFR-2 kinase inhibitors to the embryo media had profound inhibitory effects on blood vessel development (Chan *et al.*, 2002; Lee *et al.*, 2002). FGFR kinase inhibitor dramatically inhibits the regeneration of amputated fin (Poss *et al.*, 2000) and somite genesis (Sawada *et al.*, 2001; Shinya *et al.*, 2001). Inhibition of sonic hedgehog signaling by cyclopamine results in downregulation of *vegf-a* expression (Lawson *et al.*, 2002). Notch–delta signaling is regulated by proteolytic cleavage of Notch by γ-secretase (Fortini, 2002; Lai, 2002) and treatment of zebrafish embryos with a γ-secretase inhibitor blocks Notch signaling and affects embryonic neural development (Geling *et al.*, 2002).

VI. Vascular Endothelial Growth Factor

VEGF-A is a potent angiogenesis factor that appears to be a specific mitogen for EC (Conway *et al.*, 2001; Ferrara, 1999; Ferrara *et al.*, 2003; Gale and Yancopoulos, 1999; Klagsbrun and D'Amore, 1996; Klagsbrun and Moses, 1999; Yancopoulos *et al.*, 2000). However, the number of cellular targets for VEGF has been expanded recently to include cell types such as tumor and epithelial cells. The VEGF family includes the classical VEGF, now known as VEGF-A, and also VEGF-B, VEGF-C, VEGF-D, VEGF-E, and placenta growth factor (P1GF). VEGF-A is produced by numerous cell types, including tumor cells, macrophages, T cells, and smooth-muscle cells (Klagsbrun and D'Amore, 1996). VEGF-secreted by tumors plays a major role in tumor progression by inducing neovascularization of the tumor. Recently, a humanized monoclonal directed against VEGF has shown efficacy in clinical treatment of colorectal and renal tumors (Willett *et al.*, 2004).

VEGF-A is a regulator of EC differentiation, migration, proliferation, and survival (Ferrara, 1999; Ferrara *et al.*, 2003; Klagsbrun and D'Amore, 1996). VEGF-A plays an important role in embryonic vascular development as a regulator of both vasculogenesis and angiogenesis. Much of the evidence for this has come from the *Vegf-a* knockout in mice. Targeted inactivation of *Vegf-a* results in embryonic lethality between E10 and E11 (Carmeliet *et al.*, 1996; Ferrara *et al.*, 1996). It is interesting that even

5. VEGF and Blood Vessel Development   137

disruption of a single *Vegf-a* allele in mice results in embryonic lethality between E11 and E12, suggesting that the concentration of Vegf-a is strictly regulated in embryogenesis. *Vegf-a*$^{-/+}$ mice exhibit malformation in the cardiac outflow tract, a rudimentary dorsal aorta, decreased ventricular wall thickness, and greatly reduced extra-embryonic and intra-embryonic vasculature. There is also a severe reduction of hematopoietic stem cells and endothelial progenitor cells in *Vegf-a*$^{-/+}$ embryos.

The *Vegf-a* gene consists of eight exons and seven introns. Alternative splicing results in a number of VEGF isoforms of 121, 145, 165, 189, and 206 amino acids (Ferrara *et al.*, 2003; Neufeld *et al.*, 1999; Robinson and Stringer, 2001). Mouse isoforms have one less amino acid than human isoform counterparts. VEGF$_{121}$ differs from the other isoforms in that it is readily secreted, whereas the other isoforms bind cell surface heparan sulfate proteoglycans and are partially retained by the cell or are associated with the extracellular matrix (Ferrara *et al.*, 2003; Klagsbrun and D'Amore, 1996). As will be described later, VEGF$_{121}$, unlike VEGF$_{165}$, does not bind NRPs. Transgenic mice have been generated that express only one of the isoforms. Mice expressing the VEGF$_{120}$ isoform exclusively were generated by Cre/loxP-mediated removal of exons 6 and 7, exons which are encoded in the 164 and 188 amino acid isoforms. VEGF$_{120}$ mice are viable but have impaired postnatal myocardial angiogenesis, resulting in ischemic cardiomyopathy (Carmeliet *et al.*, 1999). These mice also exhibit severe branching defects in retinal blood vessels (Ruhrberg *et al.*, 2002; Stalmans *et al.*, 2002) and kidneys (Mattot *et al.*, 2002). Mice that express only VEGF$_{164}$ are normal, indicating that this isoform confers all of the guidance cues needed for vessel patterning during embryonic development (Stalmans *et al.*, 2002).

Zebrafish express *vegf-a*, predominantly *VEGF*$_{165}$ and *VEGF*$_{121}$ (Liang *et al.*, 1998). Whole-mount ISH demonstrates strong *vegf-a* expression during embryogenesis in the medial regions of the somites, the hypocord, within the anterior central nervous system, in the prospective optic stalk, in mesoderm overlapping the bilaterally located merging heart fields, and in mesoderm underlying and flanking the hindbrain (Liang *et al.*, 1998). Injection of VEGF MO diminishes blood circulation in both axial and ISV. MO-injected embryos lack expression of *fli-1* and *flk-1* in the ISV, but these genes are still expressed in axial vessels. Although there are EC present in VEGF morphants, they do not form functional axial vessels. Overall, a loss of zebrafish Vegf-a activity results in less severe phenotypes than those observed in *Vegf-a* knockout mice (Nasevicius *et al.*, 2000), but this might be attributable to duplicate *Vefg-a* genes (Bahary, personal communication).

There is evidence to suggest that sonic hedgehog (Shh) and Notch regulate zebrafish Vegf-A activity, with *shh* acting upstream of *vegf* and *notch* acting downstream of *vegf* (Lawson *et al.*, 2002). Shh is a morphogen produced by the notochord. Zebrafish *shh* mutants fail to form a dorsal aorta (Brown *et al.*, 2000), and there is diminished *vegf-a* expression in the somites (Lawson *et al.*, 2002). Inhibition of Vegf-a or Shh activity results in loss of arterial identity, such as expression of *notch 5* and *ephrinb2* in the aorta (Lawson *et al.*, 2002). Notch proteins are signaling receptors that regulate cell-fate determination in a variety of cell types. *Notch* genes are robustly expressed in the vasculature (Shawber and Kitajewski, 2004). Zebrafish studies have suggested a key role for Notch5 signaling downstream of VEGF-A. One piece of evidence is that activation of the Notch pathway in the absence of VEGF signaling can rescue arterial marker gene expression.

VEGF-C activates VEGFR-3, which is expressed predominantly in lymphatic EC. However, VEGF-C also activates VEGFR-2, which is expressed in both blood and lymphatic vessel EC (Joukov *et al.*, 1997). *Vegf-c* knockout mice die between E15.5 and E17.5 with lymphatic vessel defects and severe edema; however, the veins and arteries are normal (Karkkainen *et al.*, 2004). Proteolytic cleavage regulates the receptor binding and biological activities of VEGF-C (Joukov *et al.*, 1997). Partially processed forms of VEGF-C activate VEGFR-3, whereas the fully processed short forms are also potent stimulators of VEGFR-2. Partially processed forms of human VEGF-C and zebrafish Vegf-c, but not fully processed forms, bind to zebrafish Nrp1a (Ober *et al.*, 2004).

In contrast to *Vegf-a* and *Vegf-c* knockout mice, *Vegf-b* and *Plgf* knockout mice are viable and fertile. Targeted inactivation of *Vegf-d* and *Vegf-e* have yet to be reported. In zebrafish, Vegf-c knockdown as well as overexpression of soluble human *VEGFR3* result in defects in vasculogenesis and angiogenesis (Ober *et al.*, 2004).

VII. VEGF Receptors

There are 3 VEGF receptor tyrosine kinases that are targets for VEGF family-induced signaling, VEGFR-1 (Flt-1), VEGFR-2 (Flk-1/KDR), and VEGFR-3 (Flt-4). VEGFRs have an extracellular region with seven immunoglobulin-like domains, one transmembrane domain, and a kinase domain including a 70 amino acid insert. VEGFR-1 and VEGFR-2 are expressed in vascular EC, whereas VEGFR-3 is expressed mainly in lymphatic EC. It is now known that other cell types besides EC (e.g., neuronal cells or tumor cells) can express VEGFR-1 and -2. VEGFR1 binds VEGF-A, VEGF-B, and P1GF; VEGFR-2 binds VEGF-A, VEGF-C, and VEGF-D; VEGFR-3 binds VEGF-C and VEGF-D (Ferrara, 1999; Ferrara *et al.*, 2003; Klagsbrun *et al.*, 2002; Matsumoto and Claesson-Welsh, 2001).

The three VEGF receptors have been knocked out in mice. *Vegfr-1* knockout mice die between E8.5 and E9.5 (Fong *et al.*, 1995, 1999). EC

5. VEGF and Blood Vessel Development

are found at embryonic and extraembryonic sites, but the vessels are not organized. Thus, VEGFR-1 is needed for EC assembly into tubular structures. *Vegfr-2* knockout mice die between E8.5 and E9.5 and exhibit severe defects in vasculogenesis and hematopoiesis. They lack both mature endothelial and hematopoietic cells as a result of a failure to form blood islands. Organized blood vessels fail to develop throughout the embryo and the yolk sac (Shalaby *et al.*, 1995). *Vegfr-3* knockout mice die at E9.5. While vasculogenesis and angiogenesis occur in these mutant mice, large vessels are abnormally organized, with defective lumen formation, leading to fluid accumulation in the pericardial cavity and consequent cardiovascular failure. These findings indicate that VEGFR-3 has an essential role in the development of the embryonic cardiovascular system before the emergence of the lymphatic vessels (Dumont *et al.*, 1998; Hamada *et al.*, 2000).

Zebrafish *flk1* (*vegfr-2*) is specifically expressed in blood vessels, particularly in arteries (Fouquet *et al.*, 1997; Liao *et al.*, 1997; Sumoy *et al.*, 1997; Thompson *et al.*, 1998). The importance of VEGFR-2 for vascular development has been demonstrated by treating zebrafish with VEGFR-2 kinase inhibitors (Chan *et al.*, 2002; Lee *et al.*, 2002). Administration of VEGFR-2 kinase inhibitor at the one cell stage completely inhibits axial and ISV development, whereas administration at 24 h inhibits just the ISV (Fig. 2). This temporal sequence is consistent with the axial vessels being formed first by vasculogenesis and the ISV being formed later by angiogenesis. Significantly, the VEGFR-2 vascular phenotype can be rescued partially by activated Akt (Chan *et al.*, 2002), demonstrating that kinase inhibitor-treated embryos are good models for analyzing the signal transduction pathways that regulate blood vessel development. Recently, in a large-scale forward genetic screen, two allelic zebrafish mutants were identified that were nonsense mutations in the *flk1* gene, resulting in production of a soluble form of Flk1 due to a truncated extracellular domain (Habeck *et al.*, 2002). These mutants lack ISV and the subintestinal vein (SIV).

VIII. Neuropilins

Neuropilin (NRP) is a 130- to 140-kDa membrane glycoprotein first identified in the optic tectum of *Xenopus laevis* and subsequently in the developing brain (Fujisawa, 2002, 2004). Currently, there are two known NRP genes, designated as *NRP1* and *NRP2* (Fujisawa, 2002, 2004; Klagsbrun *et al.*, 2002; Neufeld *et al.*, 1999, 2002). NRPs have three extracellular domains (a, b, and c), a transmembrane domain, and a very short cytoplasmic domain. In addition, there is a soluble NRP (sNRP) isoform containing only the a and b domains (Gagnon *et al.*, 2000). NRP1 localizes to axons and was first identified as the receptor for Semaphorin 3A

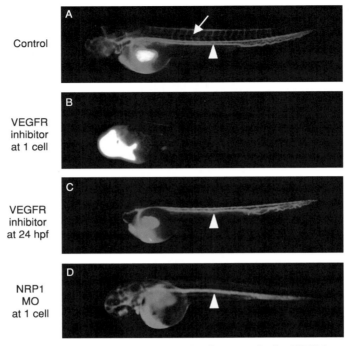

Figure 2 Vasculogenesis and Angiogenesis analyzed temporally by FITC-dextran microangiography. (A) FITC-dextran visualizes functional blood flow in axial vessels (arrowhead) and ISV (arrow). (B) Administration of VEGFR kinase inhibitor at the one-cell stage blocks blood circulation via axial vessels and ISV. (C) Administration of VEGFR kinase inhibitor at 24 hpf blocks blood circulation via ISV but not axial vessels (arrowhead). (D) Nrp1 knockdown inhibits blood circulation via ISV but not axial vessels (arrowhead). (See Color Insert.)

(originally collapsin-1), a secreted axon guidance molecule that repels axons and collapses growth cones of dorsal root ganglia (Fujisawa, 2002, 2004). More recently, NRP1 was identified as a functional receptor for $VEGF_{165}$ (Klagsbrun et al., 2002; Soker et al., 1998). NRPs do not appear to be receptor kinases but may act as coreceptors for VEGFR-2 that enhance $VEGF_{165}$ chemotactic activity (Soker et al., 1998). Binding to NRPs is isoform specific. VEGF exon 7 binds to the NRP b domain. $VEGF_{121}$ has no exon 7 domain and therefore, can not bind NRP1. In addition, NRP1 binds VEGF-B (Makinen et al., 1999), VEGF-C (Ober et al., 2004), VEGF-E (Wise et al., 1999), and P1GF2 (a splice isoform of P1GF) (Migdal et al., 1998), whereas NRP2 binds $VEGF_{165}$ (Soker et al., 1998), $VEGF_{145}$ (Gluzman-Poltorak et al., 2000), and VEGF-C (Karkkainen et al., 2001). NRP expression in EC shows a degree of specificity. During development, NRP1 is expressed by arterial EC, whereas NRP2 is expressed by venous and

5. VEGF and Blood Vessel Development

lymphatic EC (Eichmann *et al.*, 2002; Herzog *et al.*, 2001; Le Noble *et al.*, 2004; Moyon *et al.*, 2001; Mukouyama *et al.*, 2002; Yuan *et al.*, 2002).

NRPs play an important role in vascular development, as shown in a series of mouse *Nrp* knockout studies. *Nrp1* knockout mice are embryonically lethal at E12.5 to E13.5. These mice have abnormal axonal networks and also exhibit defects in yolk sac and embryonic vascular formation (Kawasaki *et al.*, 1999; Kitsukawa *et al.*, 1997). Disorganized vascular formation is observed in large and small blood vessels in yolk sac and embryos. The mouse knockout embryos have abnormal cardiovascular development, such as agenesis of the branchial arch-related great vessels, dorsal aorta, and transposition of aortic arches. Overexpression of *Nrp1* in mice also confers vascular defects (Kitsukawa *et al.*, 1995). These mice die *in utero* with morphological deformities, an excess of capillaries and blood vessels, dilation of blood vessels, hemorrhage, and malformed hearts. *Nrp2*-deficient mice show normal development of blood vessels (Chen *et al.*, 2000; Giger *et al.*, 2000) but display a severe reduction of small lymphatic vessels and capillaries (Yuan *et al.*, 2002). Arteries, veins, and the larger collecting lymphatic vessels develop normally, suggesting that NRP2, which is expressed by lymphatic EC, is selectively required for the formation of small lymphatic vessels and capillaries (Yuan *et al.*, 2002).

Double *Nrp1/Nrp2* knockout mice have a more severe vascular phenotype than either *Nrp1* or *Nrp2* null mice (Takashima *et al.*, 2002). These double mutants die very early *in utero* at E8.5 and exhibit defects in blood vessel development, including a lack of blood vessel branching in yolk sacs and no capillary formation. Furthermore, large areas of the *Nrp1/Nrp2* knockout embryos are avascular, and the mutants have marked growth retardation. The vascular phenotype of double *Nrp1/Nrp2* knockout mice resembles that of *Vegf-a* and *Vegfr-2* knockout mice (Takashima *et al.*, 2002).

Zebrafish *NRP* genes are duplicated. There are two nrp*1* (nrp*1a* and *b*) and two nrp*2* (nrp*2a* and *b*) genes (Lee *et al.*, 2002) (Bovenkamp *et al.*, 2004; Yu *et al.*, 2004.). Nrp1a and b bind $VEGF_{165}$; however, the binding of $VEGF_{165}$ to Nrp2 has not yet been shown. ISH of the spatio-temporal expression pattern of the zebrafish nrp*1* and *2* genes showed that they are expressed mainly in neuronal and vascular tissues during embryogenesis. During formation of the vascular system, nrp*1a* is first expressed by tail angioblasts toward the end of somitogenesis, and by 48 hpf, expression becomes more widespread to include the major trunk vessels and ISV. On the other hand, nrp*1b*, nrp*2a*, and nrp*2b* were not expressed in blood vessels during early embryonic development. (Bovenkamp *et al.*, 2004).

NRP1 MO-treated zebrafish have vascular defects beginning at 36 hpf, including short-circuited blood flow and a lack of circulation in the ISV, the DLAV, and the caudal vein plexus (Fig. 2). In some cases, abnormal connections (fistulas) were formed between the artery and vein, which prematurely returned blood circulation back to the heart. Blood cells accumulated posterior to the abnormal fistulas. On the other hand, even though MO was injected at the one-cell stage, the axial vessels appeared normal. Thus, these results indicate that NRP1 is a regulator of angiogenesis but not vasculogenesis. To demonstrate the interdependency of Nrp1 and Vegfa, zebrafish were treated with both NRP1 and VEGFA MO at concentrations where neither MO alone had any significant effects on blood vessel circulation. In these coinjections, all of the embryos exhibited a total loss of circulation throughout the trunk, the axial vessels, and ISV. It was concluded that VEGF and NRP1 cooperate in maintaining trunk axial and ISV function (Lee et al., 2002).

IX. Semaphorins

Of the eight classes of semaphorins (Sema), it is the class-3 Sema that bind NRPs (He and Tessier-Lavigne, 1997; Kolodkin et al., 1997). There are six class-3 Sema, A–F, with some degree of binding specificity (Chen et al., 1997, 1998; Giger et al., 1998; He and Tessier-Lavigne, 1997; Kolodkin et al., 1997; Takahashi et al., 1998). For example, NRP1 is activated by Sema3A, whereas NRP2 is activated by Sema3F. Semaphorins are 100 kDa proteins, first described as regulators of neuronal guidance that repel axons and collapse growth cones. However, semaphorins also have a vascular function. The earliest demonstration was that Sema3A is an inhibitor of EC migration and the sprouting of capillaries from aortic rings (Miao et al., 1999). It depolymerizes F-actin and induces retraction of lamelipodia. In these studies, it was also shown that $VEGF_{165}$ and Sema3A are competitive inhibitors of each other. Recently, it was found that Sema3A inhibits EC integrin activation, integrin-mediated adhesion to fibronectin and vitronectin, migration of EC, angiogenesis, and vascular remodeling (Serini et al., 2003). It has been shown in vivo that Sema3A-bound beads implanted into E4.5 forelimbs in quail cause nerves and blood vessels to deviate away from the bead (Bates et al., 2003). Besides Sema3A, Sema3F also affects blood vessels. For example, Sema3F inhibits $VEGF_{165}$ and bFGF-induced EC proliferation, EC ERK 1/2 kinase activity, and angiogenesis (Kessler et al., 2004). Overexpression of *Sema3F* in tumorigenic tumor cells resulted in a significantly lower density of blood vessels in tumors (Bielenberg and Klagsbrun, unpublished; Kessler et al., 2004).

Sema3A knockouts have been reported; however, the phenotype of the mutant mice showed variations depending on genetic background. In a 129/SV background, most of the mutant mice died from right heart failure within the first 3 days after birth (Behar et al., 1996). On the other hand, in a CD-1 background, there were vascular defects in the head and abnormal trunk blood vessels (Serini et al., 2003). At E9.5, cranial blood vessels of mutant embryos were not remodeled and instead maintained a primary capillary plexus appearance. The dorsal aorta formed normally, whereas development of the anterior cardinal vein was disrupted. Defects in the anterior cardinal vein persisted until at least E12.5. *Sema3C* knockout mice die just after birth due to aortic arch malformations and septation defects in the outflow tract of the heart (Feiner et al., 2001). *Sema3F* null mice are viable and fertile, and vascular defects have not been detected (Sahay et al., 2003).

In zebrafish, *sema3a*, *sema3d*, *sema4e*, and one unknown class-3 semaphorin have been described (Bernhardt et al., 1998; Halloran et al., 1998, 1999; Liu et al., 2004; Roos et al., 1999; Shoji et al., 1998; Xiao et al., 2003; Yee et al., 1999). Zebrafish contain two copies of the *sema3a* gene, *sema3aa* (also known as *sema1a*, *sema3a1*, or *semaZ1a*) (Shoji et al., 1998, 2003; Yee et al., 1999), and *sema3ab* (also known as *sema1b*, *sema3a2* or *semaZ1b*) (Bernhardt et al., 1998; Roos et al., 1999; Shoji et al., 2003). Angioblasts in the trunk are located ventrally and laterally to the somites. They migrate medially and dorsally along the medial surface of the somites to form the dorsal aorta just ventral to the notochord. *Sema3aa* is expressed by the dorsal and ventral regions of the somites and acts as a guidance cue that regulates the pathway of the dorsally migrating angioblasts (Shoji et al., 2003). Overexpression of *sema3aa* inhibits migration of angioblasts ventral and lateral to the somites and retards development of the dorsal aorta, resulting in severely reduced blood circulation. Thus, *sema3aa* is an inhibitor of angioblast migration that normally prevents these cells from migrating to inappropriate areas.

X. Future Perspectives

The use of zebrafish as an alternative model to mice for analyzing the developing vasculature and its regulators has greatly expanded in the last few years. Knockdown technology has identified a number of genes associated with vascular development. Transcription factors that regulate vascular gene expression are being identified. In addition, many large-scale screens are under way to identify new genes associated with vasculogenesis and angiogenesis. Mutants that result in an abnormal phenotype such as cardiovascular defects, will be useful as models of disease, for example, heart disease. Recent evidence indicates that the zebrafish also has promise as a cancer model, for example, with leukemia. Zebrafish also have the potential to rapidly screen small drugs administered to the embryo medium for adverse effects on the vasculature, for example, inhibitors of angiogenesis such as VEGFR-2, FGFR, and EGFR kinase inhibitors. These studies will provide information about efficacy, toxicity, and dosage useful for mouse tumor studies and clinical trials. The zebrafish genome is almost complete, and a protein database is being generated that will be useful for proteomics. These approaches and the ever expanding efforts of a number of laboratories, many new to the field, promise to make zebrafish a cornerstone of normal and abnormal vasculature research.

Acknowledgments

The authors thank Drs. Diane Bielenberg and Alan J. Davidson for discussion and critical review of this manuscript. We thank Melissa Mang for technical support. This work was supported by the National Institutes of Health grants CA37392 and CA45448 (M.K.).

References

Ackermann, G. E., and Paw, B. H. (2003). Zebrafish: A genetic model for vertebrate organogenesis and human disorders. *Front. Biosci.* **8,** 1227–1253.

Ahn, D. G., Ruvinsky, I., Oates, A. C., Silver, L. M., and Ho, R. K. (2000). tbx20, a new vertebrate T-box gene expressed in the cranial motor neurons and developing cardiovascular structures in zebrafish. *Mech. Dev.* **95,** 253–258.

Aird, W. C. (2002). Endothelial cell dynamics and complexity theory. *Crit. Care Med.* **30,** S180–S185.

Aird, W. C. (2003). Endothelial cell heterogeneity. *Crit. Care Med.* **31,** S221–S230.

Asahara, T., Murohara, T., Sullivan, A., Silver, M., van der Zee, R., Li, T., Witzenbichler, B., Schatteman, G., and Isner, J. M. (1997). Isolation of putative progenitor endothelial cells for angiogenesis. *Science* **275,** 964–967.

Baron, M. H. (2003). Embryonic origins of mammalian hematopoiesis. *Exp. Hemato.* **31,** 1160–1169.

Bates, D., Taylor, G. I., Minichiello, J., Farlie, P., Cichowitz, A., Watson, N., Klagsbrun, M., Mamluk, R., and Newgreen, D. F. (2003). Neurovascular congruence results from a shared patterning mechanism that utilizes Semaphorin3A and Neuropilin-1. *Dev. Biol.* **255,** 77–98.

Behar, O., Golden, J. A., Mashimo, H., Schoen, F. J., and Fishman, M. C. (1996). Semaphorin III is needed for normal patterning and growth of nerves, bones and heart. *Nature* **383,** 525–528.

Bernhardt, R. R., Goerlinger, S., Roos, M., and Schachner, M. (1998). Anterior–posterior subdivision of the somite in embryonic zebrafish: Implications for motor axon guidance. *Dev. Dyn.* **213,** 334–347.

Bovenkamp, D. E., Goishi, K., Bahary, N., Davidson, A. J., Zhou, Y., Becker, T., Becker, C. G., Zon, L. I., and Klagsbrun, M. (2004). Expression and mapping of duplicate neuropilin-1 and neuropilin-2 genes in developing zebrafish. *Gene Expr. Patterns* **4,** 361–370.

Brown, L. A., Rodaway, A. R., Schilling, T. F., Jowett, T., Ingham, P. W., Patient, R. K., and Sharrocks, A. D. (2000). Insights into early vasculogenesis revealed by expression of the ETS-domain transcription factor Fli-1 in wild-type and mutant zebrafish embryos. *Mech. Dev.* **90,** 237–252.

Carmeliet, P., Ferreira, V., Breier, G., Pollefeyt, S., Kieckens, L., Gertsenstein, M., Fahrig, M., Vandenhoeck, A., Harpal, K., Eberhardt, C., Declercq, C., Pawling, J., Moons, L., Collen, D.,

Risau, W., and Nagy, A. (1996). Abnormal blood vessel development and lethality in embryos lacking a single VEGF allele. *Nature* **380**, 435–439.
Carmeliet, P., Ng, Y. S., Nuyens, D., Theilmeier, G., Brusselmans, K., Cornelissen, I., Ehler, E., Kakkar, V. V., Stalmans, I., Mattot, V., Perriard, J. C., Dewerchin, M., Flameng, W., Nagy, A., Lupu, F., Moons, L., Collen, D., D'Amore, P. A., and Shima, D. T. (1999). Impaired myocardial angiogenesis and ischemic cardiomyopathy in mice lacking the vascular endothelial growth factor isoforms VEGF164 and VEGF188. *Nat. Med.* **5**, 495–502.
Chan, J., Bayliss, P. E., Wood, J. M., and Roberts, T. M. (2002). Dissection of angiogenic signaling in zebrafish using a chemical genetic approach. *Cancer Cell* **1**, 257–267.
Chen, H., Bagri, A., Zupicich, J. A., Zou, Y., Stoeckli, E., Pleasure, S. J., Lowenstein, D. H., Skarnes, W. C., Chedotal, A., and Tessier-Lavigne, M. (2000). Neuropilin-2 regulates the development of selective cranial and sensory nerves and hippocampal mossy fiber projections. *Neuron* **25**, 43–56.
Chen, H., Chedotal, A., He, Z., Goodman, C. S., and Tessier-Lavigne, M. (1997). Neuropilin-2, a novel member of the neuropilin family, is a high affinity receptor for the semaphorins Sema E and Sema IV but not Sema III. *Neuron* **19**, 547–559.
Chen, H., He, Z., Bagri, A., and Tessier-Lavigne, M. (1998). Semaphorin–neuropilin interactions underlying sympathetic axon responses to class III semaphorins. *Neuron* **21**, 1283–1290.
Chen, J. N., Haffter, P., Odenthal, J., Vogelsang, E., Brand, M., van Eeden, F. J., Furutani-Seiki, M., Granato, M., Hammerschmidt, M., Heisenberg, C. P., Jiang, Y. J., Kane, D. A., Kelsh, R. N., Mullins, M. C., and Nusslein-Volhard, C. (1996). Mutations affecting the cardiovascular system and other internal organs in zebrafish. *Development* **123**, 293–302.
Choi, K. (2002). The hemangioblast: A common progenitor of hematopoietic and endothelial cells. *J. Hematother. Stem Cell Res.* **11**, 91–101.
Cleaver, O., and Melton, D. A. (2003). Endothelial signaling during development. *Nat. Med.* **9**, 661–668.
Conway, E. M., Collen, D., and Carmeliet, P. (2001). Molecular mechanisms of blood vessel growth. *Cardiovasc. Res.* **49**, 507–521.
Davidson, A. J., and Zon, L. I. (2000). Turning mesoderm into blood: The formation of hematopoietic stem cells during embryogenesis. *Curr. Top. Dev. Biol.* **50**, 45–60.
Dooley, K., and Zon, L. I. (2000). Zebrafish: A model system for the study of human disease. *Curr. Opin. Genet. Dev.* **10**, 252–256.
Draper, B. W., Morcos, P. A., and Kimmel, C. B. (2001). Inhibition of zebrafish fgf8 pre-mRNA splicing with morpholino oligos: A quantifiable method for gene knockdown. *Genesis* **30**, 154–156.
Driever, W., Solnica-Krezel, L., Schier, A. F., Neuhauss, S. C., Malicki, J., Stemple, D. L., Stainier, D. Y., Zwartkruis, F., Abdelilah, S., Rangini, Z., Belak, J., and Boggs, C. (1996). A genetic screen for mutations affecting embryogenesis in zebrafish. *Development* **123**, 37–46.
Dumont, D. J., Fong, G. H., Puri, M. C., Gradwohl, G., Alitalo, K., and Breitman, M. L. (1995). Vascularization of the mouse embryo: A study of flk-1, tek, tie, and vascular endothelial growth factor expression during development. *Dev. Dyn.* **203**, 80–92.
Dumont, D. J., Jussila, L., Taipale, J., Lymboussaki, A., Mustonen, T., Pajusola, K., Breitman, M., and Alitalo, K. (1998). Cardiovascular failure in mouse embryos deficient in VEGF receptor-3. *Science* **282**, 946–949.
Eichmann, A., Pardanaud, L., Yuan, L., and Moyon, D. (2002). Vasculogenesis and the search for the hemangioblast. *J. Hematother. Stem Cell Res.* **11**, 207–214.
Ema, M., and Rossant, J. (2003). Cell fate decisions in early blood vessel formation. *Trends Cardiovasc. Med.* **13**, 254–259.

Feiner, L., Webber, A. L., Brown, C. B., Lu, M. M., Jia, L., Feinstein, P., Mombaerts, P., Epstein, J. A., and Raper, J. A. (2001). Targeted disruption of semaphorin 3C leads to persistent truncus arteriosus and aortic arch interruption. *Development* **128**, 3061–3070.

Ferrara, N. (1999). Molecular and biological properties of vascular endothelial growth factor. *J. Mol. Med.* **77**, 527–543.

Ferrara, N., Carver-Moore, K., Chen, H., Dowd, M., Lu, L., O'Shea, K. S., Powell-Braxton, L., Hillan, K. J., and Moore, M. W. (1996). Heterozygous embryonic lethality induced by targeted inactivation of the VEGF gene. *Nature* **380**, 439–442.

Ferrara, N., Gerber, H. P., and LeCouter, J. (2003). The biology of VEGF and its receptors. *Nat. Med.* **9**, 669–676.

Fina, L., Molgaard, H. V., Robertson, D., Bradley, N. J., Monaghan, P., Delia, D., Sutherland, D. R., Baker, M. A., and Greaves, M. F. (1990). Expression of the CD34 gene in vascular endothelial cells. *Blood* **75**, 2417–2426.

Fong, G. H., Rossant, J., Gertsenstein, M., and Breitman, M. L. (1995). Role of the Flt-1 receptor tyrosine kinase in regulating the assembly of vascular endothelium. *Nature* **376**, 66–70.

Fong, G. H., Zhang, L., Bryce, D. M., and Peng, J. (1999). Increased hemangioblast commitment, not vascular disorganization, is the primary defect in flt-1 knock-out mice. *Development* **126**, 3015–3025.

Fortini, M. E. (2002). Gamma-secretase-mediated proteolysis in cell-surface-receptor signalling. *Nat. Rev. Mol. Cell Biol.* **3**, 673–684.

Fouquet, B., Weinstein, B. M., Serluca, F. C., and Fishman, M. C. (1997). Vessel patterning in the embryo of the zebrafish: Guidance by notochord. *Dev. Biol.* **183**, 37–48.

Fujisawa, H. (2002). From the discovery of neuropilin to the determination of its adhesion sites. *Adv. Exp. Med. Biol.* **515**, 1–12.

Fujisawa, H. (2004). Discovery of semaphorin receptors, neuropilin and plexin, and their functions in neural development. *J. Neurobiol.* **59**, 24–33.

Gagnon, M. L., Bielenberg, D. R., Gechtman, Z., Miao, H. Q., Takashima, S., Soker, S., and Klagsbrun, M. (2000). Identification of a natural soluble neuropilin-1 that binds vascular endothelial growth factor: *In vivo* expression and antitumor activity. *Proc. Natl. Acad. Sci. USA* **97**, 2573–2578.

Gale, N. W., and Yancopoulos, G. D. (1999). Growth factors acting via endothelial cell-specific receptor tyrosine kinases: VEGFs, angiopoietins, and ephrins in vascular development. *Genes Dev.* **13**, 1055–1066.

Geling, A., Steiner, H., Willem, M., Bally-Cuif, L., and Haass, C. (2002). A gamma-secretase inhibitor blocks Notch signaling *in Vivo* and causes a severe neurogenic phenotype in zebrafish. *EMBO Rep.* **3**, 688–694.

Giger, R. J., Cloutier, J. F., Sahay, A., Prinjha, R. K., Levengood, D. V., Moore, S. E., Pickering, S., Simmons, D., Rastan, S., Walsh, F. S., Kolodkin, A. L., Ginty, D. D., and Geppert, M. (2000). Neuropilin-2 is required in vivo for selective axon guidance responses to secreted semaphorins. *Neuron* **25**, 29–41.

Giger, R. J., Urquhart, E. R., Gillespie, S. K., Levengood, D. V., Ginty, D. D., and Kolodkin, A. L. (1998). Neuropilin-2 is a receptor for semaphorin IV: Insight into the structural basis of receptor function and specificity. *Neuron* **21**, 1079–1092.

Gluzman-Poltorak, Z., Cohen, T., Herzog, Y., and Neufeld, G. (2000). Neuropilin-2 is a receptor for the vascular endothelial growth factor (VEGF) forms VEGF-145 and VEFG-165 [corrected]. *J. Biol. Chem.* **275**, 18040–18045.

Goishi, K., Lee, P., Davidson, A. J., Nishi, E., Zon, L. I., and Klagsbrun, M. (2003). Inhibition of zebrafish epidermal growth factor receptor activity results in cardiovascular defects. *Mech. Dev.* **120**, 811–822.

Grunwald, D. J., and Streisinger, G. (1992). Induction of recessive lethal and specific locus mutations in the zebrafish with ethyl nitrosourea. *Genet. Res.* **59**, 103–116.

5. VEGF and Blood Vessel Development 147

Habeck, H., Odenthal, J., Walderich, B., Maischein, H., and Schulte-Merker, S. (2002). Analysis of a zebrafish VEGF receptor mutant reveals specific disruption of angiogenesis. *Curr. Biol.* **12**, 1405–1412.

Haffter, P., Granato, M., Brand, M., Mullins, M. C., Hammerschmidt, M., Kane, D. A., Odenthal, J., van Eeden, F. J., Jiang, Y. J., Heisenberg, C. P., Kelsh, R. N., Furutani-Seiki, M., Vogelsang, E., Beuchle, D., Schach, U., Fabian, C., and Nusslein-Volhard, C. (1996). The identification of genes with unique and essential functions in the development of the zebrafish, *Danio rerio*. *Development* **123**, 1–36.

Halloran, M. C., Severance, S. M., Yee, C. S., Gemza, D. L., and Kuwada, J. Y. (1998). Molecular cloning and expression of two novel zebrafish semaphorins. *Mech. Dev.* **76**, 165–168.

Halloran, M. C., Severance, S. M., Yee, C. S., Gemza, D. L., Raper, J. A., and Kuwada, J. Y. (1999). Analysis of a Zebrafish semaphorin reveals potential functions *in vivo*. *Dev. Dyn.* **214**, 13–25.

Hamada, K., Oike, Y., Takakura, N., Ito, Y., Jussila, L., Dumont, D. J., Alitalo, K., and Suda, T. (2000). VEGF-C signaling pathways through VEGFR-2 and VEGFR-3 in vasculoangiogenesis and hematopoiesis. *Blood* **96**, 3793–3800.

Hanahan, D., and Folkman, J. (1996). Patterns and emerging mechanisms of the angiogenic switch during tumorigenesis. *Cell* **86**, 353–364.

He, Z., and Tessier-Lavigne, M. (1997). Neuropilin is a receptor for the axonal chemorepellent Semaphorin III. *Cell* **90**, 739–751.

Heasman, J. (2002). Morpholino oligos: Making sense of antisense? *Dev. Biol.* **243**, 209–214.

Herzog, Y., Kalcheim, C., Kahane, N., Reshef, R., and Neufeld, G. (2001). Differential expression of neuropilin-1 and neuropilin-2 in arteries and veins. *Mech. Dev.* **109**, 115–119.

Isogai, S., Horiguchi, M., and Weinstein, B. M. (2001). The vascular anatomy of the developing zebrafish: An atlas of embryonic and early larval development. *Dev. Biol.* **230**, 278–301.

Isogai, S., Lawson, N. D., Torrealday, S., Horiguchi, M., and Weinstein, B. M. (2003). Angiogenic network formation in the developing vertebrate trunk. *Development* **130**, 5281–5290.

Joukov, V., Sorsa, T., Kumar, V., Jeltsch, M., Claesson-Welsh, L., Cao, Y., Saksela, O., Kalkkinen, N., and Alitalo, K. (1997). Proteolytic processing regulates receptor specificity and activity of VEGF-C. *EMBO J.* **16**, 3898–3911.

Kaipainen, A., Korhonen, J., Pajusola, K., Aprelikova, O., Persico, M. G., Terman, B. I., and Alitalo, K. (1993). The related FLT4, FLT1, and KDR receptor tyrosine kinases show distinct expression patterns in human fetal endothelial cells. *J. Exp. Med.* **178**, 2077–2088.

Karkkainen, M. J., Haiko, P., Sainio, K., Partanen, J., Taipale, J., Petrova, T. V., Jeltsch, M., Jackson, D. G., Talikka, M., Rauvala, H., Betsholtz, C., and Alitalo, K. (2004). Vascular endothelial growth factor C is required for sprouting of the first lymphatic vessels from embryonic veins. *Nat. Immunol.* **5**, 74–80.

Karkkainen, M. J., Saaristo, A., Jussila, L., Karila, K. A., Lawrence, E. C., Pajusola, K., Bueler, H., Eichmann, A., Kauppinen, R., Kettunen, M. I., Yla-Herttuala, S., Finegold, D. N., Ferrell, R. E., and Alitalo, K. (2001). A model for gene therapy of human hereditary lymphedema. *Proc. Natl. Acad. Sci. USA* **98**, 12677–12682.

Kawasaki, T., Kitsukawa, T., Bekku, Y., Matsuda, Y., Sanbo, M., Yagi, T., and Fujisawa, H. (1999). A requirement for neuropilin-1 in embryonic vessel formation. *Development* **126**, 4895–4902.

Kerbel, R., and Folkman, J. (2002). Clinical translation of angiogenesis inhibitors. *Nat. Rev. Cancer* **2**, 727–739.

Kessler, O., Shraga-Heled, N., Lange, T., Gutmann-Raviv, N., Sabo, E., Baruch, L., Machluf, M., and Neufeld, G. (2004). Semaphorin-3F is an inhibitor of tumor angiogenesis. *Cancer Res.* **64**, 1008–1015.

Kitsukawa, T., Shimizu, M., Sanbo, M., Hirata, T., Taniguchi, M., Bekku, Y., Yagi, T., and Fujisawa, H. (1997). Neuropilin–semaphorin III/D-mediated chemorepulsive signals play a crucial role in peripheral nerve projection in mice. *Neuron* **19**, 995–1005.

Kitsukawa, T., Shimono, A., Kawakami, A., Kondoh, H., and Fujisawa, H. (1995). Overexpression of a membrane protein, neuropilin, in chimeric mice causes anomalies in the cardiovascular system, nervous system and limbs. *Development* **121**, 4309–4318.

Klagsbrun, M., and D'Amore, P. A. (1996). Vascular endothelial growth factor and its receptors. *Cytokine Growth Factor Rev.* **7**, 259–270.

Klagsbrun, M., and Moses, M. A. (1999). Molecular angiogenesis. *Chem. Biol.* **6**, R217–R224.

Klagsbrun, M., Takashima, S., and Mamluk, R. (2002). The role of neuropilin in vascular and tumor biology. *Adv. Exp. Med. Biol.* **515**, 33–48.

Kolodkin, A. L., Levengood, D. V., Rowe, E. G., Tai, Y. T., Giger, R. J., and Ginty, D. D. (1997). Neuropilin is a semaphorin III receptor. *Cell* **90**, 753–762.

Kwei, S., Stavrakis, G., Takahas, M., Taylor, G., Folkman, M. J., Gimbrone, M. A., Jr., and Garcia-Cardena, G. (2004). Early adaptive responses of the vascular wall during venous arterialization in mice. *Am. J. Pathol.* **164**, 81–89.

Lai, E. C. (2002). Notch cleavage: Nicastrin helps Presenilin make the final cut. *Curr. Biol.* **12**, R200–R202.

Langheinrich, U. (2003). Zebrafish: A new model on the pharmaceutical catwalk. *Bioessays* **25**, 904–912.

Lawson, N. D., Scheer, N., Pham, V. N., Kim, C. H., Chitnis, A. B., Campos-Ortega, J. A., and Weinstein, B. M. (2001). Notch signaling is required for arterial-venous differentiation during embryonic vascular development. *Development* **128**, 3675–3683.

Lawson, N. D., Vogel, A. M., and Weinstein, B. M. (2002). Sonic hedgehog and vascular endothelial growth factor act upstream of the Notch pathway during arterial endothelial differentiation. *Dev. Cell* **3**, 127–136.

Lawson, N. D., and Weinstein, B. M. (2002a). Arteries and veins: Making a difference with zebrafish. *Nat. Rev. Genet* **3**, 674–682.

Lawson, N. D., and Weinstein, B. M. (2002b). In vivo imaging of embryonic vascular development using transgenic zebrafish. *Dev. Biol.* **248**, 307–318.

Le Noble, F., Moyon, D., Pardanaud, L., Yuan, L., Djonov, V., Matthijsen, R., Breant, C., Fleury, V., and Eichmann, A. (2004). Flow regulates arterial-venous differentiation in the chick embryo yolk sac. *Development* **131**, 361–375.

Lee, P., Goishi, K., Davidson, A. J., Mannix, R., Zon, L., and Klagsbrun, M. (2002). Neuropilin-1 is required for vascular development and is a mediator of VEGF-dependent angiogenesis in zebrafish. *Proc. Natl. Acad. Sci. USA* **99**, 10470–10475.

Liang, D., Xu, X., Chin, A. J., Balasubramaniyan, N. V., Teo, M. A., Lam, T. J., Weinberg, E. S., and Ge, R. (1998). Cloning and characterization of vascular endothelial growth factor (VEGF) from zebrafish, *Danio Rerio*. *Biochim. Biophys. Acta* **1397**, 14–20.

Liao, E. C., Paw, B. H., Oates, A. C., Pratt, S. J., Postlethwait, J. H., and Zon, L. I. (1998). SCL/Tal-1 transcription factor acts downstream of cloche to specify hematopoietic and vascular progenitors in zebrafish. *Genes Dev.* **12**, 621–662.

Liao, W., Bisgrove, B. W., Sawyer, H., Hug, B., Bell, B., Peters, K., Grunwald, D. J., and Stainier, D. Y. (1997). The zebrafish gene cloche acts upstream of a flk-1 homologue to regulate endothelial cell differentiation. *Development* **124**, 381–389.

Liu, T. X., Howlett, N. G., Deng, M., Langenau, D. M., Hsu, K., Rhodes, J., Kanki, J. P., D'Andrea, A. D., and Look, A. T. (2003). Knockdown of zebrafish Fancd2 causes developmental abnormalities via p53-dependent apoptosis. *Dev. Cell* **5**, 903–914.

Liu, Y., Berndt, J., Su, F., Tawarayama, H., Shoji, W., Kuwada, J. Y., and Halloran, M. C. (2004). Semaphorin3D guides retinal axons along the dorsoventral axis of the tectum. *J. Neurosci.* **24**, 310–318.

Lyons, M. S., Bell, B., Stainier, D., and Peters, K. G. (1998). Isolation of the zebrafish homologues for the tie-1 and tie-2 endothelium-specific receptor tyrosine kinases. *Dev. Dyn.* **212,** 133–140.

MacRae, C. A., and Peterson, R. T. (2003). Zebrafish-based small molecule discovery. *Chem. Biol.* **10,** 901–908.

Makinen, T., Olofsson, B., Karpanen, T., Hellman, U., Soker, S., Klagsbrun, M., Eriksson, U., and Alitalo, K. (1999). Differential binding of vascular endothelial growth factor B splice and proteolytic isoforms to neuropilin-1. *J. Biol. Chem.* **274,** 21217–21222.

Matsumoto, T., and Claesson-Welsh, L. (2001). VEGF receptor signal transduction. *Sci. STKE* **2001,** RE21.

Mattot, V., Moons, L., Lupu, F., Chernavvsky, D., Gomez, R. A., Collen, D., and Carmeliet, P. (2002). Loss of the VEGF(164) and VEGF(188) isoforms impairs postnatal glomerular angiogenesis and renal arteriogenesis in mice. *J. Am. Soc. Nephrol.* **13,** 1548–1560.

Miao, H. Q., Soker, S., Feiner, L., Alonso, J. L., Raper, J. A., and Klagsbrun, M. (1999). Neuropilin-1 mediates collapsin-1/semaphorin III inhibition of endothelial cell motility: Functional competition of collapsin-1 and vascular endothelial growth factor-165. *J. Cell Biol.* **146,** 233–242.

Migdal, M., Huppertz, B., Tessler, S., Comforti, A., Shibuya, M., Reich, R., Baumann, H., and Neufeld, G. (1998). Neuropilin-1 is a placenta growth factor-2 receptor. *J. Biol. Chem.* **273,** 22272–22278.

Mikkola, H. K., and Orkin, S. H. (2002). The search for the hemangioblast. *J. Hematother. Stem Cell Res.* **11,** 9–17.

Milan, D. J., Peterson, T. A., Ruskin, J. N., Peterson, R. T., and MacRae, C. A. (2003). Drugs that induce repolarization abnormalities cause bradycardia in zebrafish. *Circulation* **107,** 1355–1358.

Motoike, T., Loughna, S., Perens, E., Roman, B. L., Liao, W., Chau, T. C., Richardson, C. D., Kawate, T., Kuno, J., Weinstein, B. M., Stainier, D. Y., and Sato, T. N. (2000). Universal GFP reporter for the study of vascular development. *Genesis* **28,** 75–81.

Moyon, D., Pardanaud, L., Yuan, L., Breant, C., and Eichmann, A. (2001). Plasticity of endothelial cells during arterial-venous differentiation in the avian embryo. *Development* **128,** 3359–3370.

Mukouyama, Y. S., Shin, D., Britsch, S., Taniguchi, M., and Anderson, D. J. (2002). Sensory nerves determine the pattern of arterial differentiation and blood vessel branching in the skin. *Cell* **109,** 693–705.

Nasevicius, A., and Ekker, S. C. (2000). Effective targeted gene "knockdown" in zebrafish. *Nat. Genet.* **26,** 216–220.

Nasevicius, A., Larson, J., and Ekker, S. C. (2000). Distinct requirements for zebrafish angiogenesis revealed by a VEGF-A morphant. *Yeast* **17,** 294–301.

Neufeld, G., Cohen, T., Gengrinovitch, S., and Poltorak, Z. (1999). Vascular endothelial growth factor (VEGF) and its receptors. *FASEB J.* **13,** 9–22.

Neufeld, G., Cohen, T., Shraga, N., Lange, T., Kessler, O., and Herzog, Y. (2002). The neuropilins: Multifunctional semaphorin and VEGF receptors that modulate axon guidance and angiogenesis. *Trends Cardiovasc. Med.* **12,** 13–19.

Ober, E. A., Olofsson, B., Makinen, T., Jin, S. W., Shoji, W., Koh, G. Y., Alitalo, K., and Stainier, D. Y. (2004). VEGFC is required for vascular development and endoderm morphogenesis in zebrafish. *Embo. Rep.* **5,** 78–84.

Parker, L., and Stainier, D. Y. (1999). Cell-autonomous and non-autonomous requirements for the zebrafish gene cloche in hematopoiesis. *Development* **126,** 2643–2651.

Pelster, B. (2003). Developmental plasticity in the cardiovascular system of fish, with special reference to the zebrafish. *Comp. Biochem. Physiol. Mol. Integr. Physiol.* **133,** 547–553.

Peters, K. G., De Vries, C., and Williams, L. T. (1993). Vascular endothelial growth factor receptor expression during embryogenesis and tissue repair suggests a role in endothelial differentiation and blood vessel growth. *Proc. Natl. Acad. Sci. USA* **90**, 8915–8919.

Peterson, R. T., Link, B. A., Dowling, J. E., and Schreiber, S. L. (2000). Small molecule developmental screens reveal the logic and timing of vertebrate development. *Proc. Natl. Acad. Sci. USA* **97**, 12965–12969.

Peterson, R. T., Mably, J. D., Chen, J. N., and Fishman, M. C. (2001). Convergence of distinct pathways to heart patterning revealed by the small molecule concentramide and the mutation heart-and-soul. *Curr. Biol.* **11**, 1481–1491.

Poss, K. D., Shen, J., Nechiporuk, A., McMahon, G., Thisse, B., Thisse, C., and Keating, M. T. (2000). Roles for Fgf signaling during zebrafish fin regeneration. *Dev. Biol.* **222**, 347–358.

Ransom, D. G., Haffter, P., Odenthal, J., Brownlie, A., Vogelsang, E., Kelsh, R. N., Brand, M., van Eden, F. J., Furutani-Seiki, M., Granato, M., Hammerschmidt, M., Heisenberg, C. P., Jiang, Y. J., Kane, D. A., Mullins, M. C., and Nusslein-Volhard, C. (1996). Characterization of zebrafish mutants with defects in embryonic hematopoiesis. *Development* **123**, 311–319.

Ribatti, D., Vacca, A., Nico, B., Ria, R., and Dammacco, F. (2002). Cross-talk between hematopoiesis and angiogenesis signaling pathways. *Curr. Mol. Med.* **2**, 537–543.

Ribatti, D., Vacca, A., and Presta, M. (2000). The discovery of angiogenic factors: A historical review. *Gen. Pharmaco.* **35**, 227–331.

Risau, W. (1997). Mechanisms of angiogenesis. *Nature* **386**, 671–674.

Risau, W., and Flamme, I. (1995). Vasculogenesis. *Annu. Rev. Cell Dev. Biol.* **11**, 73–91.

Robinson, C. J., and Stringer, S. E. (2001). The splice variants of vascular endothelial growth factor (VEGF) and their receptors. *J. Cell Sci.* **114**, 853–865.

Roos, M., Schachner, M., and Bernhardt, R. R. (1999). Zebrafish semaphorin Z1b inhibits growing motor axons in vivo. *Mech. Dev.* **87**, 103–117.

Ruhrberg, C., Gerhardt, H., Golding, M., Watson, R., Ioannidou, S., Fujisawa, H., Betsholtz, C., and Shima, D. T. (2002). Spatially restricted patterning cues provided by heparin-binding VEGF-A control blood vessel branching morphogenesis. *Genes Dev.* **16**, 2684–2698.

Sabin, F. R. (1917). Preliminary note on the differentiation of angioblasts and the method by which they produce blood vessels, blood plasma and red blood cells as seen in the living chick. *Anat. Rec.* **13**, 199–204.

Sahay, A., Molliver, M. E., Ginty, D. D., and Kolodkin, A. L. (2003). Semaphorin 3F is critical for development of limbic system circuitry and is required in neurons for selective CNS axon guidance events. *J. Neurosci.* **23**, 6671–6680.

Sawada, A., Shinya, M., Jiang, Y. J., Kawakami, A., Kuroiwa, A., and Takeda, H. (2001). Fgf/MAPK signalling is a crucial positional cue in somite boundary formation. *Development* **128**, 4873–4880.

Schauerte, H. E., van Eeden, F. J., Fricke, C., Odenthal, J., Strahle, U., and Haffter, P. (1998). Sonic hedgehog is not required for the induction of medial floor plate cells in the zebrafish. *Development* **125**, 2983–2993.

Serini, G., Valdembri, D., Zanivan, S., Morterra, G., Burkhardt, C., Caccavari, F., Zammataro, L., Primo, L., Tamagnone, L., Logan, M., Tessier-Lavigne, M., Taniguchi, M., Puschel, A. W., and Bussolino, F. (2003). Class 3 semaphorins control vascular morphogenesis by inhibiting integrin function. *Nature* **424**, 391–397.

Shalaby, F., Rossant, J., Yamaguchi, T. P., Gertsenstein, M., Wu, X. F., Breitman, M. L., and Schuh, A. C. (1995). Failure of blood-island formation and vasculogenesis in Flk-1-deficient mice. *Nature* **376**, 62–66.

Shawber, C. J., and Kitajewski, J. (2004). Notch function in the vasculature: Insights from zebrafish, mouse and man. *Bioessays* **26**, 225–234.

Shinya, M., Koshida, S., Sawada, A., Kuroiwa, A., and Takeda, H. (2001). Fgf signalling through MAPK cascade is required for development of the subpallial telencephalon in zebrafish embryos. *Development* **128**, 4153–4164.

Shoji, W., Isogai, S., Sato-Maeda, M., Obinata, M., and Kuwada, J. Y. (2003). Semaphorin3a1 regulates angioblast migration and vascular development in zebrafish embryos. *Development* **130**, 3227–3236.

Shoji, W., Yee, C. S., and Kuwada, J. Y. (1998). Zebrafish semaphorin Z1a collapses specific growth cones and alters their pathway *in vivo*. *Development* **125**, 1275–1283.

Smithers, L., Haddon, C., Jiang, Y. J., and Lewis, J. (2000). Sequence and embryonic expression of deltaC in the zebrafish. *Mech. Dev.* **90**, 119–123.

Soker, S., Takashima, S., Miao, H. Q., Neufeld, G., and Klagsbrun, M. (1998). Neuropilin-1 is expressed by endothelial and tumor cells as an isoform-specific receptor for vascular endothelial growth factor. *Cell* **92**, 735–745.

Stainier, D. Y., Fouquet, B., Chen, J. N., Warren, K. S., Weinstein, B. M., Meiler, S. E., Mohideen, M. A., Neuhauss, S. C., Solnica-Krezel, L., Schier, A. F., Zwartkruis, F., Stemple, D. L., Malicki, J., Driever, W., and Fishman, M. C. (1996). Mutations affecting the formation and function of the cardiovascular system in the zebrafish embryo. *Development* **123**, 285–292.

Stainier, D. Y., Weinstein, B. M., Detrich, H. W., 3rd, Zon, L. I., and Fishman, M. C. (1995). Cloche, an early acting zebrafish gene, is required by both the endothelial and hematopoietic lineages. *Development* **121**, 3141–3150.

Stalmans, I., Ng, Y. S., Rohan, R., Fruttiger, M., Bouche, A., Yuce, A., Fujisawa, H., Hermans, B., Shani, M., Jansen, S., Hicklin, D., Anderson, D. J., Gardiner, T., Hammes, H. P., Moons, L., Dewerchin, M., Collen, D., Carmeliet, P., and D'Amore, P. A. (2002). Arteriolar and venular patterning in retinas of mice selectively expressing VEGF isoforms. *J. Clin Invest.* **109**, 327–336.

Summerton, J. (1999). Morpholino antisense oligomers: The case for an RNase H-independent structural type. *Biochim. Biophys. Acta* **1489**, 141–158.

Sumoy, L., Keasey, J. B., Dittman, T. D., and Kimelman, D. (1997). A role for notochord in axial vascular development revealed by analysis of phenotype and the expression of VEGR-2 in zebrafish flh and ntl mutant embryos. *Mech. Dev.* **63**, 15–27.

Takahashi, T., Nakamura, F., Jin, Z., Kalb, R. G., and Strittmatter, S. M. (1998). Semaphorins A and E act as antagonists of neuropilin-1 and agonists of neuropilin-2 receptors. *Nat. Neurosci.* **1**, 487–493.

Takashima, S., Kitakaze, M., Asakura, M., Asanuma, H., Sanada, S., Tashiro, F., Niwa, H., Miyazaki Ji, J., Hirota, S., Kitamura, Y., Kitsukawa, T., Fujisawa, H., Klagsbrun, M., and Hori, M. (2002). Targeting of both mouse neuropilin-1 and neuropilin-2 genes severely impairs developmental yolk sac and embryonic angiogenesis. *Proc. Natl. Acad. Sci. USA* **99**, 3657–3662.

Talbot, W. S., and Hopkins, N. (2000). Zebrafish mutations and functional analysis of the vertebrate genome. *Genes Dev.* **14**, 755–762.

Tanguay, R. L., Andreasen, E., Heideman, W., and Peterson, R. E. (2000). Identification and expression of alternatively spliced aryl hydrocarbon nuclear translocator 2 (ARNT2) cDNAs from zebrafish with distinct functions. *Biochim. Biophys. Acta* **1494**, 117–128.

Thisse, C., and Zon, L. I. (2002). Organogenesis—heart and blood formation from the zebrafish point of view. *Science* **295**, 457–462.

Thompson, M. A., Ransom, D. G., Pratt, S. J., MacLennan, H., Kieran, M. W., Detrich, H. W., 3rd, Vail, B., Huber, T. L., Paw, B., Brownlie, A. J., Oates, A. C., Fritz, A., Gates, M. A., Amores, A., Bahary, N., Talbot, W. S., Her, H., Beier, D. R., Postlethwait, J. H., and Zon, L. I. (1998). The cloche and spadetail genes differentially affect hematopoiesis and vasculogenesis. *Dev. Biol.* **197**, 248–269.

Torres-Vazquez, J., Kamei, M., and Weinstein, B. M. (2003). Molecular distinction between arteries and veins. *Cell Tissue Res.* **314**, 43–59.
Traver, D., and Zon, L. I. (2002). Walking the walk: Migration and other common themes in blood and vascular development. *Cell* **108**, 731–734.
Walsh, E. C., and Stainier, D. Y. (2001). UDP-glucose dehydrogenase required for cardiac valve formation in zebrafish. *Science* **293**, 1670–1673.
Weinstein, B. M., Schier, A. F., Abdelilah, S., Malicki, J., Solnica-Krezel, L., Stemple, D. L., Stainier, D. Y., Zwartkruis, F., Driever, W., and Fishman, M. C. (1996). Hematopoietic mutations in the zebrafish. *Development* **123**, 303–309.
Weinstein, B. M., Stemple, D. L., Driever, W., and Fishman, M. C. (1995). Gridlock, a localized heritable vascular patterning defect in the zebrafish. *Nat. Med.* **1**, 1143–1147.
Willett, C. G., Boucher, Y., di Tomaso, E., Duda, D. G., Munn, L. L., Tong, R. T., Chung, D. C., Sahani, D. V., Kalva, S. P., Kozin, S. V., Mino, M., Cohen, K. S., Scadden, D. T., Hartford, A. C., Fischman, A. J., Clark, J. W., Ryan, D. P., Zhu, A. X., Blaszkowsky, L. S., Chen, H. X., Shellito, P. C., Lauwers, G. Y., and Jain, R. K. (2004). Direct evidence that the VEGF-specific antibody bevacizumab has antivascular effects in human rectal cancer. *Nat. Med.* **10**, 145–147.
Wilting, J., Christ, B., Yuan, L., and Eichmann, A. (2003). Cellular and molecular mechanisms of embryonic haemangiogenesis and lymphangiogenesis. *Naturwissenschaften* **90**, 433–448.
Wise, L. M., Veikkola, T., Mercer, A. A., Savory, L. J., Fleming, S. B., Caesar, C., Vitali, A., Makinen, T., Alitalo, K., and Stacker, S. A. (1999). Vascular endothelial growth factor (VEGF)-like protein from orf virus NZ2 binds to VEGFR2 and neuropilin-1. *Proc. Natl. Acad. Sci. USA* **96**, 3071–3076.
Xiao, T., Shoji, W., Zhou, W., Su, F., and Kuwada, J. Y. (2003). Transmembrane sema4E guides branchiomotor axons to their targets in zebrafish. *J. Neurosci.* **23**, 4190–4198.
Yamaguchi, T. P., Dumont, D. J., Conlon, R. A., Breitman, M. L., and Rossant, J. (1993). flk-1, an flt-related receptor tyrosine kinase is an early marker for endothelial cell precursors. *Development* **118**, 489–498.
Yancopoulos, G. D., Davis, S., Gale, N. W., Rudge, J. S., Wiegand, S. J., and Holash, J. (2000). Vascular-specific growth factors and blood vessel formation. *Nature* **407**, 242–248.
Yancopoulos, G. D., Klagsbrun, M., and Folkman, J. (1998). Vasculogenesis, angiogenesis, and growth factors: Ephrins enter the fray at the border. *Cell* **93**, 661–664.
Yee, C. S., Chandrasekhar, A., Halloran, M. C., Shoji, W., Warren, J. T., and Kuwada, J. Y. (1999). Molecular cloning, expression, and activity of zebrafish semaphorin Z1a. *Brain Res. Bull.* **48**, 581–593.
Yeh, J. R., and Crews, C. M. (2003). Chemical genetics: Adding to the developmental biology toolbox. *Dev. Cell* **5**, 11–19.
Yu, H. H., Houart, C., and Moens, C. B. (2004). Cloning and embryonic expression of zebrafish neuropilin genes. *Gene Expr. Patterns* **4**, 371–378.
Yuan, L., Moyon, D., Pardanaud, L., Breant, C., Karkkainen, M. J., Alitalo, K., and Eichmann, A. (2002). Abnormal lymphatic vessel development in neuropilin 2 mutant mice. *Development* **129**, 4797–4806.
Zhang, J., Talbot, W. S., and Schier, A. F. (1998). Positional cloning identifies zebrafish one-eyed pinhead as a permissive EGF-related ligand required during gastrulation. *Cell* **92**, 241–251.
Zhong, T. P., Childs, S., Leu, J. P., and Fishman, M. C. (2001). Gridlock signalling pathway fashions the first embryonic artery. *Nature* **414**, 216–220.
Zhong, T. P., Rosenberg, M., Mohideen, M. A., Weinstein, B., and Fishman, M. C. (2000). gridlock, an HLH gene required for assembly of the aorta in zebrafish. *Science* **287**, 1820–1824.

6

Vascular Extracellular Matrix and Aortic Development

Cassandra M. Kelleher, Sean E. McLean, and Robert P. Mecham
Washington University School of Medicine
Department of Cell Biology and Physiology
St. Louis, Missouri 63110

I. Introduction
II. Vessel Wall Formation and Structure
III. The Vascular Extracellular Matrix
IV. Collagens
  A. Genotype–Phenotype Correlations Resulting from Mutations in the Vascular Fibrillar Collagens
V. The Elastic Fiber
  A. Elastin
  B. Fibrillin and Microfibrils
VI. Fibulins
  A. Fibulin-1
  B. Fibulin-2
  C. Fibulins-3 and -4
  D. Fibulin-5
VII. EMILIN/Multimerin Family
VIII. Fibronectin
IX. The Basement Membrane
  A. Laminins
  B. Entactin/Nidogen
X. Proteoglycans
  A. Large Proteoglycans That Form Aggregates by Interaction with Hyaluronan
  B. Small Leucine Rich Proteoglycans
XI. Matricellular Proteins
  A. Thrombospondins
  B. Tenascins
  C. SPARC (osteonectin)
XII. Correlation of Matrix Gene Expression Profile with Cytoskeletal Markers
XIII. Conclusions
  Acknowledgments
  References

I. Introduction

With the emergence of a high-pressure, pulsatile circulatory system in vertebrates came a remarkable change in blood vessel structure and function. Blood vessels no longer acted as simple tubes for channeling blood or other body fluids from a low-pressure heart. In this closed circulatory system, large arteries became an important component of proper cardiac function by serving as elastic reservoirs, enabling the arterial tree to undergo large-volume changes with little change in pressure. Without elastic vessels, the tremendous surge of pressure as blood ejected from the heart would inhibit the heart from emptying, and the pressure in the vessels would fall so rapidly that the heart could not refill. Furthermore, distension of the elastic arterial wall by blood pushed from the heart is translated into kinetic energy when the arterial wall contracts, which helps move the blood down the vascular tree. The change that brought about this critical step in the evolution of higher organisms was the emergence of a vascular wall containing cells specialized in the production and organization of an extracellular matrix (ECM) uniquely designed to provide elastic recoil.

In addition to providing the structural and mechanical properties required for vessel function, the ECM provides instructional signals that induce, define, and stabilize smooth muscle phenotypes. There are many examples of ECM molecules playing critical roles in the regulation of gene expression by interacting with specific matrix receptors on cells and by binding and storing growth factors that influence cellular function. This reciprocal instructive interaction between the cell and its ECM is important in directing the developmental transitions that occur in embryogenesis, postnatal development, and in response to injury. How vascular cells interpret these regulatory signals is a major area of research today.

This review will discuss the ECM molecules made by vessel wall cells during vascular development, with the primary focus on the developing mouse aorta. Several excellent reviews have summarized our current understanding of smooth muscle cell phenotypes based on expression of cytoskeletal and other marker proteins (Glukhova and Koteliansky, 1995; Hungerford et al., 1996; Owens, 1995). There are also numerous ultrastructural studies documenting the architecture of the developing vessel wall (Albert, 1972; Berry et al., 1972; Gerrity and Cliff, 1975; Haust et al., 1965; Karrer, 1961; Paule, 1963; Pease and Paule, 1960; Thyberg et al., 1979), although most of these studies have been in animals other than mouse. The morphogenesis of the aortic wall in the rat, however, has been well investigated (Berry et al., 1972; Cliff, 1967; Gerrity and Cliff, 1975; Nakamura, 1988; Paule, 1963; Pease and Paule, 1960) and shows many similarities with mouse wall structure (Davis, 1993; Karrer, 1961). For the interested reader, extensive information on the vascular smooth muscle cell and a still timely discussion of questions and issues driving research in vascular biology can be found in a monograph by Schwartz and Mecham (1995).

II. Vessel Wall Formation and Structure

While the role of endothelial cells in the formation of the vascular primordia is beginning to be well understood (Carmeliet, 2000; Drake *et al.*, 1998; Rossant and Howard, 2002), surprisingly little is known about how vessels acquire their coat of smooth muscle cells that make up the vessel wall. Presumptive vascular smooth muscle cells (VSMCs) form from the surrounding mesenchyme and/or cardiac neural crest in response to soluble factors secreted by endothelial cells. The angiopoietin/Tie receptor pathway (Dumont *et al.*, 1995; Sato *et al.*, 1993) is clearly a major player in early stages of this process, but questions remain about what other factors guide smooth muscle differentiation through the various stages of vessel wall formation. Complicating our understanding of the VSMC is the cellular heterogeneity (Frid *et al.*, 1994; Gittenberger-de Groot *et al.*, 1999) and phenotypic plasticity (Schwartz and Mecham, 1995) observed during embryogenesis and vessel maturation. As the vessel wall matures, the SMCs go through multiple overlapping phenotypic transitions, characterized broadly by cellular proliferation, matrix production, and the assembly of an appropriate contractile apparatus within the cell cytoplasm. In medium and large vessels, the major function of the SMC is to synthesize and organize the unique extracellular matrix responsible for the mechanical properties of the wall. Unlike cells in the small muscular and resistance vessels, the smooth muscle cells of the elastic conducting vessels contribute little to the static mechanical properties of the wall. Hence, their ability to produce ECM can be considered to be their "differentiated" phenotype. Because the formation of a functional extracellular matrix must occur in an organized sequence, the "matrix phenotype" is changing throughout the entire period of vessel wall development. As pointed out by Little and colleagues (Drake *et al.*, 1998; Hungerford *et al.*, 1996), the expression pattern of ECM proteins may be a better indicator of VSMC differentiation status than the presence or absence of intracellular markers.

The general histological form of the large blood vessels includes three compartments: the *tunica intima*, consisting of a single layer of endothelial cells that sit directly on the internal elastic lamina (IEL); the *tunica media*, consisting of concentric layers of smooth muscle cells between sheets of elastin (the elastic laminae); and the *tunica adventitia*, made up of myofibroblasts that produce mainly collagen fibers. Within the medial layer, the collagen and elastin fibers are arranged to form a "two phase" system, in which circumferentially aligned collagen fibers of high tensile strength and elastic modulus bear most of the stressing force at and above physiologic blood pressure. Elastin, which is distensible and has a low tensile strength, functions primarily as an elastic reservoir and distributes stress evenly throughout the wall and onto collagen fibers (Berry et al., 1972; Gerrity and Cliff, 1975; Wolinsky and Glagov, 1967). The number of lamellar units (generally defined as the elastic lamella and adjacent smooth muscle cells) in a vascular segment is related linearly to tensional forces within the wall (Clark and Glagov, 1985; Leung et al., 1977; Wolinsky and Glagov, 1967), with the greatest number of elastic layers occurring in the larger, more proximal vessels that experience the highest wall stress.

A role for hemodynamics in vessel wall development (Folkow, 1983; Langille, 1996) and in modulating elastin production (Faury et al., 2003; Keeley and Alatawi, 1991; Keeley and Johnson, 1986) has been suggested from numerous studies of vascular remodeling in response to altered pressure and flow. In the developing chick coronary artery, for example, SMC recruitment from undifferentiated mesenchyme does not occur until the connection to the aorta is made and actual blood flow through these vessels has begun (Bergwerff et al., 1996). When the vessel wall is forming, SMC differentiation, lamellar number, and elastin content coordinately increase with the gradual rise in blood pressure until the proper number of lamellar units are organized (Nakamura, 1988; Roach, 1983). The relatively constant tension per lamellar unit and their uniformity of composition, regardless of species, indicate that the proportion of collagen, elastin, and SMCs in the media is optimal for the stresses to which the aorta is subjected (Wolinsky and Glagov, 1967).

III. The Vascular Extracellular Matrix

In addition to the structural matrix proteins (collagen, elastin, proteoglycans, etc.), vascular cells must produce matrix macromolecules that are important for cell movement, polarization, and anchorage. These molecules, which include adhesive glycoproteins such as fibronectin, basement membrane components, and the matricellular proteins that modulate cell–matrix interactions, provide important informational signals to cells that can influence gene expression and cellular function. To identify the types of matrix proteins produced by SMCs and to compare their expression pattern with other known markers of SMC differentiation, we performed large-scale gene expression analysis on developing mouse aorta using oligonucleotide microarrays (MU74Av2 chip from Affymetrix). Our dataset begins at embryonic day 12 and extends through 6 months of age in the adult mouse. Details of

6. Vascular Matrix and Aortic Development

the array procedures, sample selection and preparation, and data analysis can be found in our original publication (McLean et al., 2004).

The array data identify four major patterns for matrix gene expression. The first and most prevalent begins around day 14 shortly after mesenchymal cells recruited to the vessel wall organize into layers that closely approximate the number that will be found in the mature tissue. This expression pattern, which we will call the *matrix pattern*, consists of a major increase in matrix protein expression at embryonic day 14 followed by a steady rise through the first 7–14 days after birth. This is followed by a decrease in expression over 2–3 months to low levels that persist in the adult. Most of the structural matrix proteins follow this pattern. The second most prevalent pattern was one of consistent expression throughout the time series and was typical of basement membrane components, fibronectin, most integrins, and some matrix metalloproteinases. The third pattern consists of high expression levels in the embryonic–fetal period followed by decreased expression postnatally. The final and least populated pattern was low expression throughout development with an increase in the adult period.

Our expression data are in agreement with the appearance of structural matrix proteins in the vessel wall as assessed by ultrastructural studies (Nakamura, 1988). Figure 1 compares the vessel wall of the developing mouse aorta at embryonic days 12, 14, and 18. At day 12, there are few discernable collagen or elastin fibers in the extracellular space. By day 14, however, collagen fibers and small patches of elastin are beginning to form in the juxtacellular space. By day 18, the elastic lamellae and mature collagen fibers are clearly evident. These micrographs illustrate the tremendous rate

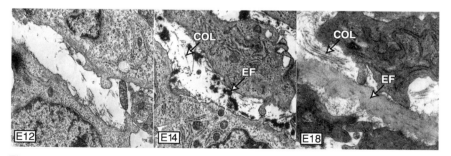

Figure 1 Electron micrograph of mouse ascending aorta at embryonic days 12, 14, and 18. Sparse collagen or elastic fibers are evident in the extracellular space at E12. At E14 there is a major increase in structural matrix protein expression, which is evident as small patches of elastic fibers (EF) and individual collagen fibers (COL). By E18, the small elastic bundles have coalesced into larger fibers that define the elastic lamellae. Bundles of mature collagen fibers are also evident. These micrographs illustrate the remarkable quantity of extracellular matrix that is deposited into the vessel wall in a relatively short period of development.

of structural matrix protein synthesis that begins around embryonic day 14. The sections that follow contain an overview of the expression pattern for the major classes of vascular ECM. Where possible, we will discuss vascular phenotypes associated with specific mutations or resulting from gene targeting experiments that produce loss of function mutations in mice. Table I summarizes the known mouse phenotypes resulting from deletion of the structural matrix genes discussed in this report.

IV. Collagens

Collagens are ubiquitous ECM proteins that impart a structural framework to tissues (Mecham, 1998). All collagens have a triple-helical domain that is composed of repeats where glycine occupies every third position in the sequence (Gly-X-Y). Three individual collagen proteins, called α chains, associate to form a righthanded triple helix. The three chains can be identical or consist of two or three different α chains. In all, 17 different collagens were identified in the developing mouse aorta by microarray analysis, with collagens I, III, IV, V, and VI having the highest expression levels. Present in lesser amounts were collagens VII, VIII, IX, X, XI, XIV, XV, XVIII, and XIX. Collagens II, XII, XIII, and XVII had low expression levels. Collagens type I, II, III, and V are fibril-forming collagens that assemble into striated fibers of varying diameter and are usually the most abundant collagens in tissues. Type VI, also a fibrillar collagen, forms a beaded filament. Collagens IV, VIII, and X are members of the network-forming collagen family and

Table I Known Vascular Phenotypes Resulting From Deletion of Structural Extracellular Matrix Proteins in Mice

	Null phenotype
Collagen 1A1	Embryonically lethal—vessel rupture E12–E14
Collagen 3A1	Decreased viability, vessel rupture in viable adults
Collagen 6A1	Decreased capillary lumen size, enlarged endothelial cells
Elastin	Perinatally lethal—aortic occlusion by VSMC
EMILIN-1	Alterations in elastic fibers, changes in morphology and anchorage of endothelial cells and SMCs to elastic lamellae
Fibrillin-1	Vascular aneurysms
Fibulin-1	Narrow capillary lumen, perineural and skin hemorrhages
Fibulin-5	Large vessel tortuosity, disrupted lamellar structure
Fibronectin	Embryonically lethal—no endothelial tube formation
Laminin $\alpha 4$	Microvascular degeneration
Thrombospondin-1	Increased vascularity in healing wounds
Thrombospondin-2	Increased vascularity (small and midsized vessels)

create "basket weave–like" structures through associations between their helical and non-helical domains. Type IV collagen is the major structural protein of basement membranes. Collagens IX, XIV, and XIX are FACIT collagens that attach to the surface of fibril-forming collagens but do not form fibers themselves. Collagen XIII is a collagen with a transmembrane domain that resides in adhesive structures of cells and has been implicated in cell adhesion. Collagens XV and XVIII are closely related nonfibrillar collagens found associated with basement membranes. Type XV is thought to help anchor cells to the basement membrane, and the C-terminal fragment of type XVIII collagen, called endostatin, is a potent inhibitor of angiogenesis and endothelial cell migration.

Expression analysis by gene array of collagens I, III, and VI, the major fibrillar collagens in the aorta, showed a pattern typical of the matrix expression pattern previously described (Fig. 2). A major increase in expression beginning at embryonic day 14 was followed by high expression through postnatal day 10. Expression then decreased relatively rapidly over several weeks and continued to fall gradually into the adult period. Collagens XV and XVIII were the only collagens to show increased expression in adult animals.

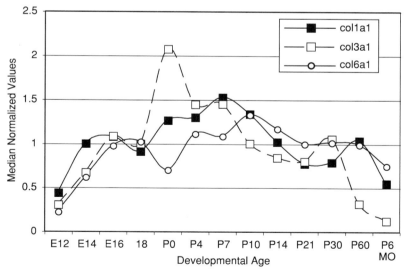

Figure 2 Expression for $\alpha 1$ chains of types I, III, and VI collagens as determined by oligonucleotide microarray (median normalized values). These values have been confirmed by real-time PCR. All three collagen chains show a major increase in expression beginning at E14 and continuing through postnatal days 7 to 10. Thereafter, expression decreases into the adult period. This pattern is typical of most of the structural matrix proteins that make up the aortic wall.

Collagens can interact with cells via integrin receptors, specifically $\alpha 1\beta 1$. In the human aorta, $\alpha 1\beta 1$ integrins are expressed as early as 10 weeks gestation (Glukhova and Koteliansky, 1995). Collagen types I, II, III, and IV act as ECM ligands of $\alpha 1\beta 1$. Expression of activated $\beta 1$ integrins has been speculated to facilitate maintainence of VSMCs in a contractile phenotype. Possibly, therefore, collagens may play a role in maintaining the integrity of the vessel wall not only by structural mechanisms, but by stabilizing the VSMC phenotype.

A. Genotype–Phenotype Correlations Resulting from Mutations in the Vascular Fibrillar Collagens

1. Collagen Type I

Immunohistochemical studies have localized type I collagen to the intimal, medial, and advential layers of the fetal bovine aorta, and in the adult human, in the media and intima (Howard and Macarak, 1989; Voss and Rauterberg, 1986). In the media, type I collagen is distributed around VSMCs in close contact to elastic lamallae (Dingemans et al., 2000; McCullagh et al., 1980).

Type I collagen is known to be important for blood vessel stability as mice homozygous for targeted interruption of collA1 die between E12 and E14 from vessel rupture (Lohler et al., 1984). Human diseases that result from mutations in the col1 gene have a wide array of phenotypes. It is interesting that none of the human phenotypes resulting from col1 gene mutations are vascular in nature. This may be a result of the embryonic lethality of vascular phenotypes caused by col1 mutations, although there is little data from the human literature to support or refute this speculation. Type I collagen is also a large structural component of bone. The phenotypes resulting from col1 mutations, osteogenesis imperfecta (OI) and Ehlers–Danlos Syndrome Type VIIA and B, are a result of altered expression levels or mutated col1 protein found in bone (Byers et al., 1997; Gajko-Galicka, 2002; Raff et al., 2000). One study of aortic tissue from human fetuses with the lethal form of OI showed that the vessels had smaller collagen fibrils and disordered elastic lamellae (Pasquali-Ronchetti et al., 1986). No follow-up studies have been reported on the vascular consequences of col1 mutations in humans.

2. Collagen Type III

Type III collagen is found associated with the intimal, medial, and adventitial layers of the developing bovine aorta and in the intima and media of adult human aortas (Howard and Macarak, 1989; Voss and Rauterberg,

1986). Collagen III has been localized in dense deposits in close proximity to the elastic laminae in human aortic media (McCullagh *et al.*, 1980). In electronmicrographic studies, collagen type III colocalizes with types I and V collagen in the areas adjacent to the elastic lamellae (Dingemans *et al.*, 2000).

In humans, mutations in the col3A1 gene result in Ehlers–Danlos syndrome (EDS) type IV (Pope *et al.*, 1977; Schwarze *et al.*, 2001). The vascular phenotype found in this disease includes fragility of blood vessels and a propensity toward large-vessel aneurysm and rupture. Most mutations in col3A1 that lead to EDS-IV are single amino acid substitutions in the Gly-X-Y repeat sequences of the gene. However, functional haploinsufficiency for col3A1 caused by nonsense-mediated decay of mutant mRNA has also been shown to cause the disease (Schwarze *et al.*, 2001). It is interesting that mice heterozygous for col3A1 were normal. The homozygous null mouse generated by targeted disruption of col3A1, however, had a 90% perinatal mortality, with those surviving having a short life span and a phenotype similar to EDS-IV in humans (Liu *et al.*, 1997). Death in the adult null mice was due to vascular rupture. Perhaps the most significant finding from the creation of the col3A1 mutant mice came from the ultrastructural analysis of the collagen fibers themselves. It appears that type III collagen helps to regulate the diameter of collagen type I fibrils. With the loss of type III collagen, the type I fibril diameters were larger and inconsistent, and the total number of fibrils was decreased by one-third when compared to wild-type animals. No difference in the number or alignment of VSMCs was noted in the null animals (Liu *et al.*, 1997).

3. Collagen V

Collagen V has been localized immunohistochemically to the media of human arteries and has also been seen in the basement membrane (discussed later) surrounding VSMCs of the media (McCullagh *et al.*, 1980; Voss and Rauterberg, 1986). The expression patterns of col5A1 and col5A2 parallel that of col3A1, peaking at P0 and decreasing over the postnatal time points (McLean *et al.*, 2004).

Mutations in the human col5A1 gene that result in functional haploinsufficiency cause types I and II, or classical EDS. Phenotypically, these patients exhibit hyperextensible skin and joints, easy bruising, and abnormal scarring. Similarly to mutations in the col3A1 gene, these mutations lead to dysregulation of the size of collagen type I fibrils (Bouma *et al.*, 2001; Schwarze *et al.*, 2000; Wenstrup *et al.*, 2000). There is no described vascular phenotype in classical EDS.

4. Collagen VI

The pattern of type VI collagen expression in the mouse aorta is similar to type I, but the changes in expression levels are not as dramatic. Expression increases slowly during embryonic time points, dips sharply at P0, and recovers to peak at P7 days. From 1 week of life, the expression levels then again decline.

Collagen VI is found in association with Fibrilin-1 in oxytalan fibers in the media of the human aortae. These fibers were shown by immuno-electron microscopy to connect elastic lamallae to the basement membrane of VSMCs. Some of these fibers also ran along the surface of VSMCs, putting collagen type VI in the closest proximity to the cell of all of the fibrillar collagens (Dingemans et al., 2000).

In humans, Bethlem myopathy and Ullrich's disease results from deficiency of collagen VI. It is interesting that mutations in the col6A2 gene leading to premature termination codons and functional haploinsufficiency cause both the dominantly inherited Bethlem myopathy and the recessive Ullrich's disease. Mutations in col6A1 and col6A3 also cause Bethlem myopathy. Patients with both disorders have joint contractures and muscle weakness with progressive muscle wasting. The mouse model of Bethlem myopathy was made by targeted disruption of the col6A1 gene. Null animals had a phenotype that closely paralleled the human Bethlem myopathy (Bonaldo et al., 1998).

Until recently, there was no vascular phenotype known in patients with mutations in collagen VI. Niiyama et al. (2003) have shown that the capillaries in muscle from patients with Ullrich's disease have narrow lumens, enlarged endothelial cell nuclei, and fenestrations. Although it does not appear that muscular arteries were examined in these patients, the phenotype was not realized until electron microscopic studies on the capillaries were undertaken. This leaves open the possibility that mutations in type VI collagen may cause a subtle change in VSMC development or function that will only be detected after more detailed investigation.

V. The Elastic Fiber

A. Elastin

The elastic fiber is a multicomponent structure whose main protein is elastin. In contrast to the genetic diversity evident in the collagen gene family, elastin is encoded by only one gene. Tropoelastin, the monomeric gene product, contains alternating domains of hydrophobic amino acids that contribute to the protein's elastic properties and sequences that contain lysine residues that will serve to cross-link the protein into a functional polymer (elastin).

6. Vascular Matrix and Aortic Development

The carboxy-terminal region of the protein contains an important assembly domain such that mutations resulting in its deletion or modification account for two classes of human diseases (discussed later). In the extracellular space, lysine residues within tropoelastin are specifically modified to form covalent crosslinks by one or more lysyl oxidases, a multigene family consisting of five members. These enzymes are also responsible for cross-linking collagen molecules.

Similar to what was found for type I collagen, elastin expression in the mouse aorta is lowest in the embryonic period and increases (with a dip at P0) until P14, then decreases rapidly to low levels in the adult (Fig. 3). Expression of lysyl oxidase-like (LOX-1) showed a similar pattern to elastin and collagen, whereas expression of lysyl oxidase-like (LOX-2) protein was low and unchanging (not shown).

Three inherited diseases, supravalvular aortic stenosis (SVAS), Williams syndrome (WS), and autosomal dominant cutis laxa (ADCL), have been linked to mutations in the elastin gene (Ewart et al., 1993; Olson et al., 1993; Tassabehji et al., 1998). Loss of function mutations in one allele of the elastin gene are responsible for supravalvular aortic stenosis (SVAS) in humans, a congenital narrowing of large arteries occurring sporadically, as a familial condition with autosomal dominant inheritance (Eisenberg et al., 1964) or as a condition of Williams' syndrome (Morris, 1998; Morris and Mervis, 2000). The severity and onset of this disease is variable and, if untreated, can lead to heart failure, myocardial infarction, and death. Pathologic studies of patients affected by SVAS have shown hypertrophy and hyperplasia of smooth muscle cells in affected vessels, fragmentation or dissolution of elastic lamellae, and changes in extracellular matrix composition. Causative mutations include large elastin intragenic deletions, crossover events, and point mutations that include nonsense, frameshift, and splice site mutations (Li et al., 1997; Metcalfe et al., 2000; Milewicz et al., 2000; Urbán et al., 2000).

A second disease linked to mutations in elastin is autosomal dominant cutis laxa (ADCL) characterized by lax, inelastic skin and, in many instances, internal manifestations that can include pulmonary artery stenosis, aneurysms, emphysema, bronchiectasis, and hernias (Milewicz et al., 2000; Tassabehji et al., 1998). The molecular basis of cutis laxa is not known with certainty, although the recent identification of frameshift mutations in the 3' end of the elastin coding region suggest either a dominant-negative or gain-of-function mechanism (Tassabehji et al., 1998; Zhang et al., 1999). It is now clear, however, that elastin mutations are not the exclusive cause of the disease. Two reports have shown that mutations in fibulin-5 (discussed later) lead to the cutis laxa phenotype in humans and mice (Loeys et al., 2002; Markova et al., 2003).

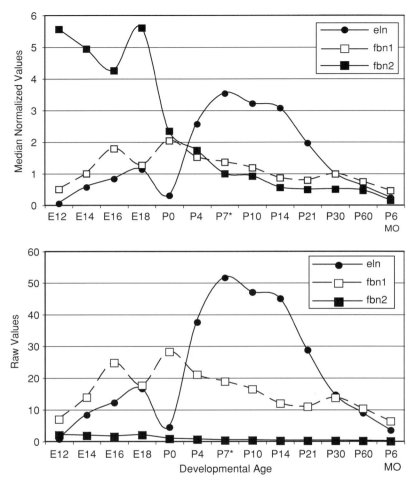

Figure 3 Expression pattern for elastin (eln) and fibrillins (fbn)-1 and -2 as determined by oligonucleotide microarray. These values have been confirmed by real-time PCR. The top panel shows the mean normalized values, which highlights the expression pattern for each individual gene over the time course. The bottom panel shows the raw expression values, which provides an estimate of the relative expression levels for the different genes. It is interesting to note that whereas fibrillin-2 is decreasing in the embryonic period, fibrillin-1 is increasing. A comparison of the unnormalized data shows that even when expression of fibrillin-2 is at its highest, it is still appreciably lower in terms of absolute amount than fibrillin-1 or elastin. This suggests that fibrillin-1 is the major fibrillin in aortic microfibrils, with fibrillin-2 playing a minor role.

B. Fibrillin and Microfibrils

The second major structural component of elastic fibers is the microfibril. The structural building blocks of these long linear fibers are the fibrillin molecules. Several microfibril-associated proteins have also been described, but their importance to microfibril structure and functions is not yet clear. The best characterized microfibril-associated proteins are the latent TGF-β-binding proteins (LTBP 1-4), microfibril-associated glycoproteins (MAGP-1 and -2), and members of the fibulin family. A list of these and other proteins can be found in a recent review by Kielty *et al.* (2002).

The human genome contains three fibrillins, but fibrillin-3 appears to have been inactivated in the mouse genome due to chromosome rearrangements (Corson *et al.*, 2004). These 350-kDa glycoproteins are highly homologous, with modular structures consisting of repeating calcium-binding epidermal growth factor (EGF)-like domains interspersed between 8-cysteine domains similar to those found in the latent transforming growth factor-β-binding protein (LTBP) family (Handford *et al.*, 2000). It has long been assumed that microfibrils provide a scaffold or template for elastin assembly by binding and aligning tropoelastin monomers so that lysine-containing regions are in register for cross-linking. Recent studies from fibrillin knockout mice (Bunton *et al.*, 2001; Pereira *et al.*, 1997), however, and our own studies of elastin assembly (Kozel *et al.*, 2003, 2004) are beginning to call this function into question.

There is accumulating evidence that a prominent function of fibrillin-containing microfibrils is to bind and sequester growth factors into the ECM (Neptune *et al.*, 2003). In addition, all three fibrillins contain the integrin-binding RGD sequence, and fibrillin-1 and fibrillin-2 have been shown to interact with $\alpha v\beta 3$ and $\alpha 5\beta 1$ integrins as well as with cell-surface heparan sulfate proteoglycan (D'Arrigo *et al.*, 1998; Pfaff *et al.*, 1996; Ritty *et al.*, 2003; Sakamoto *et al.*, 1996). Thus, the fibrillins have the potential of providing instructive signals to cells either indirectly through the binding of growth factors or directly by interacting with signaling receptors on the cell surface.

In the developing mouse aorta, fibrillin-1 has an expression pattern similar to elastin, except peak expression occurs at P0 (Fig. 3). Expression of fibrillin-2, on the other hand, is markedly different from fibrillin-1 and elastin. Fib-2 expression is highest in the early embryonic period and then decreases almost linearly throughout maturation. A comparison of the unnormalized data shows that even when expression of fibrillin-2 is at its highest, it is still appreciably lower in terms of absolute amount than fibrillin-1 or elastin (Fig. 3). This suggests that fibrillin-1 is the major fibrillin in aortic microfibrils with fibrillin-2 playing a minor role.

Several genetic diseases have also been linked to mutations in microfibrillar proteins. These include Marfan syndrome, which is linked to mutations in fibrillin-1 and congenital contractural arachnodactyly, linked to mutations in fibrillin-2 (reviewed in Dietz and Mecham, 2000; Milewicz et al., 2000). Marfan syndrome is inherited as an autosomal dominant disease with high penetrance and, among other traits, results in vascular defects that lead to proximal dilatation of the aorta and aneurysm formation with risk of dissection. To date, greater than 200 mutations responsible for the syndrome have been identified. It was originally thought that the pathogenesis of Marfan syndrome occurred through a dominant negative-type mechanism resulting from mutations within the fibrillin-1 gene that affected the structural properties of microfibrils and elastic fibers. Recent studies of mice deficient in fibrillin-1, however, suggest that the inability of mutant fibrillins to sequester growth factors may be a major contributing factor to disease pathogenesis (Neptune et al., 2003).

VI. Fibulins

The fibulins are a family of ECM proteins with five members (Argraves et al., 2003; Chu and Tsuda, 2004; Timpl et al., 2003). The amino terminal region of fibulin-1 and -2 consists of an anaphylatoxin-type structure; the midportion contains multiple calcium-binding EGF-like repeats, and the C-terminus contains a motif similar to the fibrillins. Fibulins-3, -4, and -5 lack the anaphylatoxin-type domain. The fibulins are often found in association with elastin fibers and are also known to bind to multiple components of the ECM and basement membrane.

A. Fibulin-1

Fibulin-1 expression has been documented from early time points in mouse development preceding organogenesis (E10). Expression patterns of mRNA in the mouse aorta increase during the time of organ development from E12–E16, fall sharply, and then peak again at P2, with a subsequent gentle rise into adulthood (McLean et al., 2004). Protein expression has been detected in the heart endocardial cushions and aortic valve as well as in the wall of elastic arteries. Protein localization studies have shown fibulin-1 to associate most closely with the amorphous elastin component of the elastic fiber (Roark et al., 1995). Fibulin-1 has many protein interactions with other ECM proteins, including fibrinogen, fibronectin, tropoelastin, and components of the basement membrane (Timpl et al., 2003). Homozygous null mice for fibulin-1 have significant perinatal mortality due to malformations in the lungs, kidneys, and vasculature. Null embryos demonstrate hemorrhage in perineural tissue and muscle, and petechial bleeding in the skin. They also have abnormalities in capillary development, evidenced by enlarged and irregular capillary lumens. It is interesting that although endothelial cell morphology and capillary integrity was affected in these animals, there seems to be no functional alteration of elastic or muscular arteries. It was suggested that fibulin-1 does not act as an adhesion molecule in cell–ECM interactions, but instead may regulate macromolecular organization or perhaps be involved in cell signaling from the luminal surface, as fibulin-1 levels in circulating plasma are quite high. Fibulin-1 null mice did not have prolonged bleeding or coagulation times, suggesting that the circulating fibulin-1 is more likely to play a role in an alternate physiologic pathway such as cell signaling. (Kostka *et al.*, 2001). A stable chromosomal translocation that interferes with expression of only the D splice variant of fibulin-1 has been implicated in synpolydactly in humans (Debeer *et al.*, 2002). No other human diseases involving mutated or decreased expression of fibulin-1 protein have been documented.

B. Fibulin-2

Expression of fibulin-2 has been documented during early embryonic development. At E8.5, fibulin-2 can be seen in the basement membrane of the developing neural tube and aorta. By E9.5, however, there is a dramatic upregulation of fibulin-2 expression at sites of mesenchymal cell differentiation in the area of the aortic outflow tract and endocardial cushions. As the aortic arch vessels mature, fibulin-2 expression becomes exclusive to mesenchymal cell components that go on to become VSMCs (Tsuda *et al.*, 2001). In the later stages of aortic development, the mRNA expression pattern mimics that of fibulin-1 but at higher levels. Fibulin-2 interacts with many of the same ECM proteins as fibulin-1, with binding to perlecan and fibrillin-1 being specific to fibulin-2 (Timpl *et al.*, 2003). At the time of writing, a fibulin-2 knockout mouse has not been reported, and no human diseases have been definitively linked to mutations in fibulin-2.

C. Fibulin-3 and -4

Little is known about fibulin-3 and -4 with respect to expression patterns or interactions with ECM components. Likewise, data from null mice are unavailable. A potential role for fibulin-3 in the development of macular degeneration has been postulated (Stone *et al.*, 1999). No known human vascular diseases have been attributed to mutations in fibulin-3 or -4.

D. Fibulin-5

Fibulin-5 is known to co-localize with elastic fibers, and its expression is strong in elastic arteries, skin, and lung. Fibulin-5 was found to be produced by VSMC of the developing murine aorta by *in situ* hybridization as early as at E11.5. Levels are dramatically down regulated in adult vessels, except where there is ongoing angiogenesis, such as the uterus, and in injured vessels (Kowal *et al.*, 1999). Fibulin-5 binds to integrins $\alpha v \beta 3$, $\alpha v \beta 5$, and $\alpha 9 \beta 1$ via its N-terminal domain and interacts with tropoelastin in a calcium-dependent fashion (Nakamura *et al.*, 2002; Yanagisawa *et al.*, 2002). These interactions potentially act to link the cell to the ECM and may have structural, signaling, and stabilizing effects in the vessel wall. The fibulin-5 null mouse shows defective elastic fibers with disrupted laminae and abnormal elastin aggregates in the aortic wall. Defects lead to tortuous vessels and changes in vessel mechanics. There are also defects in lungs and in skin, leading to a cutis laxa phenotype in the fibulin-5 null mice. These animals do not have evidence of aneurysm or dissection, indicating that fibulin-5 is important during the development of large vessels but is less important with regard to vascular stability. No specific effects on the VSMCs were investigated in these mice (Nakamura *et al.*, 2002; Yanagisawa *et al.*, 2002). In humans, two mutations in fibulin-5 have been linked to autosomal recessive cutis laxa (Loeys *et al.*, 2002; Markova *et al.*, 2003).

VII. EMILIN/Multimerin Family

The EMILIN/Multimerin family comprises four proteins (EMILIN 1–3 and multimerin) with common structural domains. At the amino terminus is an EMI domain (a cysteine-rich sequence of $\sim$80 amino acids), a large central region thought to facilitate coiled-coil structures, and a carboxyl-terminal region homologous to the globular domain of C1q that directs the formation of trimers (Colombatti *et al.*, 2000). The function of the EMILINs is largely unknown, although there is accumulating evidence for a role for EMILIN-1 in elastogenesis and cell adhesion (Bressan *et al.*, 1993; Zanetti *et al.*, 2004). Mice deficient in EMILIN-1 exhibit structural alterations of elastic lamellae of elastic arteries. Additional alterations were observed in cell morphology and anchorage of endothelial and smooth muscle cells to elastic lamellae (Zanetti *et al.*, 2004). EMILIN-1 is the most widely expressed gene of the family and is found in the blood vessel wall as well as in most other tissues. EMILIN-2, -3, and multimerin are also expressed in the aorta, although expression of EMILIN-3 and multimerin is restricted to endothelial cells (Braghetta *et al.*, 2004). The expression pattern for the EMILIN family members is similar for other structural matrix genes: highest in the fetal and neonatal period and lowest in the adult.

VIII. Fibronectin

Numerous studies have illustrated the importance of fibronectin to vessel formation, particularly in the early embryonic periods (Francis et al., 2002; Glukhova and Koteliansky, 1995; Risau and Lemmon, 1988). In the embryonic chicken, the early vasculature is rich in fibronectin but relatively devoid of basement membrane or structural matrix proteins (Risau and Lemmon, 1988). Our expression profile data suggest the same is true in the developing mouse aorta, where fibronectin expression is high and relatively constant throughout development and into the adult period. Fibronectin plays an important role in facilitating cell movement during early migratory events in cell wall formation. During aortic morphogenesis, the pattern of fibronectin alternative splicing changes in ways that are dependent on cell type and functional state (Glukhova et al., 1990). Its continued expression after vessel maturation suggests an ongoing role in vessel homeostasis.

Gene knockout studies have shown that fibronectin is essential for the organization of heart and blood vessels (George et al., 1993). In the absence of fibronectin, aortic endothelial cells do not organize into tubes, and, as a consequence, blood vessels do not form in the vitelline yolk sac. In addition, ablation of the fibronectin receptors $\alpha 5$ integrin in mice and the $\alpha v \beta 3$ integrin in chickens results in vascular defects and early embryonic lethality.

IX. The Basement Membrane

Along with type IV collagen and entactin/nidogen, the laminins are the major structural elements of the basement membrane (also referred to as basal lamina) (Ekblom and Timpl, 1996). The molecular architecture of these matrices results from specific binding interactions among the various components. Type IV collagen chains that assemble into a covalently stabilized polygonal network form the structural skeleton. Laminin self-assembles through terminal domain interactions to form a second polymer network. Entactin/nidogen binds laminin near its center and interacts with type IV collagen, bridging the two (Mayer and Timpl, 1994). A large heparan sulfate proteoglycan (HS-PG), perlecan, binds laminin and type IV collagen through its GAG chains and forms dimers and oligomers through a core–protein interaction. Perlecan is important for charge-dependent molecular sieving, one of the critical functions of basement membrane. Other proteins that are often associated with basement membranes are agrin, SPARC, fibulins, fibronectin, and collagens XV, XVIII, and XIX.

One of the major functions of the basement membrane is to tether cells to the extracellular matrix through interactions with cell-surface receptors. The major receptors that recognize basement membrane macromolecules are members of the $\beta1$ integrin family. Integrin $\alpha1\beta1$, $\alpha2\beta1$, $\alpha3\beta1$, $\alpha6\beta1$, and $\alpha6\beta4$ all bind to laminin. Integrin $\alpha1\beta1$ and $\alpha2\beta1$ also bind to type IV collagen, and $\beta1$ integrins bind entactin/nidogen-2, perlecan, and agrin. Another major receptor is α-dystroglycan, which binds to perlecan, agrin, and several laminins. Laminins and perlecan can also interact with heparan sulfate-containing moieties on the cell surface, such as syndecans and lipid sulfatides.

A. Laminins

The laminins are modular proteins with domains that interact with both cells and ECM (Ekblom and Timpl, 1996). They constitute a family of basement membrane glycoproteins that affect cell proliferation, migration, and differentiation. Fifteen different laminins have been identified, each containing an α, β, and γ chain. Electron microscopy has revealed that all laminins have a crosslike shape with three short arms and one rodlike long arm, a shape well suited for mediating interactions between sites on cells and components of the ECM (Maurer and Engel, 1996; Yurchenco et al., 2004). The rodlike regions separating the globular units of the short arms are made up of repeating EGF-like domains. The long arm is formed by all three component chains folding into an α-helical coiled-coil structure and is the only domain composed of multiple chains. It is terminated by a large globular domain composed of five homologous subdomains formed by the C-terminal region of the α chain.

In the late embryonic and fetal periods, our mouse expression data (McLean et al., 2004) suggest that the predominant aortic laminin chains are $\alpha4$, $\beta1$, and $\gamma1$, which correspond to laminin-8. Laminin $\alpha1$ expression is highest at our first time point (e12) and then drops slightly between embryonic days 12–14, consistent with previous observations that laminin-1 ($\alpha1\beta1\gamma1$) is expressed earliest during embryogenesis (Li et al., 2003; Smyth et al., 1999). In the postnatal period, there are marked increases in expression of the laminin $\alpha5$ and $\beta2$ chains. Thus, after birth, laminin-9 and laminin-10, as well as laminin-8, contribute to basement membrane structure. These findings are in agreement with immunolocalization studies showing that laminin-8 is widely distributed in vascular tissues (Petajaniemi et al., 2002) and that laminin $\alpha5$ appears during the postnatal period (Sorokin et al., 1997). Laminin $\alpha3$ shows constant, but low, expression over the entire data series, whereas expression of laminin $\alpha2$ is intermediate between the low and high expressers, suggesting the presence of some laminin-2 or laminin-4.

6. Vascular Matrix and Aortic Development

Ultrastructural studies of developing aorta show little discernable basement membrane until about midgestation, when the period of intense ECM production begins following cell polarization and tissue organization. At this point, cells in the vessel wall begin to orient in distinct layers and a discontinuous basement membrane begins to develop around individual smooth muscle cells. Basement membrane proteins have also been shown to provide the substratum for endothelial cells, although this structure in elastic vessels is frequently associated directly with the internal elastic lamina, so it is difficult to discern by electron microscopy.

Gene targeting studies have shown that deletion of the laminin $\gamma 1$ chain results in embryonic lethality due to failure of blastocyst development. Deletion of the $\alpha 5$ chain also results in embryonic lethality, although at a slightly later stage than what is observed for the $\gamma 1$ chain (Miner *et al.*, 1998). The only laminin knockout that expresses a vascular phenotype is associated with deletion of the $\alpha 4$ chain, which manifests as hemorrhaging, bleeding, and microvascular degeneration (Thyboll *et al.*, 2002).

Mutations in, or deletion of, other laminin chains result in muscular dystrophy, neuropathies, and epidermolysis bullosa (reviewed in Li *et al.*, 2003).

The two other main basement membrane constituents, type IV collagen and perlecan, show constant expression through vascular development and maturity. One interesting change evident in the array data, however, is a dramatic increase in expression of the collagen IV $\alpha 5$ chain and a decrease in the $\alpha 3$ chain in the postnatal period.

B. Entactin/Nidogen

Entactin, also referred to as nidogen, is a highly conserved protein in the vertebrate basement membrane that bridges the laminin and type IV collagen networks (Carlin *et al.*, 1981; Chung *et al.*, 1977; Timpl *et al.*, 1983). There are two members of the entactin family in mammals (entactin-1 and -2 or nidogen-1 and -2). Both proteins are elongated molecules composed of three globular domains (G1, G2, and G3) connected by a flexible, protease-sensitive link and a rigid rodlike domain (Fox *et al.*, 1991). Entactin-2 is enriched in endothelial basement membranes, whereas entactin-1 shows broader localization in most basement membranes. In the developing mouse aorta, entactin-1 shows a sharp increase in expression at E18 and remains high until P7, when it drops to levels that persist through the adult stages. Expression of entactin-2, in contrast, shows a sharp increase at E14 and then decreases gradually until P21, when stable expression is obtained at levels lower than those observed during the embryonic period (McLean *et al.*, 2004).

Both proteins have been knocked out in mice (Murshed *et al.*, 2000; Schymeinsky *et al.*, 2002), and in both instances the mice show no overt abnormalities and are fertile, and basement membranes appear normal by ultrastructural analysis and immunostaining. Lebel *et al.* (2003) have documented alterations in glomerular filtration in the entactin-1 null mouse, even through no major morphological alteration of the glomerular basement membrane was evident by immunoelectron microscopy. These studies call into question the role of entactin in basement membrane formation or maintenance.

X. Proteoglycans

The proteoglycans constitute a number of genetically unrelated families of multidomain proteins that have covalently attached glycosaminoglycan (GAG) chains. To date, more than 25 distinct gene products have been identified that carry at least one GAG chain (Iozzo and Murdoch, 1996). For historical reasons, proteoglycans are named based on the type of attached GAG chain(s): (1) chondroitin sulfate and dermatan sulfate, consisting of a repeating disaccharide of galactosamine and either glucuronic acid or iduronic acid; (2) heparin and heparan sulfate, consisting of a repeating disaccharide of glucosamine and either glucuronic acid or iduronic acid; and (3) keratan sulfate, consisting of a repeating disaccharide of glucosamine and galactose. Hyaluronate is also a repeating disaccharide but is not sulfated and not bound to a core protein. GAG chains are usually attached through *O*-glycosidic linkages to serine residues in the proteoglycan core protein. A characteristic feature of GAG chains is that at physiological pH they contain one to three negative charges per disaccharide due to carboxylate and sulfate groups.

A. Large Proteoglycans That Form Aggregates by Interaction with Hyaluronan

These proteoglycans interact with strands of hyaluronate to form a very-high-molecular-weight aggregate. A structural trait shared by these proteoglycans is the presence of three functional domains: a globular hyaluronan-binding domain at the N terminus, a central extended region that carries most of the GAG chains, and a modular C-terminal domain containing two EGF repeats, a C-type lectin domain, and a complement-regulatory-protein-like motif (Iozzo and Murdoch, 1996). The largest member of this family is versican (Wight, 2002), a major proteoglycan in blood vessels that is also expressed in nonvascular tissues. Aggrecan, the

6. Vascular Matrix and Aortic Development

large aggregating proteoglycan of cartilage, has a smaller core protein than versican but contains nearly three-fold more GAG chains. Two other members of this family include neurocan and brevican, both found in brain tissues.

Versican is known to have a wide variety of functions, including induction of cell adhesion, promotion of proliferation, and influencing cell migration (Wight, 2002). Versican has been localized to both the media and endothelial layers of human aortas by *in situ* hybridization and western blotting (Yao *et al.*, 1994). In the developing mouse aorta, versican mRNA expression from E12 to birth trends downward slightly from moderate levels, rises sharply to peak at P0, and then falls sharply by P4 and is maintained at low levels through P6 mo (Fig. 4). Evidence that versican is involved in vascular development comes from *in vitro* data that shows that versican is required for proliferation and migration of human aortic smooth muscle cells in response to PDGF. It is hypothesized that versican and the hyaluronan matrix affect cell adhesion and shape and, by this mechanism, affect migration and proliferation of VSMCs (Evanko *et al.*, 1999). A recent study has also shown that expression of the C-terminal region of the versican protein leads to enhanced tumor angiogenesis through a mechanism of

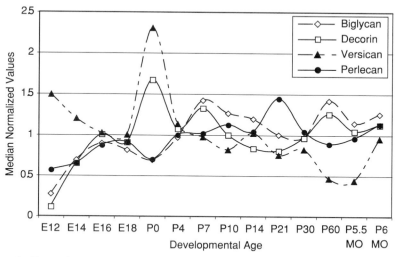

Figure 4 Expression patterns for the large proteoglycans versican and perlecan for the small proteoglycans decorin and biglycan as determined by oligonucleotide microarray. Values have not been confirmed using other RNA quantification techniques. Biglycan, decorin, and to a lesser extent, perlecan, show increased expression beginning at E14. Versican, however, shows a progressive decline in expression over the data series, suggesting an important role in the early embryonic period.

increasing endothelial cell adhesion, migration, and proliferation (Zheng et al., 2004). At present, there are no known human diseases associated with versican mutations, and no versican null mouse has been reported to date.

B. Small Leucine Rich Proteoglycans

The small leucine rich proteoglycans (SLRPs) are a family of secreted proteoglycans that can bind ECM molecules including collagen, fibronectin, and fibrillin-containing microfibrils. The SLRP family includes decorin, biglycan, fibromodulin, osteoglycin, and lumican. The Class I SLRPs, decorin and biglycan, have a U-shaped leucine-rich core domain and contain cysteine-rich clusters at the amino and carboxy terminal ends that form cysteine bonds (Ameye and Young, 2002). The concave portion of the core domain can accommodate a single collagen fibril, and both biglycan and decorin bind to and regulate collagen fibrillogenesis (Hocking et al., 1998). Both biglycan and decorin have been shown to bind TGF-β and may sequester it to the matrix. There is a suggestion that through regulation of growth factor activity, these SLRPs might regulate cell proliferation or differentiation (Riquelme et al., 2001). Biglycan localizes to all layers of the human aorta by immunohistochemical staining, whereas decorin is found only in the adventitia (Theocharis and Karamanos, 2002).

The expression profile of decorin in the mouse aorta closely parallels that of type I collagen in embryonic time points but peaks at P0. Decorin expression decreases somewhat in the postnatal time points but remains constitutively expressed at a moderate level. Biglycan, on the other hand, shows increasing expression over the embryonic time points to peak at P7 at levels similar to col1A1. Expression levels fall over the first postnatal month but rise again as the animal enters adulthood (P 5.5–6 months) (Fig. 4).

Lumican, a class II SLRP, has been localized to the outer layer of medial VSMCs and adventitia of nonatherosclorotic human coronary arteries (Onda et al., 2002). The class II SLRPs have 10 leucine-rich repeats, the same as class I, but have a unique N-terminal cysteine sequence as well as a sulfated tyrosine residue (Ameye and Young, 2002). Mouse aortic expression of lumican occurs from E12–P6 months at low to moderate levels. There is a small peak in expression levels at P0. Lumican, like decorin and biglycan, regulates collagen fibrillogenesis (Chakravarti et al., 1998). The other SLRPs have not been shown to be expressed in the vascular wall.

Studies of mice null for decorin, biglycan, and lumican show phenotypes in bone, tendon, and skin. Specifically, collagen fibril diameter and organization are dysregulated, with different fibril diameters and organizations dependent on the tissue type examined. No changes in blood vessel structure or stability have been reported. It is possible that collagen fibrillogenesis is dysregulated in the blood vessel as well, but that the change is not significant enough to cause vessel rupture or other gross phenotype. The expression profiles of the collagens and SLRPs in the developing mouse aorta certainly suggest that the SLRPs may function in a role similar to that in skin, bone, and cartilage collagen fibrillogenesis. Further investigation into such questions may help to elucidate the role of the SLRPs in vascular development.

XI. Matricellular Proteins

The term *matricellular* has been applied to a group of extracellular proteins that function by binding to matrix proteins and to cell surface receptors but do not contribute to the structural integrity of the ECM (Bornstein and Sage, 2002). Proposed members of this group include the thrombospondins, members of the tenascin protein family, SPARC/osteonectin, and osteopontin. These proteins are frequently called "antiadhesive proteins" because of their ability to induce rounding and partial detachment of some cells *in vitro*. Their ability to interact with many different matrix proteins and cell surface receptors may explain their complex range of biological functions.

A. Thrombospondins

The thrombospondins (TSP) are a family of secreted glycoproteins found in the ECM. The family has five members in vertebrates divided into two subgroups, A and B. Thrombospondins in subgroup A (-1 and -2) are secreted as disulfide-bonded trimers, whereas those in group B (-3, -4, and -5) are secreted as pentamers (Bornstein, 1995; Lawler, 2002).

Nearly all of the TSPs have been localized to the vessel wall. *In situ* hybridization using probes to TSP-1, -2, and -3 showed that only TSP-2 was present in large vessels in the developing murine embryo. TSP-2 transcripts were seen at E11 in the dorsal aorta and could be seen in other large vessels as well as areas with ongoing angiogenesis throughout embryogenesis (Iruela-Arispe *et al.*, 1993). TSP-1 transcripts were seen at later developmental stages (E16–E18) associated with capillaries but not elastic or muscular arteries. Immunohistochemical studies of TSP-1, however, have localized TSP-1 protein to the luminal portion of human blood vessels (Wight *et al.*, 1985). No TSP-3 could be detected in vascular structures in the developing mouse by *in situ* hybridization (Iruela-Arispe *et al.*, 1993). Gene array data shows a similar pattern as the *in situ* data, with TSP-2 expression higher than TSP-1 or TSP-3 in the mouse aorta. Levels of mRNA of TSP-2 rise dramatically through embryogenesis and remain high during the first week of postnatal life, falling steeply thereafter. Levels of TSP-1 rise modestly around E16–P0, then decrease to low levels. TSP-3 levels in the gene array data from mouse aorta show a modest embryonic expression with a dip at P0 and recovery with slight increase over the postnatal time points to P6 months (Fig. 5). Recently, TSP-5 (cartilage oligomeric matrix protein, COMP) has been localized to the media of adult human vessels by immunohistochemistry (Riessen et al., 2001). RT-PCR was used to verify the expression of TSP-4 in adult human coronary artery VSMC and EC cultures (Stenina et al., 2003).

Thrombospondins-1 and -2 have been studied extensively and are known to interact with various matrix elements, TGFβ-1 and MMP-2 (see Adams, 2001 for a thorough TSP review). Both are also known inhibitors of angiogenesis. Their role in blood vessel development has been further elucidated by the generation of knockout mice. TSP-1 null mice do not have a vascular phenotype until injured, at which time wounds become hypervascularized (Agah et al., 2002; Stenina et al., 2003). TSP-2 null mice have increased numbers of small and medium-sized blood vessels and prolonged bleeding times (Kyriakides et al., 1998). Collagen fibrillogenesis was also abnormal in the TSP-2 null mice (Kyriakides et al., 1998). It is interesting that deficiency in either TSP-1, TSP-2, or both does not cause morphologic changes or decreased integrity of large elastic or muscular arteries (Agah et al., 2002). It can be postulated, then, that the TSPs may play a role in angiogenic processes but not in the vasculogenic process through which many of the large vessels are formed during embryonic development.

There is a small but growing literature on TSP-4 and -5 in relation to VSMCs and blood vessels. To date, this consists of descriptive studies localizing TSP-4 and -5 to cells of the vascular wall, and for TSP-5 evidence that it affects VSMC migration and adhesion (Riessen et al., 2001). No *in vivo* data, knockout mice, or known human genetic diseases have been presented to show a relation between TSP-4 or -5 and blood vessel development.

B. Tenascins

The tenascins constitute a gene family consisting of four members: tenascins-C,-R,-X, and -Y (Chiquet-Ehrismann and Chiquet, 2003; Jones and Jones, 2000a, b). Tenascin-C (early names include GMEM, cytotactin, J1, hexabrachion, and neuronectin) was the first form discovered and exists as a hexamer of disulfide-bonded subunits. Each subunit consists of a cysteine-rich N-terminal domain involved in oligomerization, EGF-like repeats, fibronectin type III–like repeats, and a fibrinogen-like globular domain. The number of fibronectin type III–like repeats varies as a result of alternative splicing. Like TSP, tenascin-C has diverse biological effects when applied to cells. Both stimulation and inhibition of cellular proliferation

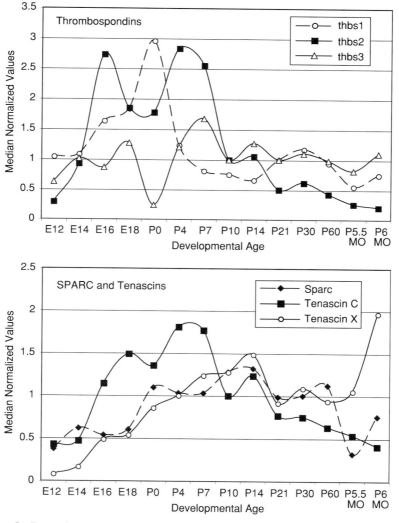

Figure 5 Expression patterns for matricellular proteins as determined by oligonucleotide. Values have not been confirmed using other RNA quantification techniques. The top panel shows median normalized expression values for thrombospondin (thbs)-1, -2, and -3. The bottom panel is similar data for SPARC and tenascins C and X.

have been observed in response to tenascin-C. In terms of cell adhesion, some cells do attach to tenascin, but weakly. In most instances, tenascin does not allow cell adhesion, and it can even inhibit cell attachment to other matrix proteins such as fibronectin and laminin. The finding that tenascin-C contains a defined cell attachment site suggests that separate domains that override the attachment domains affect the overall antiadhesive properties of the glycoprotein. Expression of tenascin-X was low in the developing mouse aorta until embryonic day 16. Expression then increased gradually through P10 and then declined slowly, only to increase again in the adult period. Tenascin-C showed an expression pattern typical of other matrix proteins, with increased expression beginning at E14, with levels gradually decreasing, beginning at P7 (Fig. 5).

No vascular defects or diseases have been linked to mutations within any of the tenascin genes, nor do knockout animals display a vascular phenotype (Mackie and Tucker, 1999; Mao *et al.*, 2002). A recessive form of the Ehlers–Danlos syndrome has recently been linked to tenascin-X deficiency, but the vascular system is not affected in these patients (Schalkwijk *et al.*, 2001).

C. SPARC (osteonectin)

Secreted protein acidic and rich in cysteine (SPARC), also known as osteonecitn or BM-40, is the glycosylated product of a single, highly conserved gene. The protein has three domains: the N-terminus is an acidic region that binds calcium and is thought to be involved in bone ossification; the second is a follistatin-like domain that binds copper and contains sequences that cause cell proliferation and stimulates angiogenesis; and the C-terminus is an extracellular calcium binding domain. SPARC, like the other matricellular proteins, binds to many ECM proteins, including collagens I–V and VIII, as well as to PDGF and VEGF (Yan and Sage, 1999).

SPARC is expressed in the early embryo (E.8.5) in the heart and placental tissues, and is found in bone, skin, and teeth at high levels in the newborn mouse (Holland *et al.*, 1987). In the developing mouse aorta, SPARC expression increases during embryogenesis and postnatal development to peak at high levels around 1 week after birth. Levels of mRNA then fall into adulthood (Fig. 5). Expression of SPARC in the adult animal mainly is limited to rapidly proliferating tissues and areas undergoing wound healing. Moderate levels have, however, been detected immunohistochemically in human VSMCs and endothelial cells (Porter *et al.*, 1995). SPARC binds to a broad range of ECM proteins and has a variety of actions, including modification of growth factor activity and matrix metalloproteinase expression, and effects on cell shape and adhesion (see Brekken and Sage, 2001, for review). The SPARC null mouse shows early development of cataracts, osteopenia, increased fat, abnormal collagen fibrilogenesis, and accelerated wound healing. No vascular abnormalities have been reported in these mice. Evidence of SPARC involvement in the vasculogenic process is not currently available in the literature and appears to be limited at best, given the lack of vascular phenotype in the SPARC null mouse. It is interesting, however, that endothelial cells in blood vessels forming by angiogenesis in the chick chorioallantoic membrane express high levels of SPARC, suggesting that the molecular control of these two processes of vessel formation is indeed quite distinct (Iruela-Arispe *et al.*, 1995). It is intriguing to speculate that the molecular control of vessel development differs dramatically from the vasculogenic to angiogenic processes. Evidence such as that presented for SPARC and thrombospondin are just beginning to elucidate these differences.

XII. Correlation of Matrix Gene Expression Profile with Cytoskeletal Markers

The most commonly used markers for smooth muscle cell identification are smooth muscle-specific isoforms of contractile proteins. Changes in cytoskeletal organization occur as cells within the vessel wall mature, so characterization of the contractile proteins expressed by these cells provides a useful way of following their phenotypic transitions. The nature of the contractile proteins as well as their expression pattern in the developing arterial wall have been extensively reviewed (Glukhova and Koteliansky, 1995; Hungerford *et al.*, 1996, 1997; Owens, 1995; Schwartz and Mecham, 1995). The relative expression pattern for smooth muscle α-actin, calponin, smooth muscle myosin heavy chain, transgelin (SM22-α), and smoothelin in the developing mouse aorta essentially agreed with published results for these proteins in other organisms (see discussion in Glukhova and Koteliansky, 1995). All of these proteins were present at E12 and throughout the time series of our gene array analysis, with changes in expression occurring at different stages of aortic development.

Alpha-smooth muscle actin showed the earliest change in expression, characterized by a continual increase from E12 up to 14 days after birth, followed by relatively constant expression through the adult period. Cytoskeletal proteins that showed a continual increase in expression over the entire data set were smoothelin and smooth-muscle myosin heavy chain (SM1). It is interesting that the increase in expression of these proteins lagged behind that seen for α-smooth muscle actin by about 2 days and occurred concurrent with increases in expression of most of the structural matrix genes (between E14 and E16). Increases in expression of vinculin and SM22-α were also concurrent with the onset of matrix production, with vinculin showing constant expression from E16 through the adult period and expression of SM22-α constant after E16 but increasing in the late adult vessel.

XIII. Conclusions

The control of vessel wall formation involves the complex interaction of a multitude of signaling events and structural developments. From the earliest hint of an endothelial tube network, extracellular matrix molecules are important to this process. Gene expression analysis of the developing aorta provides evidence for a dramatic phenotypic switch in smooth muscle cells beginning at embryonic day 14, characterized by a major increase in structural matrix protein production. Over a period of ~20 days, the cells in the vascular wall deposit greater than 90% of the matrix proteins required to impart the mechanical properties the vessel will retain into the adult period. But the ECM is not just cables, elastic bands, and glue. In addition to its structural properties, each matrix macromolecule has the potential to provide informational signals to cells, either through direct interaction with cell-surface receptors or through the binding and sequestration of growth factors. The concept that the ECM can influence cell proliferation, migration, and phenotypic stabilization must be considered when assessing the role of the ECM in a developing tissue. Understanding how these molecules all work together, both as structural components and as signaling moieties that direct VSMC differentiation and tissue maturation, will be critical to understanding the biology (and pathology) of the vessel wall.

Acknowledgments

The original work sited in this review was funded by grants to R.P.M. from the National Institutes of Health (HL53325, HL62295, HL71960). C.M.K. was supported by Pediatric Cardiology Training Grant T32 HL07873. We would like to thank Terese Hall for expert editorial assistance and Dr. Thomas Mariani and Brigham Mecham at Harvard University for assistance with the microarray analysis. We also thank Russel Knutsen and Marilyn Levy for expert electron microscopy.

References

Adams, J. C. (2001). Thrombospondins: Multifunctional regulators of cell interactions. *Annu. Rev. Cell Dev. Biol.* **17,** 25–51.

Agah, A., Kyriakides, T. R., Lawler, J., and Bornstein, P. (2002). The lack of thrombospondin-1 (TSP1) dictates the course of wound healing in double-TSP1/TSP2-null mice. *Am. J. Pathol.* **161,** 831–839.

Albert, E. N. (1972). Developing elastin tissue. An electron microscopic study. *Am. J. Pathol.* **69,** 89–102.

Ameye, L., and Young, M. F. (2002). Mice deficient in small leucine-rich proteoglycans: Novel *in vivo* models for osteoporosis, osteoarthritis, Ehlers-Danlos syndrome, muscular dystrophy, and corneal disease. *Glycobiology* **12,** 107R–116R.

6. Vascular Matrix and Aortic Development

Argraves, W. S., Greene, L. M., Cooley, M. A., and Gallagher, W. M. (2003). Fibulins: Physiological and disease perspectives. *EMBO Rep.* **4,** 1127–1131.

Bergwerff, M., DeRuiter, M. C., Poelmann, R. E., and Gittenberger-deGroot, A. C. (1996). Onset of elastogenesis and downregulation of smooth muscle actin as distinguishing phenomena in artery differentiation in the chick embryo. *Anat. Embryol.* **194,** 545–557.

Berry, C. L., Looker, T., and Germain, J. (1972). The growth and development of the rat aorta. I. Morphological aspects. *J. Anat.* **113,** 1–16.

Bonaldo, P., Braghetta, P., Zanetti, M., Piccolo, S., Volpin, D., and Bressan, G. M. (1998). Collagen VI deficiency induces early onset myopathy in the mouse: An animal model for Bethlem myopathy. *Hum. Mol. Genet.* **7,** 2135–2140.

Bornstein, P. (1995). Diversity of function is inherent in matricellular proteins: An appraisal of thrombospondin 1. *J. Cell. Biol.* **130,** 503–506.

Bornstein, P., and Sage, E. H. (2002). Matricellular proteins: Extracellular modulators of cell function. *Curr. Opin. Cell. Biol.* **14,** 608–616.

Bouma, P., Cabral, W. A., Cole, W. G., and Marini, J. C. (2001). COL5A1 exon 14 splice acceptor mutation causes a functional null allele, haploinsufficiency of alpha 1(V) and abnormal heterotypic interstitial fibrils in Ehlers-Danlos syndrome II. *J. Biol. Chem.* **276,** 13356–13364.

Braghetta, P., Ferrari, A., De Gemmis, P., Zanetti, M., Volpin, D., Bonaldo, P., and Bressan, G. M. (2004). Overlapping, complementary and site-specific expression pattern of genes of the EMILIN/Multimerin family. *Matrix Biol.* **22,** 549–556.

Brekken, R. A., and Sage, E. H. (2001). SPARC, a matricellular protein: At the crossroads of cell-matrix communication. *Matrix Biol.* **19,** 816–827.

Bressan, G. M., Daga-Gordini, D., Colombatti, A., Castellani, I., Marigo, V., and Volpin, D. (1993). Emilin, a component of elastic fibers preferentially located at the elastin-microfibrils interface. *J. Cell Biol.* **121,** 201–212.

Bunton, T. E., Biery, N. J., Myers, L., Gayraud, B., Ramirez, F., and Dietz, H. C. (2001). Phenotypic alteration of vascular smooth muscle cells precedes elastolysis in a mouse model of Marfan syndrome. *Circ. Res.* **88,** 37–43.

Byers, P. H., Duvic, M., Atkinson, M., Robinow, M., Smith, L. T., Krane, S. M., Greally, M. T., Ludman, M., Matalon, R., Pauker, S., Quanbeck, D., and Schwarze, U. (1997). Ehlers-Danlos syndrome type VIIA and VIIB result from splice-junction mutations or genomic deletions that involve exon 6 in the COL1A1 and COL1A2 genes of type I collagen. *Am. J. Med. Genet.* **72,** 94–105.

Carlin, B., Jaffe, R., Bender, B., and Chung, A. E. (1981). Entactin, a novel basal lamina-associated sulfated glycoprotein. *J. Biol. Chem.* **256,** 5209–5214.

Carmeliet, P. (2000). Mechanisms of angiogenesis and arteriogenesis. *Nat. Med.* **6,** 389–395.

Chakravarti, S., Magnuson, T., Lass, J. H., Jepsen, K. J., LaMantia, C., and Carroll, H. (1998). Lumican regulates collagen fibril assembly: Skin fragility and corneal opacity in the absence of lumican. *J. Cell Biol.* **141,** 1277–1286.

Chiquet-Ehrismann, R., and Chiquet, M. (2003). Tenascins: Regulation and putative functions during pathological stress. *J. Pathol.* **200,** 488–499.

Chu, M. L., and Tsuda, T. (2004). Fibulins in development and heritable disease. *Birth Defects Res. Part C Embryo Today* **72,** 25–36.

Chung, A. E., Freeman, I. L., and Braginski, J. E. (1977). A novel extracellular membrane elaborated by a mouse embryonal carcinoma-derived cell line. *Biochem. Biophys. Res. Commun.* **79,** 859–868.

Clark, J. M., and Glagov, S. (1985). Transmural organization of the arterial media. The lamellar unit revisited. *Arteriosclerosis* **5,** 19–34.

Cliff, W. J. (1967). The aortic tunica media in growing rats studied with the electron microscope. *Lab. Invest.* **17,** 599–615.

Colombatti, A., Doliana, R., Bot, S., Canton, A., Mongiat, M., Mungiguerra, G., Paron-Cilli, S., and Spessotto, P. (2000). The EMILIN protein family. *Matrix Biol.* **19**, 289–301.

Corson, G. M., Charbonneau, N. L., Keene, D. R., and Sakai, L. Y. (2004). Differential expression of fibrillin-3 adds to microfibril variety in human and avian, but not rodent, connective tissues. *Genomics* **83**, 461–472.

D'Arrigo, C., Buri, S., Withers, A. P., Dobson, H., Black, C., and Boxer, M. (1998). TGF-beta1 binding protein-like modules of fibrillin-1 and -2 mediate integrin-dependent cell adhesion. *Connect. Tiss. Res.* **37**, 29–47.

Davis, E. C. (1993). Smooth muscle cell to elastic lamina connections in developing mouse aorta. Role in aortic medial. *Lab. Invest.* **68**, 89–99.

Debeer, P., Schoenmakers, E. F., Twal, W. O., Argraves, W. S., De Smet, L., Fryns, J. P., and Van De Ven, W. J. (2002). The fibulin-1 gene (FBLN1) is disrupted in a t(12;22) associated with a complex type of synpolydactyly. *J. Med. Genet.* **39**, 98–104.

Dietz, H. C., and Mecham, R. P. (2000). Mouse models of genetic diseases resulting from mutations in elastic fiber proteins. *Matrix Biol.* **19**, 481–482.

Dingemans, K. P., Teeling, P., Lagendijk, J. H., and Becker, A. E. (2000). Extracellular matrix of the human aortic media: An ultrastructural histochemical and immunohistochemical study of the adult aortic media. *Anat. Rec.* **258**, 1–14.

Drake, C. J., Hungerford, J. E., and Little, C. D. (1998). Morphogenesis of the first blood vessels. *Ann. NY Acad. Sci.* **857**, 155–179.

Dumont, D. J., Fong, G.-H., Puri, M. C., Gradwohl, G., Alitalo, K., and Breitman, M. L. (1995). Vascularization of the mouse embryo: A study of flk-1, tek, tie, and vascular endothelial growth factor expression during development. *Develop. Dynam.* **203**, 80–92.

Eisenberg, R., Young, D., Jacobson, B., and Boito, A. (1964). Familial supravavular aortic stenosis. *Am. J. Dis. Child.* **108**, 341–347.

Ekblom, P., and Timpl, R. (1996). The Laminins. Harwold Academic Publishers, New York.

Evanko, S. P., Angello, J. C., and Wight, T. N. (1999). Formation of hyaluronan- and versican-rich pericellular matrix is required for proliferation and migration of vascular smooth muscle cells. *Arterioscler. Thromb. Vasc. Biol.* **19**, 1004–1013.

Ewart, A. K., Morris, C. A., Ensing, G. J., Loker, J., Moore, C., Leppert, M., and Keating, M. (1993). A human vascular disorder, supravalvular aortic stenosis, maps to chromosome 7. *Proc. Natl. Acad. Sci. USA* **90**, 3226–3230.

Faury, G., Pezet, M., Knutsen, R. H., Boyle, W. A., Hexamer, S. P., McLean, S. E., Minkes, R. K., Blumer, K. J., Kovacs, A., Kelly, D. P., Li, D. Y., Starcher, B., and Mecham, R. P. (2003). Developmental adaptation of the mouse cardiovascular system to elastin haploinsufficiency. *J. Clin. Invest.* **112**, 1419–1428.

Folkow, B. (1983). 'Structural autoregulation'–the local adaptation of vascular beds to chronic changes in pressure. *Ciba Foundation Symp.* **100**, 56–79.

Fox, J. W., Mayer, U., Nischt, R., Aumailley, M., Reinhardt, D., Wiedemann, H., Mann, K., Timpl, R., Krieg, T., Engel, J., *et al.* (1991). Recombinant nidogen consists of three globular domains and mediates binding of laminin to collagen type IV. *EMBO J.* **10**, 3137–3146.

Francis, S. E., Goh, K. L., Hodivala-Dilke, K., Bader, B. L., Stark, M., Davidson, D., and Hynes, R. O. (2002). Central roles of alpha5beta1 integrin and fibronectin in vascular development in mouse embryos and embryoid bodies. *Arterioscler. Thromb. Vasc. Biol.* **22**, 927–933.

Frid, M. G., Moiseeva, E. P., and Stenmark, K. R. (1994). Multiple phenotypically distinct smooth muscle cell populations exist in the adult and developing bovine pulmonary arterial media *in vivo*. *Circ. Res.* **75**, 669–681.

Gajko-Galicka, A. (2002). Mutations in type I collagen genes resulting in osteogenesis imperfecta in humans. *Acta Biochim. Pol.* **49**, 433–441.

George, E. L., Georges-Labouesse, E. N., Patel-King, R. S., Rayburn, H., and Hynes, R. O. (1993). Defects in mesoderm, neural tube and vascular development in mouse embryos lacking fibronectin. *Development* **119**, 1079–1091.

Gerrity, R. G., and Cliff, W. J. (1975). The aortic tunica media of the developing rat. I. Quantitative stereologic and biochemical analysis. *Lab. Invest.* **32**, 585–600.

Gittenberger-de Groot, A. C., DeRuiter, M. C., Bergwerff, M., and Poelmann, R. E. (1999). Smooth muscle cell origin and its relation to heterogeneity in development and disease. *Arterioscler. Thromb. Vasc. Biol.* **19**, 1589–1594.

Glukhova, M. A., Frid, M. G., Shekhonin, B. V., Balabanov, Y. V., and Koteliansky, V. E. (1990). Expression of fibronectin variants in vascular and visceral smooth muscle cells in development. *Dev. Biol.* **141**, 193–202.

Glukhova, M. A., and Koteliansky, V. E. (1995). Integrins, Cytoskeletal and Extracellular Matrix Proteins in Developing Smooth Muscle Cells of Human Aorta. In "The Vascular Smooth Muscle Cell: Molecular and Biological Responses to the Extracellular Matrix" (S.M. Schwartz and R.P. Mecham, Eds.), pp. 37–79. Academic Press, San Diego.

Handford, P. A., Downing, A. K., Reinhardt, D. P., and Sakai, L. Y. (2000). Fibrillin: From domain structure to supramolecular assembly. *Matrix Biol.* **19**, 457–470.

Haust, M. D., More, R. H., Benscome, S. A., and Balis, J. U. (1965). Elastogenesis in human aorta: An electron microscopic study. *Exp. Mol. Pathol.* **4**, 508–524.

Hocking, A. M., Shinomura, T., and McQuillan, D. J. (1998). Leucine-rich repeat glycoproteins of the extracellular matrix. *Matrix Biol.* **17**, 1–19.

Holland, P. W., Harper, S. J., McVey, J. H., and Hogan, B. L. (1987). *In Vivo* expression of mRNA for the Ca++ binding protein SPARC (osteonectin) revealed by *in situ* hybridization. *J. Cell Biol.* **105**, 473–482.

Howard, P. S., and Macarak, E. J. (1989). Localization of collagen types in regional segments of the fetal bovine aorta. *Lab. Invest.* **61**, 548–555.

Hungerford, J. E., Hoeffler, J. P., Bowers, C. W., Dahm, L. M., Flachetto, R., Shabanowitz, J., Hunt, D. F., and Little, C. D. (1997). Identification of a novel marker for primordial smooth muscle and its differential expression pattern in contractile vs. noncontractile cells. *J. Cell Biol.* **137**, 925–937.

Hungerford, J. E., Owens, G. K., Argraves, W. S., and Little, C. D. (1996). Development of the aortic vessel wall as defined by vascular smooth muscle and extracellular matrix markers. *Develop. Biol.* **178**, 375–392.

Iozzo, R. V., and Murdoch, A. D. (1996). Proteoglycans of the extracellular environment: Clues from the gene and protein side offer novel perspectives in molecular diversity and function. *FASEB J.* **10**, 598–614.

Iruela-Arispe, M. L., Lane, T. F., Redmond, D., Reilly, M., Bolender, R. P., Kavanagh, T. J., and Sage, E. H. (1995). Expression of SPARC during development of the chicken chorioallantoic membrane: Evidence for regulated proteolysis *in vivo*. *Mol. Biol. Cell* **6**, 327–343.

Iruela-Arispe, M. L., Liska, D. J., Sage, E. H., and Bornstein, P. (1993). Differential expression of thrombospondin 1, 2, and 3 during murine development. *Dev. Dyn.* **197**, 40–56.

Jones, F. S., and Jones, P. L. (2000a). The tenascin family of ECM glycoproteins: Structure, function, and regulation during embryonic development and tissue remodeling. *Dev. Dyn.* **218**, 235–259.

Jones, P. L., and Jones, F. S. (2000b). Tenascin-C in development and disease: Gene regulation and cell function. *Matrix Biol.* **19**, 581–596.

Karrer, H. E. (1961). An electron microscope study of the aorta in young and in aging mice. *J. Ultrastruct. Res.* **5**, 1–17.

Keeley, F. W., and Alatawi, A. (1991). Response of aortic elastin synthesis and accumulation to developing hypertension and the inhibitory effect of colchicine on this response. *Lab. Invest.* **64**, 499–507.

Keeley, F. W., and Johnson, D. J. (1986). The effect of developing hypertension on the synthesis and accumulation of elastin in the aorta of the rat. *Biochem. Cell Biol.* **64**, 38–43.

Kielty, C. M., Sherratt, M. J., and Shuttleworth, C. A. (2002). Elastic fibres. *J. Cell Sci.* **115**, 2817–2828.

Kostka, G., Giltay, R., Bloch, W., Addicks, K., Timpl, R., Fassler, R., and Chu, M. L. (2001). Perinatal lethality and endothelial cell abnormalities in several vessel compartments of fibulin-1-deficient mice. *Mol. Cell. Biol.* **21**, 7025–7034.

Kowal, R. C., Richardson, J. A., Miano, J. M., and Olson, E. N. (1999). EVEC, a novel epidermal growth factor-like repeat-containing protein upregulated in embryonic and diseased adult vasculature. *Circ. Res.* **84**, 1166–1176.

Kozel, B. A., Ciliberto, C. H., and Mecham, R. P. (2004). Deposition of tropoelastin into the extracellular matrix requires a competent elastic fiber scaffold but not live cells. *Matrix Biol.* **23**, 23–34.

Kozel, B. A., Wachi, H., Davis, E. C., and Mecham, R. P. (2003). Domains in tropoelastin that mediate elastin deposition *in vitro* and *in vivo*. *J. Biol. Chem.* **278**, 18491–18498.

Kyriakides, T. R., Zhu, Y. H., Smith, L. T., Bain, S. D., Yang, Z., Lin, M. T., Danielson, K. G., Iozzo, R. V., LaMarca, M., McKinney, C. E., Ginns, E. I., and Bornstein, P. (1998). Mice that lack thrombospondin 2 display connective tissue abnormalities that are associated with disordered collagen fibrillogenesis, an increased vascular density, and a bleeding diathesis. *J. Cell Biol.* **140**, 419–430.

Langille, B. L. (1996). Arterial remodeling: Relation to hemodynamics. *Can. J. Physiol. Pharmacol.* **74**, 834–841.

Lawler, J. (2002). Thrombospondin-1 as an endogenous inhibitor of angiogenesis and tumor growth. *J. Cell Mol. Med.* **6**, 1–12.

Lebel, S. P., Chen, Y., Gingras, D., Chung, A. E., and Bendayan, M. (2003). Morphofunctional studies of the glomerular wall in mice lacking entactin-1. *J. Histochem. Cytochem.* **51**, 1467–1478.

Leung, D. Y. M., Glagov, S., and Mathews, M. B. (1977). Elastin and collagen accumulation in rabbit ascending aorta and pulmonary trunk during postnatal growth. Correlation of cellular synthetic response with medial tension. *Circ. Res.* **41**, 316–323.

Li, S., Edgar, D., Fassler, R., Wadsworth, W., and Yurchenco, P. D. (2003). The role of laminin in embryonic cell polarization and tissue organization. *Dev. Cell* **4**, 613–624.

Li, W., Nellaiappan, K., Strassmaier, T., Graham, L., Thomas, K. M., and Kagan, H. M. (1997). Localization and activity of lysyl oxidase within nuclei of fibrogenic cells. *Proc. Natl. Acad. Sci. USA* **25**, 12817–12822.

Liu, X., Wu, H., Byrne, M., Krane, S., and Jaenisch, R. (1997). Type III collagen is crucial for collagen I fibrillogenesis and for normal cardiovascular development. *Proc. Natl. Acad. Sci. USA* **94**, 1852–1856.

Loeys, B., Van Maldergem, L., Mortier, G., Coucke, P., Gerniers, S., Naeyaert, J. M., and De Paepe, A. (2002). Homozygosity for a missense mutation in fibulin-5 (FBLN5) results in a severe form of cutis laxa. *Hum. Mol. Genet.* **11**, 2113–2118.

Lohler, J., Timpl, R., and Jaenisch, R. (1984). Embryonic lethal mutation in mouse collagen I gene causes rupture of blood vessels and is associated with erythropoietic and mesenchymal cell death. *Cell* **38**, 597–607.

Mackie, E. J., and Tucker, R. P. (1999). The tenascin-C knockout revisited. *J. Cell Sci.* **112**(Pt 22), 3847–3853.

Mao, J. R., Taylor, G., Dean, W. B., Wagner, D. R., Afzal, V., Lotz, J. C., Rubin, E. M., and Bristow, J. (2002). Tenascin-X deficiency mimics Ehlers-Danlos syndrome in mice through alteration of collagen deposition. *Nat. Genet.* **30**, 421–425.

Markova, D., Zou, Y., Ringpfeil, F., Sasaki, T., Kostka, G., Timpl, R., Uitto, J., and Chu, M. L. (2003). Genetic heterogeneity of cutis laxa: A heterozygous tandem duplication within the fibulin-5 (FBLN5) gene. *Am. J. Hum. Genet.* **72**, 998–1004.

Maurer, P., and Engel, J. (1996). Structure of laminins and their chain assembly. In "The Laminin" (P. Ekblom and R. Timpl, Eds.), pp. 27–49. Harwood Academic Publishers, New York.

Mayer, U., and Timpl, R. (1994). Nidogen: A versatile binding protein of basement membranes. In "Extracellular Matrix Assembly" (P. D. Yurchenco, D. E. Birk, and R. P. Mecham, Eds.), pp. 318–416. Academic Press, San Diego.

McCullagh, K. G., Duance, V. C., and Bishop, K. A. (1980). The distribution of collagen types I, III and V (AB) in normal and atherosclerotic human aorta. J. Pathol. **130**, 45–55.

McLean, S. E., Mecham, B. H., Mariani, T. J., Corry, S., Ciliberto, C. H., and Mecham, R. P. (2004). Submitted for publication.

Mecham, R. P. (1998). Overview of Extracellular Matrix. In "Current Protocols in Cell Biology" pp. 10.1.1–10.1.13. John Wiley & Sons, Inc.

Metcalfe, K., Rucka, A. K., Smoot, L., Hofstadler, G., Tuzler, G., McKeown, P., Siu, V., Rauch, A., Dean, J., Dennis, N., Ellis, I., Reardon, W., Cytrynbaum, C., Osborne, L., Yates, J. R., Read, A. P., Donnai, D., and Tassabehji, M. (2000). Elastin: Mutational spectrum in supravalvular aortic stenosis. Eur. J. Hum. Genet. **8**, 955–963.

Milewicz, D. M., Urbán, Z., and Boyd, C. D. (2000). Genetic disorders of the elastic fiber system. Matrix Biol. **19**, 471–480.

Miner, J. H., Cunningham, J., and Sanes, J. R. (1998). Roles for laminin in embryogenesis: Exencephaly, syndactyly, and placentopathy in mice lacking the laminin alpha5 chain. J. Cell Biol. **143**, 1713–1723.

Morris, C. A. (1998). Genetic aspects of supravalvular aortic stenosis. Current Opin. Cardiol. **13**, 214–219.

Morris, C. A., and Mervis, C. B. (2000). Roles for laminin in embryogenesis: Exencephaly, syndactyly, and placentopathy in mice lacking the laminin alpha5 chain. Annu. Rev. Genomics Hum. Genet. **1**, 461–484.

Murshed, M., Smyth, N., Miosge, N., Karolat, J., Krieg, T., Paulsson, M., and Nischt, R. (2000). The absence of nidogen 1 does not affect murine basement membrane formation. Mol. Cell. Biol. **20**, 7007–7012.

Nakamura, H. (1988). Electron microscopic study of the prenatal development of the thoracic aorta in the rat. Am. J. Anat. **181**, 406–418.

Nakamura, T., Lozano, P. R., Ikeda, Y., Iwanaga, Y., Hinek, A., Minamisawa, S., Cheng, C. F., Kobuke, K., Dalton, N., Takada, Y., Tashiro, K., Ross, Jr., J., Honjo, T., and Chien, K. R. (2002). Fibulin-5/DANCE is essential for elastogenesis in vivo. Nature **415**, 171–175.

Neptune, E. R., Frischmeyer, P. A., Arking, D. E., Myers, L., Bunton, T. E., Gayraud, B., Ramirez, F., Sakai, L. Y., and Dietz, H. C. (2003). Dysregulation of TGF-beta activation contributes to pathogenesis in Marfan syndrome. Nat. Genet. **33**, 407–411.

Niiyama, T., Higuchi, I., Hashiguchi, T., Suehara, M., Uchida, Y., Horikiri, T., Shiraishi, T., Saitou, A., Hu, J., Nakagawa, M., Arimura, K., and Osame, M. (2003). Capillary changes in skeletal muscle of patients with Ullrich's disease with collagen VI deficiency. Acta Neuropathol. (Berl.) **106**, 137–142.

Olson, T. M., Michels, V. V., Lindor, N. M., Pastores, G. M., Weber, J. L., Schaid, D. J., Driscoll, D. J., Feldt, R. H., and Thibodeau, S. N. (1993). Autosomal dominant supravalvular aortic stenosis: Localization to chromosome 7. Hum. Mol. Genet. **2**, 869–873.

Onda, M., Ishiwata, T., Kawahara, K., Wang, R., Naito, Z., and Sugisaki, Y. (2002). Expression of lumican in thickened intima and smooth muscle cells in human coronary atherosclerosis. Exp. Mol. Pathol. **72**, 142–149.

Owens, G. K. (1995). Regulation of differentiation of vascular smooth muscle cells. Physiol Rev. 1995 Jul;75(3): 487–517. Review. Physiol. Rev. **75**, 487–517.

Pasquali-Ronchetti, I., Quaglino, D., Baccarani-Contri, M., Tenconi, R., Bressan, G. M., and Volpin, D. (1986). Aortic elastin abnormalities in osteogenesis imperfecta type II. Coll. Relat. Res. **6**, 409–421.

Paule, W. J. (1963). Electron microscopy of the newborn rat aorta. *J. Ultrastruct. Res.* **8**, 219–235.
Pease, D. C., and Paule, W. J. (1960). Electron microscopy of elastic arteries; the thoracic aorta of the rat. *J. Ultrastruct. Res.* **3**, 469–483.
Pereira, L., Andrikopoulos, K., Tian, J., Lee, S. Y., Keene, D. R., Ono, R., Reinhardt, D. P., Sakai, L. Y., Biery, N. J., Bunton, T., Dietz, H. C., and Ramirez, F. (1997). Targetting of the gene encoding fibrillin-1 recapitulates the vascular aspect of Marfan syndrome. *Nature genetics* **17**, 218–222.
Petajaniemi, N., Korhonen, M., Kortesmaa, J., Tryggvason, K., Sekiguchi, K., Fujiwara, H., Sorokin, L., Thornell, L. E., Wondimu, Z., Assefa, D., Patarroyo, M., and Virtanen, I. (2002). Localization of laminin alpha4-chain in developing and adult human tissues. *J. Histochem. Cytochem.* **50**, 1113–1130.
Pfaff, M., Reinhardt, D. P., Sakai, L. Y., and Timpl, R. (1996). Cell adhesion and integrin binding to recombinant human fibrillin-1. *FEBS Letters* **384**, 247–250.
Pope, F. M., Martin, G. R., and McKusick, V. A. (1977). Inheritance of Ehlers-Danlos type IV syndrome. *J. Med. Genet.* **14**, 200–204.
Porter, P. L., Sage, E. H., Lane, T. F., Funk, S. E., and Gown, A. M. (1995). Distribution of SPARC in normal and neoplastic human tissue. *J. Histochem. Cytochem.* **43**, 791–800.
Raff, M. L., Craigen, W. J., Smith, L. T., Keene, D. R., and Byers, P. H. (2000). Partial COL1A2 gene duplication produces features of osteogenesis imperfecta and Ehlers-Danlos syndrome type VII. *Hum. Genet.* **106**, 19–28.
Riessen, R., Fenchel, M., Chen, H., Axel, D. I., Karsch, K. R., and Lawler, J. (2001). Cartilage oligomeric matrix protein (thrombospondin-5) is expressed by human vascular smooth muscle cells. *Arterioscler. Thromb. Vasc. Biol.* **21**, 47–54.
Riquelme, C., Larrain, J., Schonherr, E., Henriquez, J. P., Kresse, H., and Brandan, E. (2001). Antisense inhibition of decorin expression in myoblasts decreases cell responsiveness to transforming growth factor beta and accelerates skeletal muscle differentiation. *J. Biol. Chem.* **276**, 3589–3596.
Risau, W., and Lemmon, V. (1988). Changes in the vascular extracellular matrix during embryonic vasculogenesis and angiogenesis. *Dev. Biol.* **125**, 441–450.
Ritty, T. M., Broekelmann, T. J., Werneck, C. C., and Mecham, R. P. (2003). Fibrillin-1 and -2 contain heparin-binding sites important for matrix deposition and that support cell attachment. *Biochem. J.* **375**, 425–432.
Roach, M. (1983). The pattern of elastin in the aorta and large arteries of mammals. *Ciba Foundation Symp.* **100**, 37–55.
Roark, E. F., Keene, D. R., Haudenschild, C. C., Godyna, S., Little, C. D., and Argraves, W. S. (1995). The association of human fibulin-1 with elastic fibers: An immunohistological, ultrastructural, and RNA study. *J. Histochem. Cytochem.* **43**, 401–411.
Rossant, J., and Howard, L. (2002). Signaling pathways in vascular development. *Annu. Rev. Cell. Dev. Biol.* **18**, 541–573.
Sakamoto, H., Broekelmann, T., Cheresh, D. A., Ramirez, F., Rosenbloom, J., and Mecham, R. P. (1996). Cell-type specific recognition of RGD- and non-RGD-containing cell binding domains in fibrillin-1. *J. Biol. Chem.* **271**, 4916–4922.
Sato, T. N., Qin, Y., Kozak, C. A., and Audus, K. L. (1993). Tie-1 and tie-2 define another class of putative receptor tyrosine kinase genes expressed in early embryonic vascular system. *Proc. Natl. Acad. Sci. USA* **90**, 9355–9358.
Schalkwijk, J., Zweers, M. C., Steijlen, P. M., Dean, W. B., Taylor, G., van Vlijmen, I. M., van Haren, B., Miller, W. L., and Bristow, J. (2001). A recessive form of the Ehlers-Danlos syndrome caused by tenascin-X deficiency. *N. Engl. J. Med.* **345**, 1167–1175.
Schwartz, S. M., and Mecham, R. P. (1995). The Vascular Smooth Muscle Cell: Molecular and Biological Responses to the Extracellular Matrix. pp. 410. Academic Press, San Diego.

Schwarze, U., Atkinson, M., Hoffman, G. G., Greenspan, D. S., and Byers, P. H. (2000). Null alleles of the COL5A1 gene of type V collagen are a cause of the classical forms of Ehlers-Danlos syndrome (types I and II). *Am. J. Hum. Genet.* **66**, 1757–1765.

Schwarze, U., Schievink, W. I., Petty, E., Jaff, M. R., Babovic-Vuksanovic, D., Cherry, K. J., Pepin, M., and Byers, P. H. (2001). Haploinsufficiency for one COL3A1 allele of type III procollagen results in a phenotype similar to the vascular form of Ehlers-Danlos syndrome, Ehlers-Danlos syndrome type IV. *Am. J. Hum. Genet.* **69**, 989–1001.

Schymeinsky, J., Nedbal, S., Miosge, N., Poschl, E., Rao, C., Beier, D. R., Skarnes, W. C., Timpl, R., and Bader, B. L. (2002). Gene structure and functional analysis of the mouse nidogen-2 gene: Nidogen-2 is not essential for basement membrane formation in mice. *Mol. Cell. Biol.* **22**, 6820–6830.

Smyth, N., Vatansever, H. S., Murray, P., Meyer, M., Frie, C., Paulsson, M., and Edgar, D. (1999). Absence of basement membranes after targeting the LAMC1 gene results in embryonic lethality due to failure of endoderm differentiation. *J. Cell. Biol.* **144**, 151–160.

Sorokin, L. M., Pausch, F., Frieser, M., Kroger, S., Ohage, E., and Deutzmann, R. (1997). Developmental regulation of the laminin alpha5 chain suggests a role in epithelial and endothelial cell maturation. *Dev. Biol.* **189**, 285–300.

Stenina, O. I., Desai, S. Y., Krukovets, I., Kight, K., Janigro, D., Topol, E. J., and Plow, E. F. (2003). Thrombospondin-4 and its variants: Expression and differential effects on endothelial cells. *Circulation* **108**, 1514–1519.

Stone, E. M., Lotery, A. J., Munier, F. L., Heon, E., Piguet, B., Guymer, R. H., Vandenburgh, K., Cousin, P., Nishimura, D., Swiderski, R. E., Silvestri, G., Mackey, D. A., Hageman, G. S., Bird, A. C., Sheffield, V. C., and Schorderet, D. F. (1999). A single EFEMP1 mutation associated with both Malattia Leventinese and Doyne honeycomb retinal dystrophy. *Nat. Genet.* **22**, 199–202.

Tassabehji, M., Metcalfe, K., Hurst, J., Ashcroft, G. S., Kielty, C., Wilmot, C., Donnai, D., Read, A. P., and Jones, C. J. P. (1998). An elastin gene mutation producing abnormal tropoelastin and abnormal elastic fibres in a patient with autosomal dominant cutis laxa. *Hum. Molec. Gen.* **7**, 1021–1028.

Theocharis, A. D., and Karamanos, N. K. (2002). Decreased biglycan expression and differential decorin localization in human abdominal aortic aneurysms. *Atherosclerosis* **165**, 221–230.

Thyberg, J., Hinek, A., Nilsson, J., and Friberg, U. (1979). Electron microscopic and cytochemical studies of rat aorta. Intracellular vesicles containing elastin- and collagen-like material. *Histochem. J.* **11**, 1–17.

Thyboll, J., Kortesmaa, J., Cao, R., Soininen, R., Wang, L., Iivanainen, A., Sorokin, L., Risling, M., Cao, Y., and Tryggvason, K. (2002). Deletion of the laminin alpha4 chain leads to impaired microvessel maturation. *Mol. Cell. Biol.* **22**, 1194–202.

Timpl, R., Dziadek, M., Fujiwara, S., Nowack, H., and Wick, G. (1983). Nidogen: A new, self-aggregating basement membrane protein. *Eur. J. Biochem.* **137**, 455–465.

Timpl, R., Sasaki, T., Kostka, G., and Chu, M. L. (2003). Fibulins: A versatile family of extracellular matrix proteins. *Nat. Rev. Mol. Cell. Biol.* **4**, 479–489.

Tsuda, T., Wang, H., Timpl, R., and Chu, M. L. (2001). Fibulin-2 expression marks transformed mesenchymal cells in developing cardiac valves, aortic arch vessels, and coronary vessels. *Dev. Dyn.* **222**, 89–100.

Urbán, Z., Michels, V. V., Thibodeau, S. N., Davis, E. C., Bonnefont, J.-P., Munnich, A., Eyskens, B., Gewillig, M., Devriendt, K., and Boyd, C. D. (2000). Isolated supravalvular aortic stenosis: Functional haploinsufficiency of the elastin gene as a result of nonsense-mediated decay. *Hum. Genet.* **106**, 577–588.

Voss, B., and Rauterberg, J. (1986). Localization of collagen types I, III, IV and V, fibronectin and laminin in human arteries by the indirect immunofluorescence method. *Pathol. Res. Pract.* **181**, 568–575.

Wenstrup, R. J., Florer, J. B., Willing, M. C., Giunta, C., Steinmann, B., Young, F., Susic, M., and Cole, W. G. (2000). COL5A1 haploinsufficiency is a common molecular mechanism underlying the classical form of EDS. *Am. J. Hum. Genet.* **66,** 1766–1776.

Wight, T. N. (2002). Versican: A versatile extracellular matrix proteoglycan in cell biology. *Curr. Opin. Cell Biol.* **14,** 617–623.

Wight, T. N., Raugi, G. J., Mumby, S. M., and Bornstein, P. (1985). Light microscopic immunolocation of thrombospondin in human tissues. *J. Histochem. Cytochem.* **33,** 295–302.

Wolinsky, H., and Glagov, S. (1967). A lamellar unit of aortic medial structure and function in mammals. *Circ. Res.* **20,** 99–111.

Yan, Q., and Sage, E. H. (1999). SPARC, a matricellular glycoprotein with important biological functions. *J. Histochem. Cytochem.* **47,** 1495–506.

Yanagisawa, H., Davis, E. C., Starcher, B. C., Ouchi, T., Yanagisawa, M., Richardson, J. A., and Olson, E. N. (2002). Fibulin-5 is an elastin-binding protein essential for elastic fibre development *in vivo. Nature* **415,** 168–171.

Yao, L. Y., Moody, C., Schonherr, E., Wight, T. N., and Sandell, L. J. (1994). Identification of the proteoglycan versican in aorta and smooth muscle cells by DNA sequence analysis, in situ hybridization and immunohistochemistry. *Matrix Biol.* **14,** 213–225.

Yurchenco, P. D., Amenta, P. S., and Patton, B. L. (2004). Basement membrane assembly, stability and activities observed through a developmental lens. *Matrix Biol.* **22,** 521–538.

Zanetti, M., Braghetta, P., Sabatelli, P., Mura, I., Doliana, R., Colombatti, A., Volpin, D., Bonaldo, P., and Bressan, G. M. (2004). EMILIN-1 deficiency induces elastogenesis and vascular cell defects. *Mol. Cell. Biol.* **24,** 638–650.

Zhang, M., Pierce, R. A., Wachi, H., Mecham, R. P., and Parks, W. C. (1999). An open reading frame element mediates posttranscriptional regulation of tropoelastin and responsiveness to transforming growth factor beta1. *Molec. Cell. Biol.* **19,** 7314–7326.

Zheng, P. S., Wen, J., Ang, L. C., Sheng, W., Viloria-Petit, A., Wang, Y., Wu, Y., Kerbel, R. S., and Yang, B. B. (2004). Versican/PG-M G3 domain promotes tumor growth and angiogenesis. *FASEB J.* **18,** 754–756.

7

Genetics in Zebrafish, Mice, and Humans to Dissect Congenital Heart Disease: Insights in the Role of VEGF

Diether Lambrechts and Peter Carmeliet
Flanders Interuniversity Institute for Biotechnology
Center for Transgene Technology and Gene Therapy
KU Leuven, Leuven, B-3000, Belgium

I. Conserved Body Plan Architecture in Vertebrates
II. Vascular Malformations
III. Genetic Causes of CHDs and Vascular Anomalies
   A. Chromosomal Anomalies
   B. Malformation Syndromes and Isolated Familial CHDs
   C. CHDs with Multifactorial Origin
   D. CHDs with Other Types of Inheritance
IV. The Use of Animal Models
   A. The Mouse, the Model Organism of Choice
   B. The Zebrafish Model
V. Role of VEGF in Cardiovascular Development
VI. Expression Pattern of VEGF in Mice Reveal Other Distinct Biological Functions
VII. The Role of VEGF During Heart Septation
VIII. VEGF: a Connector Between the Developing Vascular and Neuronal System
IX. A Role for VEGF in Arterial EC Specification
X. Normal Ontogenesis but Abnormal Remodeling of PAAs in Mice Lacking VEGF164
XI. Association of VEGF Gene Variations with Cardiovascular Defects in DGS
XII. Conclusion
   References

Heart development and the establishment of a functional circulatory circuit are complex biological processes in which subtle perturbations may result in catastrophic consequences of cardiovascular birth defects. Studies in model organisms, most notably the mouse and the zebrafish, have identified genes that also cause these life-threatening defects when mutated in humans. Gradually, a framework for the genetic pathway controlling these events is now beginning to emerge. However, the puzzling phenotypic variability of the cardiovascular disease phenotype in humans and the recent identification of phenotypic modifiers using model organisms indicates that other genetic loci might interact to modify the disease phenotype. To illustrate this, we review the role of vascular endothelial growth factor (VEGF) during vascular and cardiac development and stress how zebrafish and mouse genetic studies have helped us to understand the role this growth factor has in human disease, in particular in the Di-George syndrome. © 2004, Elsevier Inc.

I. Conserved Body Plan Architecture in Vertebrates

The heart is one of the first organs to develop, and its formation involves a complex series of morphological and morphogenetic events, which are each triggered by the strict temporarily and spatially regulated expression of numerous signaling molecules and transcription factors. Because the process of heart development occurs through an evolutionary conserved program that is common to all vertebrates, information about the genes and their mechanisms of action can be extrapolated from small animal models, most notably the chick, the zebrafish, the frog, and the mouse, to human heart development (Srivastava and Olson, 2000).

In mice and humans, heart development starts with the specification of anterior lateral plate mesoderm-derived cells to cardiomyocytes, which converge and fuse along the ventral midline of the embryo to form a single heart tube that spontaneously beats. This straight heart tube is patterned in an anterior–posterior polarity in which the prospective tissues of the aortic sac, outflow tract (conotruncus), ventricles, and atria are present (Fig. 1) and consist of an external myocardial and internal endocardial layer (reviewed in Towbin and Belmont, 2000). Under the control of a number of genes that are associated with right–left asymmetry, the linear heart tube then forms a rightward loop, thereby creating left–right polarity and bringing the embryonic heart structures into position to engage in septation (reviewed in Burdine and Schier, 2000; Capdevila et al., 2000). Septation encompasses the division of the heart into chambers and the formation of valves; these processes involve the initial swelling of extracellular matrix or cardiac jelly between the inner endothelium and outer myocardium, after which a subset of endothelial cells (ECs) undergo endothelial/mesenchymal transformation to produce mesenchyme cells that form the valves and septa (reviewed in Anderson et al., 2003a,b; Moorman et al., 2003; Pierpont et al., 2000). Ultimately, septation leads to the formation of a four-chambered heart structure with separate venous and arterial poles and with atrioventricular and semilunar valves to ensure unidirectional blood flow. In the zebrafish, heart development continues only until cardiac looping, after which the adult configuration is obtained. Consequently, the zebrafish heart consists of a tubular structure that is divided into four serially connected chambers: sinus venosus, atrium, ventricle, and bulbus arteriosus (Fig. 1). It pumps venous blood from caudal venous connections to a single cranial arterial connection that is connected to the gill arches where oxygenation occurs and

7. Genetics and VEGF

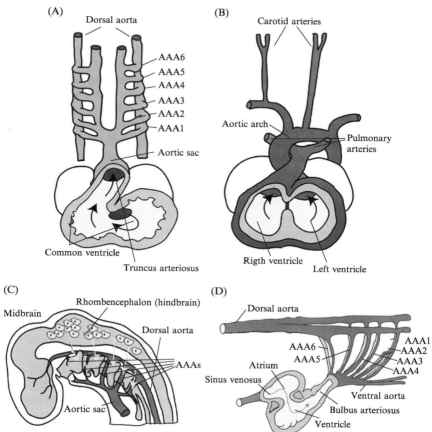

Figure 1 Development of the heart and pharyngeal arch system (PAA) in mice and zebrafish. (A) The developing heart and the PAA system, before the remodeling of the arch artery system and before the formation of the aortico-pulmonary septum, with a common ventricle, a single outflow vessel called truncus arteriosus, and the aortic sac, which gives rise to six bilaterally symmetric vessels known as PAAs; (B) the mature heart configuration, in which the truncus arteriosus has become divided into an aortic and pulmonary trunk: the aortic trunk arises from the left ventricle, the pulmonary trunk from the right ventricle. The aortic arch system has extensively remodeled into the mature aortic arch and proximal pulmonary arteries; (C) lateral view of a murine embryo of 10.5 days old, showing the pharyngeal arches (1–6) and their corresponding arch arteries. Neural crest cells (yellow arrows) migrate from the neural folds in the rhombencephalon or hindbrain to the adjacent pharyngeal arches. The crest cells that fill the first arch arise from the caudal midbrain and rhombomeres 1 and 2 of the hindbrain; those that populate the second arch emerge primarily from rhombomere 4; and, finally, the more caudal arches are filled by crest cells generated by rhombomeres 6 and 7; (D) schematic overview of the heart and blood vascular systems of an early zebrafish embryo. At 48-h postfertilization, the zebrafish heart consists of a smooth-walled tube partitioned into all four segments with the definite structure of sinus venosus, atrium, ventricle, and bulbus arteriosus, which gives rise to six bilaterally symmetric vessels known as aortic arch arteries (AAA) that drain into the dorsal aorta. (See Color Insert.)

from where it is distributed to the rest of the body (Hu *et al.*, 2000; Stainier, 2001; Thisse and Zon, 2002).

Also, the development of the aortic arch system, which is the equivalent of the pharyngeal arch artery (PAA) system in mice and humans, is conserved between fish and mammals. The development of the pharyngeal arches (PAs) is a complex process and involves the contribution of different embryonic cell types, including ectodermal, mesodermal, endodermal, and neural crest cells (NCCs, reviewed in Creazzo *et al.*, 1998; Graham and Smith, 2001; Maschhoff and Baldwin, 2000). The latter originate from the caudal hindbrain and migrate into the PAs, where they play a pivotal role in patterning and subsequent remodeling of the PAs into the organs that develop from it. Prior to these remodeling events, the aortic sac is located rostral to the conotruncus and gives rise to six bilaterally symmetric vessels known as the PAAs (Fig. 1). Each PAA traverses a PA before joining one of the paired dorsal aortae. The NCCs that derive from the posterior rhombencephalon or hindbrain (rombomeres 6 and 7) migrate into the caudal PAs (third, fourth, and sixth) where they contribute to the vascular development of the great arteries (Fig. 1c). Some of these NCCs continue to migrate into the developing conotruncal endocardial cushions of the outflow tract to play an essential role in the organization of cardiac outflow tract septation. Proper migration of NCCs is thus required to ensure the correct configuration of the cardiac outflow tract and aortic arch. But also, most of the craniofacial bones and cartilaginous structures, such as the mandible, maxilla, thyroid cartilage, and middle ear elements, are derived from NCCs. These NCCs are referred to as the cranial neural crest, which is the cartilaginous component of the PAs. The nervous component of the PAs, which also derives from the cranial neural crest, gives rise to the cranial nerves, while the parathyroid glands and thymus originate from the pouch between the third and the fourth PAs. Hence, NCC migratory defects result in a typical spectrum of conotruncal septation defects in the heart, craniofacial, thymic, thyroid, and parathyroid defects. Although the gross remodeling of the caudal four PAAs that occurs in mammals does not take place in fish, as these vessels remain largely intact as gill arteries, much of the genetic network specifying the interactions between endoderm, mesodermally derived aortic arch arteries, and neural crest in the PAs is conserved between mice and fish. Indeed, recent studies in the zebrafish support evidence that PA patterning is not completely dependent on NCCs but that its final patterning is rather the integrated result of crest-dependent and crest-independent cues (discussed later) (Graham and Smith, 2001; Piotrowski and Nusslein-Volhard, 2000). This makes the zebrafish also a well-suited model organism to study NCC migration and differentiation.

7. Genetics and VEGF

The development of blood vessels occurs via vasculogenesis and angiogenesis. The former process involves the differentiation of precursor cells or angioblasts into ECs to form a primordial vascular plexus, whereas the latter process is used to describe the entire process of how this primitive network matures into a complex network (reviewed in Carmeliet, 2000). Angiogenesis involves (1) the sprouting of new blood vessels from pre-existing vessels; (2) intussusceptive growth, whereby a preexisting vessel is split into two daughter vessels by formation of trans-capillary pillars and by invagination of surrounding pericytes and extracellular matrix; and (3) the intercalated growth of blood vessels, in which preexisting capillaries merge or additional ECs fuse into existing vessels to increase their length and diameter. The development of blood vessels in the zebrafish is largely similar to that in other vertebrates. The same primary vasculogenic vessels, such as the primary trunk vessels, the dorsal aorta, and posterior cardinal vein, which establish the initial circulatory circuits in all vertebrates, are present in the zebrafish, but also many of the later-forming zebrafish blood vessels, which are believed to form via angiogenesis, have analogues in other vertebrates (Isogai et al., 2001, 2003).

In order to establish a functional vascular network, blood vessels also must become specialized into arteries and veins, but little is still known about the molecular mechanisms of how arterial and venous vessels form parallel but distinct and separate networks (Lawson and Weinstein, 2002a). Differences in blood pressure and direction of blood flow were previously thought to be responsible for determining whether an EC ended up as an arterial or venous cell. However, recent data suggest that molecular differences between arterial and venous ECs are present before onset of circulation (Lawson et al., 2002). Fate-mapping studies in the zebrafish have shown that the majority of the angioblast precursors for the trunk artery and vein are spatially mixed in the lateral posterior mesoderm but that the progeny of each precursor becomes restricted to one of the vessels upon differentiation (Zhong et al., 2001). The notochord is known to play a role in the patterning of nearby tissues and has also been implicated in providing guidance cues for these angioblasts, which become restricted to the aorta or trunk vein. Conceivably, *floating head* and *no tail* mutant zebrafish embryos, which lack a notochord, also exhibit failures in the formation of the dorsal aorta, whereas the formation of the axial vein is less affected (Fouquet et al., 1997). This indicates that the notochord guides angioblasts in fashioning the dorsal aorta but seems to be less important for the formation of the vein. Thus, zebrafish are also appropriate organisms for studying the specification of early vascular progenitors and arterial and venous-specific patterning mechanisms (Roman and Weinstein, 2000).

II. Vascular Malformations

It is clear that the formation of the circulatory system involves a complex series of events, and that subtle perturbations of these process may have catastrophic consequences on our health in the form of congenital heart disease or vascular anomalies. Congenital heart defects (CHDs) are the most common form of birth defects and have a prevalence that has been estimated at 7–9 per 1000 births (Carlgren *et al.*, 1987; Hanna *et al.*, 1994; Mitchell *et al.*, 1971). The exact number of CHDs depends on many factors such as, for instance, the methods with which cases were ascertained, the inclusion criteria used, and the time frame during which the diagnosis was made (reviewed by Hoffman and Kaplan, 2002). Epidemiological studies, therefore, often distinguish between "severe CHDs," which include patients who are severely ill within the newborn period or early infancy and often need to undergo surgery, and "less-severe CHDs," in which a significant proportion of the patients are asymptomatic and often undergo spontaneous resolution of their defect at later age. Isolated ventricular septal defects (VSDs) are by far the most common form of CHDs and are included within the latter category. The more severe defects are, however, those impacting on the health care system and readily detectable in any good medical diagnostic

Table I

a. Severe cardiac defect
Hypoplastic left heart syndrome (HLHS)
Single ventricle
Tricuspid valve atresia
Common truncus
Interrupted aortic arch (IAA)
Pulmonary valve atresia with intact VSD
d-Transposition of great arteries (d-TGA)
Double outlet right ventricle (DORV)
Endocardial cushion defect (ECD)
Total anomalous pulmonary venous return (TAPVR)
Tetralogy of Fallot (ToF)
Ebstein's anomaly

b. Less evere cardiac defect
Ventricular septal defect (VSD)
Atrial septal defect (ASD)
Corrected transposition
Coarctation of aorta
Aortic valve stenosis
Pulmonary valve stenosis

system. Many studies have been restricted to these defects, which have yielded more consistent incidence rates, varying between 2.3–4.8 cases per 1000 births (Ferencz et al., 1985; Fyler et al., 1981; Grabitz et al., 1988; Loffredo, 2000). In Table I, severe CHDs, identified in sufficiently large numbers to permit meaningful epidemiological analyses, are listed together with the common less-severe CHDs. Likewise, vascular birth defects cause formidable medical problems and social trauma for the affected patients and family. Vascular anomalies usually affect a limited number of vessels in a restricted area of the body. Unlike vessel tumors (or hemangiomas), where ECs grow in excess, vessels in vascular malformations have normal numbers of ECs but are improperly formed and remodeled. Vascular malformations can be divided according to the type of vessel that they affect into capillary, venous, arteriovenous, lymphatic, and combined malformations. A well-known example of vascular anomalies are the capillary malformations, or "port-wine stains," which occur in 3 out of 1000 people (reviewed by Brouillard and Vikkula, 2003; Vikkula et al., 2001).

III. Genetic Causes of CHDs and Vascular Anomalies

Disease modifiers are capable of influencing the wide phenotypic spectrum of CHDs in patients with identical disease-causing alleles and in mutant mice bred into varying genetic backgrounds (Chien, 2000). The incidence of CHDs has remained relatively constant in time and varies little between populations, suggesting that there is only a small aetiological contribution from environmental factors and a larger contribution from genetic factors. This latter hypothesis is further supported by twin studies that show a greater concordance in monozygotic twins than in dizygotic twins. Nonetheless, the genetic causes for CHDs may be different in each patient (Pradat et al., 2003).

A. Chromosomal Anomalies

A number of well-recognized chromosome aneuploidies cause malformation syndromes, often associated with CHDs (e.g., Down's syndrome or trisomy 21; Pradat et al., 2003). The mechanisms by which such chromosome imbalance disturbs cardiac development are poorly understood and, sometimes, they can be far more complex than a simple gene-dosage effect. On a smaller scale, microdeletions of discrete chromosomal regions are also an important cause of syndromal CHD. In a recent study, Harris et al. (2003) reported that 18% of infants with CHDs had known chromosomal anomalies. The percentage of infants with chromosomal anomalies varied considerably between the types of cardiac defect investigated. The highest proportion was for atrioventricular septal defect, or AVSD (68%, reflecting the well-known association of endocardial cushion defects with Down's syndrome), followed by VSDs and atrial septal defects, or ASDs (18–32%). Of the infants with a single ventricle, double outlet right ventricle (DORV), pulmonary valve stenosis, or tetralogy of Fallot (ToF), about 10% had chromosomal anomalies. Although these chromosomal anomalies are relatively rare, they are particularly useful in localizing developmental genes and understanding the basis of the non-syndromatic CHDs, as it is generally assumed that haploinsufficiency of one or more genes in the deleted chromosome segment plays a pivotal role in causing the developmental defect. Systematic approaches have now been undertaken to map phenotypic associations with chromosomal deletions and segmental polyploidy across the entire human genome (Brewer et al., 1998, 1999). A well-known example of a chromosomal deletion phenotype is the DiGeorge syndrome (DGS), which is the commonest cause of syndromal CHD after Down's syndrome. Hemizygous microdeletions of chromosome 22q11 constitute the genetic basis of DGS and cause the highly variable spectrum of life-threatening cardiovascular birth defects and craniofacial, thymic, and parathyroid anomalies (Driscoll, 1994; Scambler, 1993).

B. Malformation Syndromes and Isolated Familial CHDs

CHDs can also be an intrinsic part of a Mendelian inherited syndrome, in which a mutation in a single gene affects different developmental processes, including heart development. Again, identification of the causative genes for these syndromes might provide us with valuable insight into the genes and processes controlling cardiac development. Also, isolated forms of CHDs, such as AVSDs and laterality defects, may run in families (Gill et al., 2003). For example, septal formation defects such as ASDs and VSDs may occur in certain families and mutations in several genes, including the GATA4, TBX5, and NKX2.5 transcription factors, which have already been identified in some of these families (Basson et al., 1997; Garg et al., 2003; Li et al., 1997; Schott et al., 1998). The phenomenon whereby mutations in different genes result in a similar or identical clinical phenotype is referred to as locus heterogeneity. Recent data reveal that a multiprotein complex formed by the GATA4, TBX5, and NKX2.5 transcription factors is involved in regulating the formation of atrial and ventricular septa, thereby explaining why mutations in either of these genes may result in a similar congenital disease phenotype (Garg et al., 2003). Although most family members with mutations in NKX2.5 develop ASDs, other CHDs have also been documented at a lower penetrance in a few other family members (Benson *et al.*, 1999; Goldmuntz *et al.*, 2001; Schott *et al.*, 1998). Moreover, NKX2.5 mutations have been identified in some individuals with isolated ToF, DORV, hypoplastic left heart syndrome, and other conditions. Thus, mutations in a single gene sometimes result in a clinically variable spectrum of distinct CHDs (Elliott *et al.*, 2003; Goldmuntz *et al.*, 2001; McElhinney *et al.*, 2003). The complexity of these monogenetic types of CHDs indicates that environmental factors and most likely also additional modifier genes may determine the variable expressivity of CHDs in patients.

C. CHDs with Multifactorial Origin

In most studies of familial CHD, recurrence risks for first-degree relatives vary between 2% and 5% (Gill *et al.*, 2003; Nora and Nora, 1978). Mathematical modeling suggests that disease liability in these cases may be attributable to the interaction of several susceptibility genes with environmental factors. In this model of "multifactorial inheritance", a relatively small pool of disease alleles, present at a relatively high frequency in the healthy population, determines disease liability for CHDs (Gill *et al.*, 2003). Various examples of associations with common disease alleles also exist for other complex human disorders, for example, the APOE-ε4 allele in Alzheimer's disease or the factor V 169G→A-allele (also known as FV Leiden) in deep venous thrombosis (Wright *et al.*, 1999). These support the hypothesis that common variants indeed may hold the secret for many susceptibility diseases. The majority of CHDs is caused by multifactorial inheritance. However, despite their potential impact on the general population, very few genetic susceptibility factors have been discovered so far. Susceptibility factors, affecting the phenotypic variability of CHDs, should, however, continue to attract the attention of investigators. With respect to the increasing number of genes that cause a CHD phenotype in the mouse or zebrafish, it is not unreasonable to consider each of these genes as a potential modifier gene.

D. CHDs with Other Types of Inheritance

Some disorders might also exhibit an inheritance pattern, which falls into a category between that of Mendelial or multifactorial inheritance. Families with apparent dominant inheritance but very low penetrance have been described in families with situs abnormalities (Morelli *et al.*, 2001). Recent studies in mice support another hypothesis in which two copies of a recessive gene and one copy of a second dominant gene are needed for the full expression of a phenotype (Oh and Li, 2002). That such a mode of inheritance indeed may occur has recently been proven for the Bardet–Biedl syndrome (Badano and Katsanis, 2002; Badano *et al.*, 2003). Most vascular anomalies are sporadic, but as for CHDs, Mendelian inheritance or chromosomal abnormalities have also been observed (Vikkula *et al.*, 2001). However, to explain the focal nature of most vascular lesions, as well as the occurrence of "apparently" sporadic cases with vascular malformations, a paradominant pattern of inheritance was proposed (Steijlen and van Steensel, 1999). According to this concept, individuals heterozygous for a particular mutation would be phenotypically normal but the mutant allele would be transmitted imperceptibly through many generations. The trait would only be expressed when a somatic mutation occurred in the wild-type allele at an early stage of embryogenesis, thereby giving rise to a population of cells either homozygous or hemizygous for the vascular mutation. One example of a situation in which a mutation on a second allele is needed for expression of the phenotype are mutations in the GLOMULIN gene, which are responsible for cutaneous venous malformations associated with glomus cells (Brouillard *et al.*, 2002).

IV. The Use of Animal Models

Now that the human genome has been sequenced, the next challenge is to unravel how these genes function in normal health and, even more medically relevant, how they contribute to disorders (Lander *et al.*, 2001; Venter *et al.*, 2001). However, the experience so far has taught us that the genetic causes of CHDs may be surprisingly complex (previously discussed) and, therefore, the identification of CHD genes using human genetics alone may still pose inevitable difficulties. In recent years, characterization of mutations affecting cardiovascular development in animal models has, however, identified many of the signaling molecules and transcriptional regulators required for heart formation. With the availability of complete genome sequences for humans and several model organisms, a powerful system by which to dissect vertebrate gene function has thus emerged, making it possible to identify human genes through their orthologues in other species. Also, the identification of genes acting downstream in hierarchies of complex disease has become available by the ability to profile gene expression in the healthy and diseased model organisms. Therefore, comparative analysis of genes involved in gene networks of different species will enable human disease genes to be identified, while vice versa identification of candidate genes through human genetics can also be verified through additional biological proofs in model organisms.

7. Genetics and VEGF

A. The Mouse, the Model Organism of Choice

Functions of a gene can be best studied in its natural context. To this end, engineering gene mutations in mouse embryonic stem (ES) cells and generating mice harboring the corresponding genetic changes have been proven tremendously valuable. For human inherited disorders, the natural context is defined by the interaction between three factors that determine sensitivity to disease: (1) the nature of the defect in the responsible gene, (2) the context in which the defective gene operates (i.e., the genetic background and the modifier genes present), and (3) the environmental influences. In this respect, genetically well-defined inbred mouse strains offer a significant advantage over other animal models. Indeed, the mouse and human genomes share considerable homology, and mice develop many diseases similar to those observed in human beings, including, for example, the wide spectrum of cardiac defects in humans. In addition, the availability, short gestation period, large litter sizes, and well-defined genetic map of numerous inbred mouse strains, as well as the ability to control the environment in which the animals are reared, all establish the inbred mouse as an ideal choice for these types of study. Furthermore, the required experimental tools, including monoclonal antibodies, recombinant cytokines, and transgenic, gene-knockout and gene-transfer technologies, are available for the mouse.

All these available technologies have made mouse-gene-targeting studies an ideal tool for the functional validation of target genes in the development of disease, as well as for the evaluation of therapeutic strategies (reviewed in Capecchi, 2001; Evans, 2001; Goldstein, 2001; Smithies, 2001). In addition, powerful experimental tools have recently become available to rapidly assign functions to entire sets of genes within the genome. In particular, gene-trap vector-induced mutagenesis of ES cells and high-throughput automated transgenesis, using targeting vectors based on bacterial artificial chromosomes (BACs), appear to become fundamental steps in the future of functional genomics. Vectors used in gene-trapping are specifically designed to integrate in intragenic regions of the genome and contain a promoterless reporter that, when integrated downstream of a gene's regulatory sequence, becomes expressed as a fusion transcript between the endogenous gene and the reporter at levels that mimic endogenous gene expression at the inserted locus. After aggregation of the ES clone, analysis of the trapped gene expression pattern in the mouse embryo and identification of the trapped gene by 5' RACE technology identifies candidate genes for further study in the mouse (reviewed in Leighton *et al.*, 2001; Mitchell *et al.*, 2001; Zambrowicz *et al.*, 1998). Moreover, the availability of techniques to differentiate ES cells into different cell lineages allows for the preselection of genes in which integration has occurred. An example is jumonji (Jmj), which is a gene that was cloned in a gene-trap screen to identify genes important for heart development.

Phenotypic analysis of mice homozygous for the Jmj mutation revealed the presence of heart malformations, including VSD, DORV, and dilated atria in these mice, indicating that this experimental strategy is also promising when analyzing cardiac development (Lee *et al.*, 2000).

Recently, the possibility to subclone or modify DNA from plasmids and bacterial artificial chromosomes (BAC) without the need for restriction enzymes or DNA ligases has become available through homologous recombination in *Escherichia coli*. This new form of chromosome engineering, termed recombineering, can be used to construct targeting vectors and is much more efficient and less time-consuming than the conventional cloning techniques (Copeland *et al.*, 2001; Narayanan *et al.*, 1999; Yu *et al.*, 2000; Zhang *et al.*, 1998). In addition, the use of BAC vectors to target genes in ES cells, combined with a loss-of-native gene assay, which easily identifies correctly targeted ES cells by detecting loss of gene copies in these cells, seems to be very promising (Testa *et al.*, 2003; Valenzuela *et al.*, 2003; Yang and Seed, 2003). Furthermore, conventional targeting vectors require isogenic homologous arms for optimal targeting efficiency, while BAC vectors do not seem to require this isogenicity, presumably because of their very long homology arms. The availability of a high-throughput approach that allows the rapid generation of gene-targeted vectors in combination with automated screening of correctly targeted ES cells and the rapid generation of mice bearing almost any genetic alteration in the gene of interest will certainly accelerate mouse genetic studies of gene function by an order of magnitude in the postgenome era (Valenzuela *et al.*, 2003).

B. The Zebrafish Model

Conventional gene-targeting techniques in the mouse remain slow, labor intensive, and costly. Therefore, more rapid, efficient, and high-throughput functional genomic technologies are required. The zebrafish is an ideal model to rapidly study a large number of genes and to evaluate their interaction in pathways (Dodd *et al.*, 2000; Thisse and Zon, 2002). In the zebrafish, most of the cardiovascular system has developed by 3–4 days postfertilization (dpf) and can be easily visualized in the transparent embryos, which develop *ex utero* (Childs *et al.*, 2002; Crosier *et al.*, 2002; Lawson and Weinstein, 2002b). Deviations from normal contractility or morphology can be seen in real time in the intact living embryo under a microscope. Likewise, the presence of an effective circulation may also be directly assessed by observing the flow of erythrocytes pumped from the heart to the vessels of the head and trunk. Because of its small size and relatively low metabolism, which allows for the diffusion of oxygen and nutrition, zebrafish can survive for several days with a compromised cardiac function, while a similar phenotype in the mouse would result in early lethality and reabsorption (Stainier et al., 1996). Thus, the ease of scoring cardiac and circulatory failure in zebrafish makes them ideal for screening for abnormalities of heart function. Furthermore, the zebrafish has tremendous potential as a genetic model due to its relatively small size, which allows for the maintenance of large numbers of animals, relatively short generation time (2–3 months), and production of over 200 embryos per female per week. Mutations, which affect cardiac function in the zebrafish, have also great relevance to human disease. For this reason, large-scale mutagenesis screens, focusing on cardiac and vascular development, were initiated almost one decade ago (Chen et al., 1996; Stainier et al., 1996). Positional cloning efforts are now beginning to identify the mutated genes. Importantly, the zebrafish can also be easily and efficiently genetically manipulated by using conventional transgenesis to overexpress transgenes or by using morpholino oligomers to down regulate the expression of target genes (Nasevicius and Ekker, 2000). Thus, the zebrafish allows one to set up a genetic pre-screen and to identify rapidly true cardiovascular targets. Because of the relative ease by which combinatorial studies can be performed, it may also be particularly interesting to study the interaction between subtle gene defects that cause polygenic disorders in zebrafish.

To illustrate how valuable mouse and zebrafish genetic studies can be used to dissect the molecular mechanisms and pathways by which genes operate, and to determine how they might contribute to human genetic disorders, the multi-potent function of vascular endothelial growth factor (VEGF) during cardiovascular development will now be discussed.

V. Role of VEGF in Cardiovascular Development

Compelling evidence indicates that VEGF is one of the major regulators during the processes of blood vessel and heart formation (Carmeliet, 2000). The predominant role of VEGF can be best illustrated by the fact that loss of even a single allele in the mouse results in haploinsufficiency, with early embryonic lethality due to severe vascular defects (Carmeliet et al., 1996; Ferrara et al., 1996). Heterozygous VEGF-deficient mice were characterized by a poorly developed dorsal aorta, in which ECs lined a much smaller lumen than normal. In addition, sprouting of the vessels in the intersomitic regions, in the head mesenchyme and in the neuroepithelium, was reduced, and connections of the large blood vessels with the heart outflow tract region appeared abnormal. Blood vessel development was more affected in homozygous VEGF-deficient embryos than in heterozygous embryos, suggesting a tight gene dose-dependent relationship. Significant VEGF dose-dependency is also suggested by studies involving the overexpression of VEGF. In mice, overexpression of VEGF causes embryonic lethality at embryonic day (E) 12.5–14 due to malformations that primarily affected heart septation (e.g., defective ventricular septation and outflow tract abnormalities) and the development of both the myocardium and coronary vasculature (Miquerol et al., 2000). Two VEGF receptors belonging to the tyrosine-kinase receptor family that bind VEGF homologues with different specificity have been identified: Vegf-receptor-1 (VEGFR-1) or fms-related tyrosine kinase (Flt-1) and VEGFR-2 or fetal-liver kinase (Flk-1) (de Vries et al., 1992; Terman et al., 1992). Flt-1 and Flk-1 have distinct temporo-spatial expression patterns during embryogenesis (Millauer et al., 1993) and fulfill different functions in vivo (Fong et al., 1995; Shalaby et al., 1995). In response to activation by VEGF, Flk-1 stimulates EC proliferation, migration, and survival. Flt-1, on the other hand, binds VEGF as well as its homologues VEGF-B and placental growth factor (PLGF) (Neufeld et al., 1999), and activation of Flt-1 by Plgf stimulates angiogenesis, but only in pathological conditions (Luttun et al., 2002). Gene targeting studies in mice have revealed that both Flt-1 and Flk-1 are essential for the development of the embryonic vasculature in mice. Deficiency of Flk-1 resulted in embryonic lethality around 10 days of gestation due to abnormal vascular development (Shalaby et al., 1995). Histological examination revealed complete absence of organized blood vessels and necrosis. Flt-1-deficient mice were also lethal at around 10 days of gestation, due to the presence of numerous differentiated ECs that, however, were abnormally large and fused and failed to form an organized vascular network (Fong et al., 1995). This suggests a possible role for Flt-1 in contact inhibition of EC growth or in endothelial assembly between EC precursors.

Mouse genetic studies of the role of VEGF and its receptors in later stages of embryonic development, as well as during physiological or pathological conditions, have been significantly complicated by the dominant, haploinsufficient nature of VEGF-deficient mice. The absence of systemic vascular defects in all these transgenic mouse models is probably due to the fact that the heart is earlier sensitive to elevation of VEGF than the vasculature. As zebrafish have only minimal requirements for a functioning circulatory system during early development, they might serve as an outstanding alternative model system to study VEGF deficiency at further stages of development (Nasevicius et al., 2000). Morpholino-based targeted gene knockdown of VEGF in zebrafish results in morphant embryos that present with little or no functioning vasculature, pericardial oedema, the absence of red blood cells, and a slight reduction in neural tube and overall body size. At high-dose injections of VEGF morpholino, a primitive vascular structure could still be detected in the heart and yolk, whereas the remaining parts of the embryo blood vessels either failed to form at all or contained no functional connections to the heart (Nasevicius et al., 2000). At lower doses, most embryos had normal heart, yolk, and head blood vessels but exhibited abnormalities in the axial or inter-segmental blood vessels (Nasevicius et al., 2000). Molecular analysis using the endothelial markers Fli-1 and Flk-1 at 1 day of development demonstrated distinct requirements for VEGF between axial and intersegmental vascular structure specification: VEGF was not required for the initial establishment of axial vasculature patterning, whereas the development of intersegmental vasculature was completely dependent on VEGF signaling (Nasevicius et al., 2000). However, both axial and intersegmental vasculature failed to function in VEGF-morphant embryos. Overexpression studies of VEGF in the zebrafish further established that the expression of EC marker genes Flk-1, but also that of the blood marker gene Gata-1, were increased, indicating that the zebrafish VEGF gene acts not only upstream of Flk-1 but that it may also play a role upstream of Gata-1 to stimulate the formation of blood lineage (Liang et al., 2001). In conclusion, VEGF appears to be an essential growth factor during the formation of the heart and the large vessels that connect with the heart, but also during the vasculogenic and angiogenic processes within other vessels in the body. Furthermore, zebrafish studies on VEGF have now revealed that VEGF also plays an essential role in the artery versus vein specification (discussed later).

VI. Expression Patterns of VEGF in Mice Reveal Other Distinct Biological Functions

VEGF is produced by cells in close vicinity to the developing ECs, for example, in the visceral endoderm cells in the yolk sac or in the ventricular zone of the developing brain (Breier et al., 1995). The VEGF receptors have distinct but overlapping temporo-spatial expression patterns during embryogenesis (Millauer et al., 1993). Initially, both receptors become expressed in the mesenchyme of the yolk sac, and in ECs of the dorsal aorta and of the perineural plexus. Subsequently, by 11.5–17.5 days of gestation, they are detected in most capillaries of the developing organs, such as the intersomitic regions, the lung, thymus, intestines, and brain (Dumont et al., 1995; Yamaguchi et al., 1993). It is interesting that explicit hot spots of VEGF expression are also present in the developing structures of the heart or in the embryonic structures that contribute to the developing heart. In situ hybridization analysis revealed that by day E9.5, VEGF is uniformly expressed throughout the myocardial tube. Shortly thereafter, however, when chamber septation has been initiated to form a distinct ventricle and atrium (E10.5), VEGF expression is up regulated in two diametrically opposed regions of the myocardium, which precisely mark the sites of atrioventricular canal formation (Dor et al., 2001). Intense VEGF expression is also detectable in periendothelial cells in the aortic sac at E10.5, in myocardial cells of the cardiac outflow tract at E10.5 and E12.5, and in the muscular portion of the nascent ventricular septum at E12.5 (Dor et al., 2001). Intense localized *in situ* hybridization signal are also detected at E10 in the endoderm of the fourth pharyngeal arch, overlaying the developing fourth arch artery, while at E10.5 a hot spot of VEGF expression in the pouch endoderm between the fourth and developing sixth pharyngeal arch precedes remodeling of the sixth arch artery (Stalmans et al., 2003). The fact that VEGF is specifically up regulated in the atrioventricular (AV) field of the heart tube as well as in the fourth and sixth pharyngeal arches suggests that VEGF also plays a role in septational and outflow tract remodeling events of the developing heart. Besides its function as an angiogenic molecule, VEGF has also other functions, as, for example, its role on the survival of neurons (Storkebaum and Carmeliet, 2004), which are not included within the scope of this review.

VII. The Role of VEGF During Heart Septation

The endocardial cushions are precursors of the AV valves and a portion of the AV septum. Endocardial cushion formation occurs via epithelial-mesenchymal transformation (EMT), in which a subpopulation of ECs within the endocardial layer adjacent to the atrioventricular canal down regulate cell adhesion molecules, separate from the endocardium, and transform into migratory mesenchymal cells that invade into the underlying cardiac jelly (Eisenberg and Markwald, 1995). Conditional over-expression of VEGF exclusively in the myocardium inhibits the process of EMT transformation and leads to improper septation between atria and ventricles and to impaired development of the cardiac outflow tract (Dor et al., 2001). One of the most potent stimulators of VEGF expression is hypoxia (Ikeda et al., 1995). The fact that hypoxia induces high levels of VEGF on the one hand, and that surplus VEGF impairs proper development of endocardial cushions on the other hand, suggests that VEGF might be a crucial factor linking gestational ischemia and congenital heart defects. Indeed, in an *ex vivo* model of EMT transformation, hypoxia completely inhibits mesenchymal transformation, while addition of soluble Flt-1, which acts as a dominant negative inhibitor of VEGF signaling through sequestration of the growth factor, fully restores mesenchymal transformation and gel invasion. This indeed indicates that VEGF is a key mediator of hypoxia-induced defects in the formation of endocardial cushions and, by extrapolation, suggests that VEGF might mediate additional teratogenic effects of hypoxia on heart development.

Further studies are required to address the signaling pathway downstream of VEGF during cardiac cushion formation. However, one possible mediator of VEGF signaling in the endocardium is the transcription factor NFATc1, whose nuclear translocation in ECs is induced by VEGF. It is interesting that mice deficient in NF-ATc also die *in utero* due to valvuloseptal defects (de la Pompa *et al.*, 1998; Ranger *et al.*, 1998). Defects resulting from aberrant endocardial cushion formation, such as atrioventricular septal defects, are also strongly associated with maternal diabetes (Loffredo *et al.*, 2001). Hyperglycemia, as hypoxia, can lead to changes in the levels of VEGF expression (Pinter *et al.*, 2001). In the developing fetus, reductions in VEGF levels may occur in response to hyperglycemia (Pinter *et al.*, 2001). Recent studies have suggested a VEGF-dependent mechanism for the occurrence of maternal diabetes-induced septal defects: High-glucose levels would cause a reduction in myocardial-expressed VEGF, resulting in transient changes of tyrosine phosphorylation in cell adhesion molecules such as PECAM-1 (Enciso *et al.*, 2003). Prominent expression of PECAM-1 has been observed during the earliest stages of vasculogenesis, whereas changes in PECAM-1 tyrosine phosphorylation have been associated with EC migration, vasculogenesis, and angiogenesis both *in vitro* and *in vivo*. Thus, persistence of PECAM-1 expression would lead to persistent adhesion between ECs, preventing disassociation of these cells from the endocardium and consequently reducing the number of migrating EC cells. Overall, the observation that increased as well as reduced levels of VEGF lead to cardiac septation defects illustrates the necessity for exquisitely tight regulation of VEGF expression for normal cardiac development.

VIII. VEGF: a Connector Between the Developing Vascular and Neuronal System

Recent genetic insights show that both the neural and vascular system are guided by common signals, which instruct extending axons or blood vessels to make connections within the complex environment of appearing and disappearing structures of the developing embryo (Carmeliet, 2003). The process of axon or vessel guidance is dependent on the presence of filopodia, which are the sensory structures at the edge of extending axons or vessels that integrate repulsive and attractive cues through a diverse set of receptor complexes. Neuropilin receptors are expressed both on neural and vascular cells, and they can bind both semaphorins, which belong to a large family of cell-associated and secreted repulsive guidance cues and VEGF family members (Neufeld *et al.*, 2002). Semaphorine3A (SEMA3A) and VEGF act antagonistically: SEMA3A repels the filopodia, whereas VEGF attracts them (Bagnard et al., 2001; Carmeliet, 2003; Miao et al., 1999). Likewise, NRP-2 is necessary for SEMA3F-mediated filopodia repulsion (Takahashi et al., 1998), whereas NRP-2 also binds to VEGF (Gluzman-Poltorak et al., 2000). Generation of NRP-1 and NRP-2-deficient mice and NRP-1 over-expression mice have provided insights into the in vivo functions of these receptors. NRP-1-deficient mice die around E10.5–E12.5 (Kawasaki et al., 1999). Besides multiple nervous system defects caused by disorganized projection of nerve fibers, they also exhibit many vascular abnormalities, including partial impairment of neural vascularization, transposition of great vessels, disorganized and insufficient development of vascular networks in the yolk sac, and defects in heart development. NRP-2-deficient mice, on the other hand, exhibit multiple nervous system defects and have a paucity of lymphatic vessels (Chen et al., 2000; Giger et al., 2000; Yuan et al., 2002). By intercrossing these deficient mice with ligand-specific gene-targeted mice, that is, mice expressing an NRP-1 receptor that only binds to VEGF but not to members of the semaphorin family, it became evident that the interaction between semaphorins and NRP-1 is required for axonal path finding in the central nervous system and for proper heart development, whereas it appears to be much less crucial during vascular development (Gu et al., 2003). NRP-1 is expressed in various cell types that contribute to the development of the cardiovascular system, including cardiac NCCs and ECs. The cardiovascular anomalies in NRP-1-deficient mice may therefore be attributable at least in part to the inappropriate migration of NCCs. It is interesting that at least one member of the semaphorin class, SEMA3C, is known to repulse NCCs in vitro and delineates territories for non-permissive neural crest migration in vivo (Eickholt et al., 1999). However, also VEGF-NRP-1 interactions may mediate multiple processes during development of the heart. Endothelial-specific NRP-1 knockout mice also exhibit abnormalities in septation of the cardiac outflow tract as revealed by the high occurrence of a persistent truncus arteriosus (PTA) and VSDs (Gu et al., 2003). Thus, at least two ligands, SEMA3C and VEGF, each act through NRP-1 to coordinate cardiac outflow tract septation. Zebrafish genetic studies revealed that NRP-1 is also involved during vascular development (Lee et al., 2002). Embryos microinjected with NRP-1 antisense morpholino oligonucleotides exhibited vascular defects including impaired circulation in the intersegmental vessels, whereas circulation via trunk axial vessels was not affected. Coinjection of NRP and zebrafish VEGF morpholinos at concentrations that individually did not inhibit blood vessel development significantly yielded a severe aberrant vascular phenotype, including loss of both trunk axial and intersegmental vessel circulation. These results demonstrate that zebrafish VEGF activity is NRP-1 dependent.

IX. A Role for VEGF in Arterial EC Specification

Increasing evidence suggests that VEGF also plays a key role in artery versus vein specification in vertebrates. Although Flt-1 and Flk-1 become expressed on nearly all ECs, NRP-1 and NRP-2 are differentially expressed on arteries, veins, and lymphatic vessels, respectively (Herzog et al., 2001). When released from Schwann cells, VEGF also induces the arterial specification of vessels tracking alongside these nerves (Mukouyama et al., 2002). Moreover, VEGF determines arterial EC specification after birth in the heart and retina; especially, the matrix-binding VEGF188 isoform (discussed later) is crucial for arterial development as transgenic mice expressing only VEGF188 isoform exhibit vascular defects, which are characterized by normal venous but severely impaired arterial development (Stalmans et al., 2002). In zebrafish, knockdown of VEGF leads to specific arterial differentiation defects and abnormal expression of arterial expression markers. Since activation of the NOTCH pathway in the absence of VEGF signaling is able to rescue abnormal arterial marker expression patterns, it has been proposed that NOTCH transduces VEGF-induced arterial-specifying EC signals (Lawson et al., 2002). Also, members of the Hedgehog family of proteins play an important role during the development of the embryonic vasculature. In the zebrafish mutant *sonic youth (sy)*, which encodes the zebrafish homolog of sonic hedgehog (SHH), formation of the dorsal aorta and the posterior cardinal vein is impaired (Chen et al., 1996). Consistent with these findings, embryos with a morpholino-targeted reduction of SHH-expression fail to undergo arterial differention, as evidenced by the expression of artery-specific markers such as EPHRIN2 (Lawson et al., 2002). Remarkably, exogeneous addition of VEGF in *sy* mutants is sufficient to restore the vascular expression patterns of EPHRIN2, and, therefore, VEGF VEGF appears to act downstream of SHH to induce arterial differentiation (Lawson et al., 2002). Thus, novel insights from zebrafish are assembling the following genetic hierarchy of molecules involved in artery and vein specification, in which SHH expressed in the notochord induces the expression of VEGF, which then acts on adjacent angioblasts to switch NOTCH expression and activate downstream artery-specific genes.

X. Normal Ontogenesis but Abnormal Remodeling of PAAs in Mice Lacking VEGF164

It is now well established that alternative exon splicing of a single VEGF gene results in the generation of four different molecular species, having, respectively, 121, 165, 189, and 206 amino acids following signal sequence cleavage (reviewed in Carmeliet and Collen, 1999; Ferrara *et al.*, 2003). The murine VEGF isoforms are shorter than the human VEGF isoforms by one amino acid (VEGF120, VEGF164, and VEGF188). The addition of the 44 amino acid–long peptide encoded by exon 7 of the VEGF gene distinguishes VEGF165 from VEGF121 and confers a heparin-binding ability on VEGF165. Notably, VEGF121 has more than 100-fold reduced EC mitogenic potency than VEGF165. VEGF189 and VEGF206 contain the peptides encoded by exons 6 and 7 and display an even higher affinity to heparin and heparan-sulfates than does VEGF165. VEGF189 is sequestered on heparan-sulfate proteoglycans of cell surfaces and in the extra-cellular matrix, and, it is therefore considered to have a reduced bioavailability than VEGF165 *in vivo*. Due to their difference in solubility and distinct bioactivity, the different VEGF isoforms are thought to provide a spatial gradient of patterning information during blood vessel formation. The VEGF121 isoform is soluble and diffuses randomly from its source, providing a chaotic pattern for vessel growth. The VEGF165 isoform is also soluble, yet it binds sufficiently to the extracellular matrix to provide more restricted matrix-associated guidance cues. Although secreted, the VEGF189 isoform is not soluble and remains attached to the extracellular matrix, close to the site of VEGF production. As such, VEGF isoforms establish a VEGF-gradient or a "trail" along which outgrowing vessels are attracted to the region of VEGF production (ischemic region).

To unravel the differential roles of the distinct VEGF-isoforms, transgenic mice, expressing only one VEGF-isoform (VEGF120, VEGF164 and VEGF188 mice), were generated (Carmeliet *et al.*, 1999). Compared to VEGF knockout mice, mice engineered to express a single VEGF isoform develop further in gestation but exhibit distinct phenotypes. VEGF164 mice are normal, indicating that the VEGF164 isoform is sufficient to assemble a functional vascular network. In contrast, 55% of the mice expressing only the VEGF188 isoform die between E9.5 and E13, whereas the remaining mice survive for prolonged periods after birth. VEGF120 mice develop to term: 50% of the mice become sick immediately after birth and die within the first few hours after birth, while the remaining mice are healthy at birth but succumb to ischemic heart disease within two weeks after birth (Carmeliet *et al.*, 1999). When studying the cause of birth defects in mice that lack the VEGF164 isoform, it became clear that both VEGF120 and VEGF188 mice die from a wide spectrum of congenital cardiac anomalies, reminiscent of those found in patients with PAA remodeling defects (Stalmans *et al.*, 2003).

While numerous molecules have been implicated in the early assembly of ECs into a primitive vascular network, it is less well understood which molecules control the extensive remodeling into a complex functional vasculature (Carmeliet, 2000). The PAAs are a prototypical vascular network where coordinated remodeling is required to establish the aortic arch vessels.

7. Genetics and VEGF

Life-threatening defects of aortic arch remodeling are one of the most common congenital cardiovascular disorders, affecting 1 in 4000 infants (Devriendt et al., 1998). Such vascular malformations can exist in isolation but often occur in combination with craniofacial dysmorphism and thymic and parathyroid abnormalities in the DiGeorge syndrome (DGS). Most DGS patients are hemizygous for a microdeletion of $\sim$300 kilobases on human chromosome 22q11, and haploinsufficiency of one or more genes in this region was originally believed to underlie the phenotype (Lindsay, 2001). However, since nonoverlapping deletions have been mapped in this region in patients with DGS, not a single but a combination of candidate genes within or, possibly, outside this region could be responsible for the syndrome (Lindsay, 2001). Recent transgenic studies have implicated the T-box transcription factor (TBX1) in DGS (Jerome and Papaioannou, 2001; Lindsay et al., 2001; Merscher et al., 2001), and the recent identification of TBX1 mutations in the rare DGS patients without the 22q11 deletion has further confirmed the involvement of TBX1 (Yagi et al., 2003). However, unlike most DGS patients (in whom a single TBX1 allele is lacking), both TBX1 alleles must be deleted in mice to reproduce most of the disease spectrum (Jerome and Papaioannou, 2001; Lindsay et al., 2001; Merscher et al., 2001). Moreover, DGS in humans and in TBX1-deficient mice is characterized by an extreme phenotypic variability and influenced by genetic background (Lindsay, 2001; Schinke and Izumo, 1999). Thus, multiple genes (including genes outside the 22q11 region) and environmental influences likely contribute to the complex disease spectrum of DGS (Epstein, 2001; Lindsay, 2001; Scambler, 1999; Schinke and Izumo, 1999).

NCCs play a major role in the development of the PAs (Graham and Smith, 2001), and defects in these cells result in the phenotypic abnormalities in the derivative structures such as the cardiac outflow tract. It is therefore believed that DGS was caused by a deficient contribution of NCCs to pharyngeal development. However, recent evidence indicates that the endoderm, overlying the arch arteries, plays an important role in aortic arch remodeling by releasing yet unidentified inductive signals to the nearby arch vessels. Since NCCs are normal in the TBX1-deficient mouse model for DGS, but their migration seems to be misdirected due to the lack of sufficient guidance cues from the pouch endoderm, one of these signals may be TBX1. Indeed, also in the zebrafish mutant *van gogh*, which is caused by a mutation in TBX1 (Piotrowski et al., 2003), the neural crest derived cartilages of the arches are disorganized, often fusing with each other or not forming at all. This is not due to defects in the neural crest but is rather caused by alterations in the endoderm, which fails to segment in these mutants (Piotrowski and Nusslein-Volhard, 2000). One of the directing signals may thus be the TBX1 gene (Fig. 2).

The absence of the VEGF164 isoform also causes birth defects in mice that were strikingly similar to those seen in DGS patients; for example, a

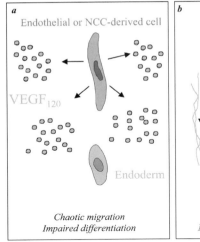 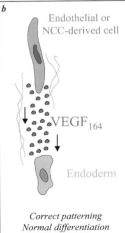 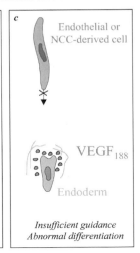

Figure 2 Schematic model of the hypothetical role of the different VEGF isoforms. VEGF isoform provide distinct guidance and differentiation signals for the remodeling of the pharyngeal arch arteries. In all panels, VEGF is produced by the pharyngeal arch endoderm, while the endothelial cells of the pharyngeal arch arteries or the neural crest cell (NCC)–derived periendothelial cells around the arteries are the VEGF-responsive target cells. Since VEGF is produced by the endoderm at the periphery of the arch and the pharyngeal arch arteries are located centrally in the arch, VEGF must diffuse over a finite distance to reach its target cells. VEGF could be necessary for providing spatial guidance cues as well as essential differentiation signals. (*a*) The VEGF120 isoform is too soluble and therefore diffuses randomly in a chaotic pattern and over too long distances, resulting in a lack of focused instructive spatial guidance and insufficient differentiation cues. (*b*) The VEGF164 isoform is soluble, yet binds sufficiently to the extracellular matrix so that it can reach its target cell but still provides spatial guidance through its association with the matrix and differentiation signals via binding to neuropilin-1. (*c*) The VEGF188 isoform is secreted but not soluble, as it is attached to the extracellular matrix. Since this isoform cannot reach its target cell, insufficient spatial guidance and differentiation signals are generated. Thus, either excessive or insufficient matrix binding results in imbalanced vascular remodeling. Similar mechanisms might be operational in other DiGeorge syndrome–affected organs, although the VEGF-responsive cell type may vary. (See Color Insert.)

ToF, a PTA, a hypoplasia of the pulmonary trunk, or an isolated VSD were frequently observed in VEGF120 and VEGF188 neonates (Stalmans *et al.*, 2003). In addition, VEGF120 mice had facial deformities, respiratory problems due to tracheal compression by an ectopic thymus, fewer circulating lymphocytes, and mislocated or even absent parathyroid glands (Stalmans *et al.*, 2003). This suggests that VEGF could also be one of these signals released by the endoderm. Indeed, hotspots of VEGF expression were detected in the endoderm just prior to the onset of remodeling of the individual arch arteries. Consistent with the latter findings, we were also unable to document a gross impairment of NCC migration, accumulation, or differentiation *in vitro* and *in vivo*. NRP-1, on the other hand, was predominantly expressed in ECs of the pharyngeal arch arteries and, at a lower level, also in the mesenchyme around the pharyngeal arch arteries. Both VEGF164 and NRP-1 were in fact coordinatedly expressed at all DGS-predilection sites (i.e., pharyngeal arch arteries, heart, thymus, craniofacial skeleton), suggesting that NRP-1 plays a role in transmitting the instructive VEGF164 cues. For a VEGF isoform to provide instructive signals, it should reach the vessel, which is centrally located in the pharyngeal arch. Thus, the VEGF isoform should be diffusible, yet interact with guidance cues in the extracellular matrix. The VEGF164 isoform fits all these criteria. In contrast, the short VEGF120 isoform may be too soluble, and therefore diffuse in a chaotic manner or overshoot its vascular target, while the matrix-bound VEGF188 isoform may not be releasable at all and fail to reach the vessel (Fig. 3). Interestingly, SEMA3C, another NRP-1-ligand (Takahashi *et al.*, 1998), was coexpressed with VEGF and NRP-1 in the remodeling arch arteries and outflow tract, and mice lacking SEMA3C exhibit isolated vascular malformations (Feiner *et al.*, 2001). SEMA3A, which is a close relative of SEMA3A, is known to inhibit EC migration by antagonizing binding of VEGF164 to NRP-1 (Miao *et al.*, 1999). If SEMA3C would behave similarly, it could displace VEGF164 from NRP-1 and thereby regulate the levels of its instructive guidance or differentiation signals. Perhaps such a precise temporo-spatial balance between SEMA3C and VEGF164 signals, molecularly integrated at the level of NRP-1, is required for the sophisticated remodeling of the pharyngeal arches.

However, since the VEGF gene does not map to the deleted 22q11 region in DGS patients, these mouse genetic findings raised the question of whether VEGF might act in the genetic pathway of a critical gene that is deleted in del22q11 patients. It could indeed be argued that mice expressing a single VEGF isoform are a laboratory model without human correlate. Quantification of the expression of VEGF and TBX1 revealed that TBX1 expression was reduced in the PAs of VEGF120 mice, but not in wild-type or VEGF164 embryos, thereby suggesting that VEGF might interact with TBX1 in PA development. To obtain additional evidence for this and to provide functional confirmation that reduced VEGF expression increases the risk of PA vessel malformations, we utilized the zebrafish model. The zebrafish orthologues of TBX1 and VEGF were expressed in the PAs beyond 1-day postfertilization (dpf). Microinjection of increasing amounts of VEGF morpholinos (60, 80, 100, 120, 200, and 300 μM solutions) dose-dependently caused defects of the head, axial, tail, sub-intestinal, and intersegmental vessels at 2 and 3 dpf, as analyzed by *in situ* hybrization for Flk1 and alkaline phosphatase, consistent with previous reports. Microinjection of TBX1 morpholinos did not affect aortic arch artery (AAA) development at 50, 100, 150, or 200 μM but at 300 μM, the sixth pharyngeal arch artery and the common

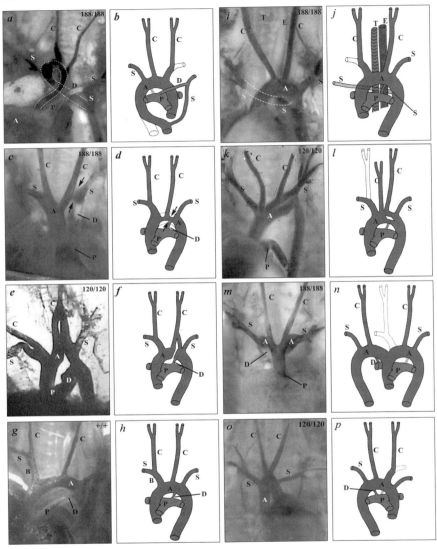

Figure 3 Aortic arch and conotruncal defects in VEGF120 and VEGF188 embryos and neonates. This shows intraventricular injection of ink or Evans blue to visualize the vascular malformations in VEGF120 and VEGF188 neonates. For each set of two panels, the left panel displays a micrograph of the thoracic vessels, while the right panel schematically illustrates the vascular malformation. (*a, b*) Normal aortic arch configuration in a wild-type neonate; (*c, d*) VEGF188 neonate with a coarctation of the aorta (indicated between arrows) and aberrant connection of the ductus arteriosus to left subclavian artery, resulting from abnormal partial regression of the left fourth arch artery. The ascending aorta is also enlarged. (*e, f*) Interruption of the aortic arch in E14.5 VEGF120 embryo, caused by abnormal regression of the left fourth

trunk, draining the sixth AAA to the midline dorsal aorta, were irregular, discontinuous, or even absent in 90% of 3 dpf embryos. Severe vascular defects were not detected in the rest of the vasculature, and intersegmental vessel flow remained normal, presumably because of the restricted expression of TBX1 in the PAs. To analyze whether VEGF and tbx interacted with each other, both morpholinos were co-injected. A low concentration of each morpholino was chosen, which, when injected separately, only minimally affected AAA development (60, 80, or 100 μM for VEGF morpholino; 200 μM for TBX1 morpholino). After co-injection of VEGF (60 μM) and TBX1 (200 μM) morpholinos, AAA development was normal in all embryos. Remarkably, however, co-injection of TBX1 morpholino (200 μM) with increasing concentrations of VEGF morpholino (80–100 μM) dose-dependently impaired AAA development, resulting in more frequent and more severe defects. Therefore, these zebrafish genetic data further underscore (1) that reduced VEGF levels, as in the mouse, predispose to aortic pharyngeal arch defects; (2) that this effect is dependent on gene-dosage; and (3) that reduced levels of VEGF enhance the penetrance of the PAA defects in a TBX1 knockdown zebrafish. With respect to these results, the zebrafish model was in particular valuable because (1) microinjection of zebrafish embryos with increasing amounts of antisense morpholino oligomers allows the gradual "knockdown" of protein translation of target genes and (2) interactions of candidate genes can be rapidly studied by simultaneous injection of multiple

---

arch artery. The blood flows from the heart into the carotid artery, and via an abnormally persisting carotid duct to the descending left aorta; (*g, h*) VEGF188 neonate (oblique view): the blood flows through the right-bending aortic arch and persistent right dorsal aorta into to the left-sided descending aorta (the left aortic arch is interrupted). The left subclavian artery and the ductus connect to the left dorsal aorta. A vascular ring around the trachea and esophagus (removed for clarity) is formed by the right-bending aorta, the right dorsal aorta, the left dorsal aorta, the ductus arteriosus, and the pulmonary trunk; (*i, j*) VEGF188 neonate: the right subclavian artery (S; traced with yellow dashed line in *a* and indicated in yellow in *b* orginates aberrantly from the descending aorta and crosses the midline behind the esophagus and trachea. This "artera lusoria" is caused by abnormal regression of the right fourth arch artery and persistence of the right distal dorsal aorta. For better visualization of the vascular defects, the ductus arteriosus and pulmonary trunk were removed; (*k, l*) E14.5 VEGF120 embryo, showing a "duplication" of the aortic arch segment between the left carotid and subclavian artery, resulting from a persistent left carotid duct and a hypoplasia of the left fourth arch artery. The ductus arteriosus is absent; (*m, n*) VEGF188 neonate: double-sided aortic arch with a right-sided ductus arteriosus, caused by abnormal persistence of the right dorsal aorta, abnormal regression of the left sixth arch artery, and persistence of the right sixth arch artery. (*o, p*) VEGF120 neonate with a right-sided aortic arch and descending aorta, a right-sided ductus arteriosus (D), and a left-sided brachiocephalic trunk. This condition is the "mirror image" of the normal aortic configuration. A (aorta); AS (aortic sac); AV (atrio-ventricular channel); B (brachio-cephalic trunk); C (carotid artery); CD (carotid duct); D (ductus arteriosus); DA (dorsal aorta); E (esophagus); LV (left ventricle); P (pulmonary trunk); PTA (persistent truncus arteriosus); S (subclavian artery); T (trachea). (See Color Insert.)

morpholinos. In conclusion, VEGF isoform-specific mouse data revealed that VEGF is capable of influencing PA patterning in an engineered mouse model, while the zebrafish genetic data demonstrated that reduced VEGF levels predispose to AAA defects in a dose-dependent way.

XI. Association of VEGF Gene Variations with Cardiovascular Defects in DGS

Because VEGF had never been implicated in DGS previously, it remained outstanding whether VEGF was also involved in patients with DGS. We therefore analyzed whether three variations of single nucleotide polymorphisms in the promoter and 5' untranslated region of VEGF, previously documented to functionally regulate VEGF expression (Lambrechts et al., 2003), might confer an increased risk for DGS. Considering the prominent congenital cardiovascular malformations in VEGF120 and VEGF188 mice, a possible association of VEGF variations with cardiovascular defects was considered. Because cardiovascular defects greatly vary in severity and penetrance among DGS subjects, only DGS subjects with significant cardiovascular defects, which could be unambiguously diagnosed echocardiographically (ToF, PTA, stenosis of the pulmonary artery and/or aorta, VSD, right-bending aorta, and abnormal origin of the subclavian artery), were included. By performing a case–control association study in a limited number of available cases, we found that the low-VEGF expressing variations were significantly more common in DGS patients with cardiovascular abnormalities than in healthy individuals or than in DGS patients without cardiovascular abnormalities. When combining these variations into haplotypes, a haplotype consisting of the low-VEGF producing alleles increased, by threefold, the risk of cardiovascular birth defects in DGS individuals with a hemizygous del22q11 deletion (Stalmans et al., 2003). In the healthy population, VEGF plasma levels were almost twofold lower in individuals carrying this "at risk" haplotype, compared to individuals with any other haplotype. In vitro expression studies subsequently revealed that VEGF expression was lowered at the level of transcription and translation (Lambrechts et al., 2003). Since the at-risk haplotype was strongly associated with reduced VEGF expression levels, we propose that a reduction of all VEGF isoforms will also reduce the expression of the VEGF165 isoform to below a critical threshold, required for proper pharyngeal arch patterning and cardiovascular development. Considering its critical gene dosage-dependent activity, VEGF also appears to be a candidate modifier gene of the cardiovascular del22q11 syndrome phenotype in humans.

Taken together, we therefore interpret the mouse, fish, and human genetic data as follows: allelic VEGF promoter variations that lower the expression of VEGF will also lower the expression of the VEGF165 isoform. Thus, when VEGF165 levels in human del22q11 fetuses carrying an at-risk haplotype fall below a critical threshold required for proper PA development, their risk of birth defects increases. The zebrafish knockdown experiments support a model whereby VEGF threshold levels are critical for TBX1-dependent PA development.

XII. Conclusion

When it was discovered that most DGS patients have a three-million-bp deletion on chromosome 22, it seemed obvious that haplo-insufficiency of one or more genes in the deleted region (22q11) was the cause. Mouse studies subsequently identified the transcription factor gene TBX1 as the prime suspect of the syndrome. The syndrome's perplexing phenotypic variability, however, suggested that haploinsufficiency or genetic variations of other genes in the 22q11 region or in genes that act in the genetic pathway of TBX1 may be responsible for the variable expressivity of the disorder. By using a combined genetic approach in mice, zebrafish, and humans, we identified the VEGF gene as the first modifier gene of the cardiovascular del22q11 syndrome phenotype in humans. However, VEGF might also be a susceptibility factor for patients with isolated pharyngeal arch remodeling defects or with other forms of CHDs. Indeed, mouse and zebrafish genetic studies have revealed that VEGF is also essential during heart septation and arterial specification. Future studies should therefore examine whether genetic variations in VEGF also increase the risk for other cardiovascular malformations or in patients displaying a multifactorial disease inheritance pattern as the genetic basis for their disease. On a more general level, these studies validate the usefulness of a combined genetic approach in mice, fish, and humans to unravel the pathogenesis of human complex traits in general, and congenital heart and vascular malformations in particular.

References

Anderson, R. H., Webb, S., Brown, N. A., Lamers, W., and Moorman, A. (2003a). Development of the heart: (2) Septation of the atriums and ventricles. *Heart* **89,** 949–958.
Anderson, R. H., Webb, S., Brown, N. A., Lamers, W., and Moorman, A. (2003b). Development of the heart: (3) formation of the ventricular outflow tracts, arterial valves, and intrapericardial arterial trunks. *Heart* **89,** 1110–1118.
Badano, J. L., and Katsanis, N. (2002). Beyond Mendel: An evolving view of human genetic disease transmission. *Nat. Rev. Genet.* **3,** 779–789.
Badano, J. L., Kim, J. C., Hoskins, B. E., Lewis, R. A., Ansley, S. J., Cutler, D. J., Castellan, C., Beales, P. L., Leroux, M. R., and Katsanis, N. (2003). Heterozygous mutations in BBS1,

BBS2 and BBS6 have a potential epistatic effect on Bardet–Biedl patients with two mutations at a second BBS locus. *Hum. Mol. Genet.* **12**, 1651–1659.

Bagnard, D., Vaillant, C., Khuth, S. T., Dufay, N., Lohrum, M., Puschel, A. W., Belin, M. F., Bolz, J., and Thomasset, N. (2001). Semaphorin 3A-vascular endothelial growth factor-165 balance mediates migration and apoptosis of neural progenitor cells by the recruitment of shared receptor. *J. Neurosci.* **21**, 3332–3341.

Basson, C. T., Bachinsky, D. R., Lin, R. C., Levi, T., Elkins, J. A., Soults, J., Grayzel, D., Kroumpouzou, E., Traill, T. A., Leblanc-Straceski, J., Renault, B., Kucherlapati, R., Seidman, J. G., and Seidman, C. E. (1997). Mutations in human TBX5 [corrected] cause limb and cardiac malformation in Holt–Oram syndrome. *Nat. Genet.* **15**, 30–35.

Benson, D. W., Silberbach, G. M., Kavanaugh-McHugh, A., Cottrill, C., Zhang, Y., Riggs, S., Smalls, O., Johnson, M. C., Watson, M. S., Seidman, J. G., Seidman, C. E., Plowden, J., and Kugler, J. D. (1999). Mutations in the cardiac transcription factor NKX2.5 affect diverse cardiac developmental pathways. *J. Clin. Invest.* **104**, 1567–1573.

Breier, G., Clauss, M., and Risau, W. (1995). Coordinate expression of vascular endothelial growth factor receptor-1 (flt-1) and its ligand suggests a paracrine regulation of murine vascular development. *Dev. Dyn.* **204**, 228–239.

Brewer, C., Holloway, S., Zawalnyski, P., Schinzel, A., and FitzPatrick, D. (1998). A chromosomal deletion map of human malformations. *Am. J. Hum. Genet.* **63**, 1153–1159.

Brewer, C., Holloway, S., Zawalnyski, P., Schinzel, A., and FitzPatrick, D. (1999). A chromosomal duplication map of malformations: Regions of suspected haplo- and triplolethality—and tolerance of segmental aneuploidy—in humans. *Am. J. Hum. Genet.* **64**, 1702–1708.

Brouillard, P., Boon, L. M., Mulliken, J. B., Enjolras, O., Ghassibe, M., Warman, M. L., Tan, O. T., Olsen, B. R., and Vikkula, M. (2002). Mutations in a novel factor, glomulin, are responsible for glomuvenous malformations ("glomangiomas"). *Am. J. Hum. Genet.* **70**, 866–874.

Brouillard, P., and Vikkula, M. (2003). Vascular malformations: Localized defects in vascular morphogenesis. *Clin. Genet.* **63**, 340–351.

Burdine, R. D., and Schier, A. F. (2000). Conserved and divergent mechanisms in left–right axis formation. *Genes Dev.* **14**, 763–776.

Capdevila, J., Vogan, K. J., Tabin, C. J., and Izpisua Belmonte, J. C. (2000). Mechanisms of left–right determination in vertebrates. *Cell* **101**, 9–21.

Capecchi, M. R. (2001). Generating mice with targeted mutations. *Nat. Med.* **7**, 1086–1090.

Carlgren, L. E., Ericson, A., and Kallen, B. (1987). Monitoring of congenital cardiac defects. *Pediatr. Cardiol.* **8**, 247–256.

Carmeliet, P. (2000). Mechanisms of angiogenesis and arteriogenesis. *Nat. Med.* **6**, 389–395.

Carmeliet, P. (2003). Blood vessels and nerves: Common signals, pathways and diseases. *Nat. Rev. Genet.* **4**, 710–720.

Carmeliet, P., and Collen, D. (1999). Role of vascular endothelial growth factor and vascular endothelial growth factor receptors in vascular development. *Curr. Top. Microbiol. Immunol.* **237**, 133–158.

Carmeliet, P., Ferreira, V., Breier, G., Pollefeyt, S., Kieckens, L., Gertsenstein, M., Fahrig, M., Vandenhoeck, A., Harpal, K., Eberhardt, C., Declercq, C., Pawling, J., Moons, L., Collen, D., Risau, W., and Nagy, A. (1996). Abnormal blood vessel development and lethality in embryos lacking a single VEGF allele. *Nature* **380**, 435–439.

Carmeliet, P., Ng, Y. S., Nuyens, D., Theilmeier, G., Brusselmans, K., Cornelissen, I., Ehler, E., Kakkar, V. V., Stalmans, I., Mattot, V., Perriard, J. C., Dewerchin, M., Flameng, W., Nagy, A., Lupu, F., Moons, L., Collen, D., D'Amore, P. A., and Shima, D. T. (1999). Impaired myocardial angiogenesis and ischemic cardiomyopathy in mice lacking the vascular endothelial growth factor isoforms VEGF164 and VEGF188. *Nat. Med.* **5**, 495–502.

Chen, H., Bagri, A., Zupicich, J. A., Zou, Y., Stoeckli, E., Pleasure, S. J., Lowenstein, D. H., Skarnes, W. C., Chedotal, A., and Tessier-Lavigne, M. (2000). Neuropilin-2 regulates the development of selective cranial and sensory nerves and hippocampal mossy fiber projections. *Neuron* **25**, 43–56.

Chen, J. N., Haffter, P., Odenthal, J., Vogelsang, E., Brand, M., van Eeden, F. J., Furutani-Seiki, M., Granato, M., Hammerschmidt, M., Heisenberg, C. P., Jiang, Y. J., Kane, D. A., Kelsh, R. N., Mullins, M. C., and Nusslein-Volhard, C. (1996). Mutations affecting the cardiovascular system and other internal organs in zebrafish. *Development* **123**, 293–302.

Chien, K. R. (2000). Genomic circuits and the integrative biology of cardiac diseases. *Nature* **407**, 227–232.

Childs, S., Chen, J. N., Garrity, D. M., and Fishman, M. C. (2002). Patterning of angiogenesis in the zebrafish embryo. *Development* **129**, 973–982.

Copeland, N. G., Jenkins, N. A., and Court, D. L. (2001). Recombineering: A Powerful new tool for mouse functional genomics. *Nat. Rev. Genet.* **2**, 769–779.

Creazzo, T. L., Godt, R. E., Leatherbury, L., Conway, S. J., and Kirby, M. L. (1998). Role of cardiac neural crest cells in cardiovascular development. *Annu. Rev. Physiol.* **60**, 267–286.

Crosier, P. S., Kalev-Zylinska, M. L., Hall, C. J., Flores, M. V., Horsfield, J. A., and Crosier, K. E. (2002). Pathways in blood and vessel development revealed through zebrafish genetics. *Int. J. Dev. Biol.* **46**, 493–502.

De la Pompa, J. L., Timmerman, L. A., Takimoto, H., Yoshida, H., Elia, A. J., Samper, E., Potter, J., Wakeham, A., Marengere, L., Langille, B. L., Crabtree, G. R., and Mak, T. W. (1998). Role of the NF-ATc transcription factor in morphogenesis of cardiac valves and septum. *Nature* **392**, 182–186.

De Vries, C., Escobedo, J. A., Ueno, H., Houck, K., Ferrara, N., and Williams, L. T. (1992). The fms-like tyrosine kinase, a receptor for vascular endothelial growth factor. *Science* **255**, 989–991.

Devriendt, K., Fryns, J. P., Mortier, G., van Thienen, M. N., and Keymolen, K. (1998). The annual incidence of DiGeorge/velocardiofacial syndrome. *J. Med. Genet.* **35**, 789–790.

Dodd, A., Curtis, P. M., Williams, L. C., and Love, D. R. (2000). Zebrafish: Bridging the gap between development and disease. *Hum. Mol. Genet.* **9**, 2443–2449.

Dor, Y., Camenisch, T. D., Itin, A., Fishman, G. I., McDonald, J. A., Carmeliet, P., and Keshet, E. (2001). A novel role for VEGF in endocardial cushion formation and its potential contribution to congenital heart defects. *Development* **128**, 1531–1538.

Driscoll, D. A. (1994). Genetic basis of DiGeorge and velocardiofacial syndromes. *Curr. Opin. Pediatr.* **6**, 702–706.

Dumont, D. J., Fong, G. H., Puri, M. C., Gradwohl, G., Alitalo, K., and Breitman, M. L. (1995). Vascularization of the mouse embryo: A study of flk-1, tek, tie, and vascular endothelial growth factor expression during development. *Dev. Dyn.* **203**, 80–92.

Eickholt, B. J., Mackenzie, S. L., Graham, A., Walsh, F. S., and Doherty, P. (1999). Evidence for collapsin-1 functioning in the control of neural crest migration in both trunk and hindbrain regions. *Development* **126**, 2181–2189.

Eisenberg, L. M., and Markwald, R. R. (1995). Molecular regulation of atrioventricular valvuloseptal morphogenesis. *Circ. Res.* **77**, 1–6.

Elliott, D. A., Kirk, E. P., Yeoh, T., Chandar, S., McKenzie, F., Taylor, P., Grossfeld, P., Fatkin, D., Jones, O., Hayes, P., Feneley, M., and Harvey, R. P. (2003). Cardiac homeobox gene NKX2-5 mutations and congenital heart disease: Associations with atrial septal defect and hypoplastic left heart syndrome. *J. Am. Coll. Cardiol.* **41**, 2072–2076.

Enciso, J. M., Gratzinger, D., Camenisch, T. D., Canosa, S., Pinter, E., and Madri, J. A. (2003). Elevated glucose inhibits VEGF-A-mediated endocardial cushion formation: Modulation by PECAM-1 and MMP-2. *J. Cell Biol.* **160**, 605–615.

Epstein, J. A. (2001). Developing models of DiGeorge syndrome. *Trends Genet.* **17**, S13–S17.
Evans, M. J. (2001). The cultural mouse. *Nat. Med.* **7**, 1081–1083.
Feiner, L., Webber, A. L., Brown, C. B., Lu, M. M., Jia, L., Feinstein, P., Mombaerts, P., Epstein, J. A., and Raper, J. A. (2001). Targeted disruption of semaphorin 3C leads to persistent truncus arteriosus and aortic arch interruption. *Development* **128**, 3061–3070.
Ferencz, C., Rubin, J. D., McCarter, R. J., Brenner, J. I., Neill, C. A., Perry, L. W., Hepner, S. I., and Downing, J. W. (1985). Congenital heart disease: Prevalence at livebirth. The Baltimore–Washington Infant Study. *Am. J. Epidemiol.* **121**, 31–36.
Ferrara, N., Carver-Moore, K., Chen, H., Dowd, M., Lu, L., O'Shea, K. S., Powell-Braxton, L., Hillan, K. J., and Moore, M. W. (1996). Heterozygous embryonic lethality induced by targeted inactivation of the VEGF gene. *Nature* **380**, 439–442.
Ferrara, N., Gerber, H. P., and LeCouter, J. (2003). The biology of VEGF and its receptors. *Nat. Med.* **9**, 669–676.
Fong, G. H., Rossant, J., Gertsenstein, M., and Breitman, M. L. (1995). Role of the Flt-1 receptor tyrosine kinase in regulating the assembly of vascular endothelium. *Nature* **376**, 66–70.
Fouquet, B., Weinstein, B. M., Serluca, F. C., and Fishman, M. C. (1997). Vessel patterning in the embryo of the zebrafish: Guidance by notochord. *Dev. Biol.* **183**, 37–48.
Fyler, D. C., Rothman, K. J., Buckley, L. P., Cohn, H. E., Hellenbrand, W. E., and Castaneda, A. (1981). The determinants of five-year survival of infants with critical congenital heart disease. *Cardiovasc. Clin.* **11**, 393–405.
Garg, V., Kathiriya, I. S., Barnes, R., Schluterman, M. K., King, I. N., Butler, C. A., Rothrock, C. R., Eapen, R. S., Hirayama-Yamada, K., Joo, K., Matsuoka, R., Cohen, J. C., and Srivastava, D. (2003). GATA4 mutations cause human congenital heart defects and reveal an interaction with TBX5. *Nature* **424**, 443–447.
Giger, R. J., Cloutier, J. F., Sahay, A., Prinjha, R. K., Levengood, D. V., Moore, S. E., Pickering, S., Simmons, D., Rastan, S., Walsh, F. S., Kolodkin, A. L., Ginty, D. D., and Geppert, M. (2000). Neuropilin-2 is required *in vivo* for selective axon guidance responses to secreted semaphorins. *Neuron* **25**, 29–41.
Gill, H. K., Splitt, M., Sharland, G. K., and Simpson, J. M. (2003). Patterns of recurrence of congenital heart disease: An analysis of 6,640 consecutive pregnancies evaluated by detailed fetal echocardiography. *J. Am. Coll. Cardiol.* **42**, 923–929.
Gluzman-Poltorak, Z., Cohen, T., Herzog, Y., and Neufeld, G. (2000). Neuropilin-2 is a receptor for the vascular endothelial growth factor (VEGF) forms VEGF-145 and VEGF-165 [corrected]. *J. Biol. Chem.* **275**, 18040–18045.
Goldmuntz, E., Geiger, E., and Benson, D. W. (2001). NKX2.5 mutations in patients with tetralogy of fallot. *Circulation* **104**, 2565–2568.
Goldstein, J. L. (2001). Laskers for 2001: Knockout mice and test-tube babies. *Nat. Med.* **7**, 1079–1080.
Grabitz, R. G., Joffres, M. R., and Collins-Nakai, R. L. (1988). Congenital heart disease: Incidence in the first year of life. The Alberta Heritage Pediatric Cardiology Program. *Am. J. Epidemiol.* **128**, 381–388.
Graham, A., and Smith, A. (2001). Patterning the pharyngeal arches. *Bioessays* **23**, 54–61.
Gu, C., Rodriguez, E. R., Reimert, D. V., Shu, T., Fritzsch, B., Richards, L. J., Kolodkin, A. L., and Ginty, D. D. (2003). Neuropilin-1 conveys semaphorin and VEGF signaling during neural and cardiovascular development. *Dev. Cell* **5**, 45–57.
Hanna, E. J., Nevin, N. C., and Nelson, J. (1994). Genetic study of congenital heart defects in Northern Ireland (1974–1978). *J. Med. Genet.* **31**, 858–863.
Harris, J. A., Francannet, C., Pradat, P., and Robert, E. (2003). The epidemiology of cardiovascular defects, part 2: A study based on data from three large registries of congenital malformations. *Pediatr. Cardiol.* **24**, 222–235.

Herzog, Y., Kalcheim, C., Kahane, N., Reshef, R., and Neufeld, G. (2001). Differential expression of neuropilin-1 and neuropilin-2 in arteries and veins. *Mech. Dev.* **109**, 115–119.

Hoffman, J. I., and Kaplan, S. (2002). The incidence of congenital heart disease. *J. Am. Coll. Cardiol.* **39**, 1890–1900.

Hu, N., Sedmera, D., Yost, H. J., and Clark, E. B. (2000). Structure and function of the developing zebrafish heart. *Anat. Rec.* **260**, 148–157.

Ikeda, E., Achen, M. G., Breier, G., and Risau, W. (1995). Hypoxia-induced transcriptional activation and increased mRNA stability of vascular endothelial growth factor in C6 glioma cells. *J. Biol. Chem.* **270**, 19761–19766.

Isogai, S., Horiguchi, M., and Weinstein, B. M. (2001). The vascular anatomy of the developing zebrafish: An atlas of embryonic and early larval development. *Dev. Biol.* **230**, 278–301.

Isogai, S., Lawson, N. D., Torrealday, S., Horiguchi, M., and Weinstein, B. M. (2003). Angiogenic network formation in the developing vertebrate trunk. *Development* **130**, 5281–5290.

Jerome, L. A., and Papaioannou, V. E. (2001). DiGeorge syndrome phenotype in mice mutant for the T-box gene, Tbx1. *Nat. Genet.* **27**, 286–291.

Kawasaki, T., Kitsukawa, T., Bekku, Y., Matsuda, Y., Sanbo, M., Yagi, T., and Fujisawa, H. (1999). A requirement for neuropilin-1 in embryonic vessel formation. *Development* **126**, 4895–4902.

Lambrechts, D., Storkebaum, E., Morimoto, M., Del-Favero, J., Desmet, F., Marklund, S. L., Wyns, S., Thijs, V., Andersson, J., van Marion, I., Al-Chalabi, A., Bornes, S., Musson, R., Hansen, V., Beckman, L., Adolfsson, R., Pall, H. S., Prats, H., Vermeire, S., Rutgeerts, P., Katayama, S., Awata, T., Leigh, N., Lang-Lazdunski, L., Dewerchin, M., Shaw, C., Moons, L., Vlietinck, R., Morrison, K. E., Robberecht, W., Van Broeckhoven, C., Collen, D., Andersen, P. M., and Carmeliet, P. (2003). VEGF is a modifier of amyotrophic lateral sclerosis in mice and humans and protects motoneurons against ischemic death. *Nat. Genet.* **34**, 383–394.

Lander, E. S., Linton, L. M., Birren, B., Nusbaum, C., Zody, M. C., Baldwin, J., Devon, K., Dewar, K., Doyle, M., FitzHugh, W., Funke, R., Gage, D., Harris, K., Heaford, A., Howland, J., Kann, L., Lehoczky, J., LeVine, R., McEwan, P., McKernan, K., Meldrim, J., Mesirov, J. P., Miranda, C., Morris, W., Naylor, J., Raymond, C., Rosetti, M., Santos, R., Sheridan, A., Sougnez, C., Stange-Thomann, N., Stojanovic, N., Subramanian, A., Wyman, D., Rogers, J., Sulston, J., Ainscough, R., Beck, S., Bentley, D., Burton, J., Clee, C., Carter, N., Coulson, A., Deadman, R., Deloukas, P., Dunham, A., Dunham, I., Durbin, R., French, L., Grafham, D., Gregory, S., Hubbard, T., Humphray, S., Hunt, A., Jones, M., Lloyd, C., McMurray, A., Matthews, L., Mercer, S., Milne, S., Mullikin, J. C., Mungall, A., Plumb, R., Ross, M., Shownkeen, R., Sims, S., Waterston, R. H., Wilson, R. K., Hillier, L. W., McPherson, J. D., Marra, M. A., Mardis, E. R., Fulton, L. A., Chinwalla, A. T., Pepin, K. H., Gish, W. R., Chissoe, S. L., Wendl, M. C., Delehaunty, K. D., Miner, T. L., Delehaunty, A., Kramer, J. B., Cook, L. L., Fulton, R. S., Johnson, D. L., Minx, P. J., Clifton, S. W., Hawkins, T., Branscomb, E., Predki, P., Richardson, P., Wenning, S., Slezak, T., Doggett, N., Cheng, J. F., Olsen, A., Lucas, S., Elkin, C., Uberbacher, E., Frazier, M., Gibbs, R. A., Muzny, D. M., Scherer, S. E., Bouck, J. B., Sodergren, E. J., Worley, K. C., Rives, C. M., Gorrell, J. H., Metzker, M. L., Naylor, S. L., Kucherlapati, R. S., Nelson, D. L., Weinstock, G. M., Sakaki, Y., Fujiyama, A., Hattori, M., Yada, T., Toyoda, A., Itoh, T., Kawagoe, C., Watanabe, H., Totoki, Y., Taylor, T., Weissenbach, J., Heilig, R., Saurin, W., Artiguenave, F., Brottier, P., Bruls, T., Pelletier, E., Robert, C., Wincker, P., Smith, D. R., Doucette-Stamm, L., Rubenfield, M., Weinstock, K., Lee, H. M., Dubois, J., Rosenthal, A., Platzer, M., Nyakatura, G., Taudien, S., Rump, A., Yang, H., Yu, J., Wang, J., Huang, G., Gu, J., Hood, L., Rowen, L., Madan, A., Qin, S., Davis, R. W., Federspiel, N. A.,

Abola, A. P., Proctor, M. J., Myers, R. M., Schmutz, J., Dickson, M., Grimwood, J., Cox, D. R., Olson, M. V., Kaul, R., Raymond, C., Shimizu, N., Kawasaki, K., Minoshima, S., Evans, G. A., Athanasiou, M., Schultz, R., Roe, B. A., Chen, F., Pan, H., Ramser, J., Lehrach, H., Reinhardt, R., McCombie, W. R., de la Bastide, M., Dedhia, N., Blocker, H., Hornischer, K., Nordsiek, G., Agarwala, R., Aravind, L., Bailey, J. A., Bateman, A., Batzoglou, S., Birney, E., Bork, P., Brown, D. G., Burge, C. B., Cerutti, L., Chen, H. C., Church, D., Clamp, M., Copley, R. R., Doerks, T., Eddy, S. R., Eichler, E. E., Furey, T. S., Galagan, J., Gilbert, J. G., Harmon, C., Hayashizaki, Y., Haussler, D., Hermjakob, H., Hokamp, K., Jang, W., Johnson, L. S., Jones, T. A., Kasif, S., Kaspryzk, A., Kennedy, S., Kent, W. J., Kitts, P., Koonin, E. V., Korf, I., Kulp, D., Lancet, D., Lowe, T. M., McLysaght, A., Mikkelsen, T., Moran, J. V., Mulder, N., Pollara, V. J., Ponting, C. P., Schuler, G., Schultz, J., Slater, G., Smit, A. F., Stupka, E., Szustakowski, J., Thierry-Mieg, D., Thierry-Mieg, J., Wagner, L., Wallis, J., Wheeler, R., Williams, A., Wolf, Y. I., Wolfe, K. H., Yang, S. P., Yeh, R. F., Collins, F., Guyer, M. S., Peterson, J., Felsenfeld, A., Wetterstrand, K. A., Patrinos, A., Morgan, M. J., Szustakowki, J., de Jong, P., Catanese, J. 'J., Osoegawa, K., Shizuya, H., Choi, S., and Chen, Y. J. (2001). Initial sequencing and analysis of the human genome. *Nature* **409,** 860–921.

Lawson, N. D., Vogel, A. M., and Weinstein, B. M. (2002). Sonic hedgehog and vascular endothelial growth factor act upstream of the Notch pathway during arterial endothelial differentiation. *Dev. Cell* **3,** 127–136.

Lawson, N. D., and Weinstein, B. M. (2002a). Arteries and veins: Making a difference with zebrafish. *Nat. Rev. Genet.* **3,** 674–682.

Lawson, N. D., and Weinstein, B. M. (2002b). *In Vivo* imaging of embryonic vascular development using transgenic zebrafish. *Dev. Biol.* **248,** 307–318.

Lee, P., Goishi, K., Davidson, A. J., Mannix, R., Zon, L., and Klagsbrun, M. (2002). Neuropilin-1 is required for vascular development and is a mediator of VEGF-dependent angiogenesis in zebrafish. *Proc. Natl. Acad. Sci. USA* **99,** 10470–10475.

Lee, Y., Song, A. J., Baker, R., Micales, B., Conway, S. J., and Lyons, G. E. (2000). Jumonji, a nuclear protein that is necessary for normal heart development. *Circ. Res.* **86,** 932–938.

Leighton, P. A., Mitchell, K. J., Goodrich, L. V., Lu, X., Pinson, K., Scherz, P., Skarnes, W. C., and Tessier-Lavigne, M. (2001). Defining brain wiring patterns and mechanisms through gene trapping in mice. *Nature* **410,** 174–179.

Li, Q. Y., Newbury-Ecob, R. A., Terrett, J. A., Wilson, D. I., Curtis, A. R., Yi, C. H., Gebuhr, T., Bullen, P. J., Robson, S. C., Strachan, T., Bonnet, D., Lyonnet, S., Young, I. D., Raeburn, J. A., Buckler, A. J., Law, D. J., and Brook, J. D. (1997). Holt-Oram syndrome is caused by mutations in TBX5, a member of the Brachyury (T) gene family. *Nat. Genet.* **15,** 21–29.

Liang, D., Chang, J. R., Chin, A. J., Smith, A., Kelly, C., Weinberg, E. S., and Ge, R. (2001). The role of vascular endothelial growth factor (VEGF) in vasculogenesis, angiogenesis, and hematopoiesis in zebrafish development. *Mech. Dev.* **108,** 29–43.

Lindsay, E. A. (2001). Chromosomal microdeletions: Dissecting del22q11 syndrome. *Nat. Rev. Genet.* **2,** 858–868.

Lindsay, E. A., Vitelli, F., Su, H., Morishima, M., Huynh, T., Pramparo, T., Jurecic, V., Ogunrinu, G., Sutherland, H. F., Scambler, P. J., Bradley, A., and Baldini, A. (2001). Tbx1 haploinsufficieny in the DiGeorge syndrome region causes aortic arch defects in mice. *Nature* **410,** 97–101.

Loffredo, C. A. (2000). Epidemiology of cardiovascular malformations: Prevalence and risk factors. *Am. J. Med. Genet.* **97,** 319–325.

Loffredo, C. A., Wilson, P. D., and Ferencz, C. (2001). Maternal diabetes: An independent risk factor for major cardiovascular malformations with increased mortality of affected infants. *Teratology* **64,** 98–106.

7. Genetics and VEGF

Luttun, A., Tjwa, M., Moons, L., Wu, Y., Angelillo-Scherrer, A., Liao, F., Nagy, J. A., Hooper, A., Priller, J., De Klerck, B., Compernolle, V., Daci, E., Bohlen, P., Dewerchin, M., Herbert, J. M., Fava, R., Matthys, P., Carmeliet, G., Collen, D., Dvorak, H. F., Hicklin, D. J., and Carmeliet, P. (2002). Revascularization of ischemic tissues by PlGF treatment, and inhibition of tumor angiogenesis, arthritis and atherosclerosis by anti-Flt1. *Nat. Med.* **8**, 831–840.

Maschhoff, K. L., and Baldwin, H. S. (2000). Molecular determinants of neural crest migration. *Am. J. Med. Genet.* **97**, 280–288.

McElhinney, D. B., Geiger, E., Blinder, J., Benson, D. W., and Goldmuntz, E. (2003). NKX2.5 mutations in patients with congenital heart disease. *J. Am. Coll. Cardiol.* **42**, 1650–1655.

Merscher, S., Funke, B., Epstein, J. A., Heyer, J., Puech, A., Lu, M. M., Xavier, R. J., Demay, M. B., Russell, R. G., Factor, S., Tokooya, K., Jore, B. S., Lopez, M., Pandita, R. K., Lia, M., Carrion, D., Xu, H., Schorle, H., Kobler, J. B., Scambler, P., Wynshaw-Boris, A., Skoultchi, A. I., Morrow, B. E., and Kucherlapati, R. (2001). TBX1 is responsible for cardiovascular defects in velo-cardio-facial/DiGeorge syndrome. *Cell* **104**, 619–629.

Miao, H. Q., Soker, S., Feiner, L., Alonso, J. L., Raper, J. A., and Klagsbrun, M. (1999). Neuropilin-1 mediates collapsin-1/semaphorin III inhibition of endothelial cell motility: Functional competition of collapsin-1 and vascular endothelial growth factor-165. *J. Cell Biol.* **146**, 233–242.

Millauer, B., Wizigmann-Voos, S., Schnurch, H., Martinez, R., Moller, N. P., Risau, W., and Ullrich, A. (1993). High affinity VEGF binding and developmental expression suggest Flk-1 as a major regulator of vasculogenesis and angiogenesis. *Cell* **72**, 835–846.

Miquerol, L., Langille, B. L., and Nagy, A. (2000). Embryonic development is disrupted by modest increases in vascular endothelial growth factor gene expression. *Development* **127**, 3941–3946.

Mitchell, K. J., Pinson, K. I., Kelly, O. G., Brennan, J., Zupicich, J., Scherz, P., Leighton, P. A., Goodrich, L. V., Lu, X., Avery, B. J., Tate, P., Dill, K., Pangilinan, E., Wakenight, P., Tessier-Lavigne, M., and Skarnes, W. C. (2001). Functional analysis of secreted and transmembrane proteins critical to mouse development. *Nat. Genet.* **28**, 241–249.

Mitchell, S. C., Korones, S. B., and Berendes, H. W. (1971). Congenital heart disease in 56,109 births. Incidence and natural history. *Circulation* **43**, 323–332.

Moorman, A., Webb, S., Brown, N. A., Lamers, W., and Anderson, R. H. (2003). Development of the heart: (1) formation of the cardiac chambers and arterial trunks. *Heart* **89**, 806–814.

Morelli, S. H., Young, L., Reid, B., Ruttenberg, H., and Bamshad, M. J. (2001). Clinical analysis of families with heart, midline, and laterality defects. *Am. J. Med. Genet.* **101**, 388–392.

Mukouyama, Y. S., Shin, D., Britsch, S., Taniguchi, M., and Anderson, D. J. (2002). Sensory nerves determine the pattern of arterial differentiation and blood vessel branching in the skin. *Cell* **109**, 693–705.

Narayanan, K., Williamson, R., Zhang, Y., Stewart, A. F., and Ioannou, P. A. (1999). Efficient and precise engineering of a 200 kb beta-globin human/bacterial artificial chromosome in *E. coli* DH10B using an inducible homologous recombination system. *Gene Ther.* **6**, 442–447.

Nasevicius, A., and Ekker, S. C. (2000). Effective targeted gene "knockdown" in zebrafish. *Nat. Genet.* **26**, 216–220.

Nasevicius, A., Larson, J., and Ekker, S. C. (2000). Distinct requirements for zebrafish angiogenesis revealed by a VEGF-A morphant. *Yeast* **17**, 294–301.

Neufeld, G., Cohen, T., Gengrinovitch, S., and Poltorak, Z. (1999). Vascular endothelial growth factor (VEGF) and its receptors. *Faseb. J.* **13**, 9–22.

Neufeld, G., Cohen, T., Shraga, N., Lange, T., Kessler, O., and Herzog, Y. (2002). The neuropilins: Multifunctional semaphorin and VEGF receptors that modulate axon guidance and angiogenesis. *Trends Cardiovasc. Med.* **12**, 13–19.

Nora, J. J., and Nora, A. H. (1978). The evolution of specific genetic and environmental counseling in congenital heart diseases. *Circulation* **57**, 205–213.

Oh, S. P., and Li, E. (2002). Gene-dosage-sensitive genetic interactions between inversus viscerum (iv), nodal, and activin type IIB receptor (ActRIIB) genes in asymmetrical patterning of the visceral organs along the left–right axis. *Dev. Dyn.* **224**, 279–290.

Pierpont, M. E., Markwald, R. R., and Lin, A. E. (2000). Genetic aspects of atrioventricular septal defects. *Am. J. Med. Genet.* **97**, 289–296.

Pinter, E., Haigh, J., Nagy, A., and Madri, J. A. (2001). Hyperglycemia-induced vasculopathy in the murine conceptus is mediated via reductions of VEGF-A expression and VEGF receptor activation. *Am. J. Pathol.* **158**, 1199–1206.

Piotrowski, T., Ahn, D. G., Schilling, T. F., Nair, S., Ruvinsky, I., Geisler, R., Rauch, G. J., Haffter, P., Zon, L. I., Zhou, Y., Foott, H., Dawid, I. B., and Ho, R. K. (2003). The zebrafish van gogh mutation disrupts tbx1, which is involved in the DiGeorge deletion syndrome in humans. *Development* **130**, 5043–5052.

Piotrowski, T., and Nusslein-Volhard, C. (2000). The endoderm plays an important role in patterning the segmented pharyngeal region in zebrafish (*Danio rerio*). *Dev. Biol.* **225**, 339–356.

Pradat, P., Francannet, C., Harris, J. A., and Robert, E. (2003). The epidemiology of cardiovascular defects, part I: A study based on data from three large registries of congenital malformations. *Pediatr. Cardiol.* **24**, 195–221.

Ranger, A. M., Grusby, M. J., Hodge, M. R., Gravallese, E. M., de la Brousse, F. C., Hoey, T., Mickanin, C., Baldwin, H. S., and Glimcher, L. H. (1998). The transcription factor NF-ATc is essential for cardiac valve formation. *Nature* **392**, 186–190.

Roman, B. L., and Weinstein, B. M. (2000). Building the vertebrate vasculature: Research is going swimmingly. *Bioessays* **22**, 882–893.

Scambler, P. J. (1993). Deletions of human chromosome 22 and associated birth defects. *Curr. Opin. Genet. Dev.* **3**, 432–437.

Scambler, P. J. (1999). Genetics. Engineering a broken heart. *Nature* **401**, 335–337.

Schinke, M., and Izumo, S. (1999). Getting to the heart of DiGeorge syndrome. *Nat. Med.* **5**, 1120–1121.

Schott, J. J., Benson, D. W., Basson, C. T., Pease, W., Silberbach, G. M., Moak, J. P., Maron, B. J., Seidman, C. E., and Seidman, J. G. (1998). Congenital heart disease caused by mutations in the transcription factor NKX2-5. *Science* **281**, 108–111.

Shalaby, F., Rossant, J., Yamaguchi, T. P., Gertsenstein, M., Wu, X. F., Breitman, M. L., and Schuh, A. C. (1995). Failure of blood-island formation and vasculogenesis in Flk-1-deficient mice. *Nature* **376**, 62–66.

Smithies, O. (2001). Forty years with homologous recombination. *Nat. Med.* **7**, 1083–1086.

Srivastava, D., and Olson, E. N. (2000). A genetic blueprint for cardiac development. *Nature* **407**, 221–226.

Stainier, D. Y. (2001). Zebrafish genetics and vertebrate heart formation. *Nat. Rev. Genet.* **2**, 39–48.

Stainier, D. Y., Fouquet, B., Chen, J. N., Warren, K. S., Weinstein, B. M., Meiler, S. E., Mohideen, M. A., Neuhauss, S. C., Solnica-Krezel, L., Schier, A. F., Zwartkruis, F., Stemple, D. L., Malicki, J., Driever, W., and Fishman, M. C. (1996). Mutations affecting the formation and function of the cardiovascular system in the zebrafish embryo. *Development* **123**, 285–292.

Stalmans, I., Lambrechts, D., De Smet, F., Jansen, S., Wang, J., Maity, S., Kneer, P., von der Ohe, M., Swillen, A., Maes, C., Gewillig, M., Molin, D. G., Hellings, P., Boetel, T., Haardt, M., Compernolle, V., Dewerchin, M., Plaisance, S., Vlietinck, R., Emanuel, B.,

Gittenberger-de Groot, A. C., Scambler, P., Morrow, B., Driscol, D. A., Moons, L., Esguerra, C. V., Carmeliet, G., Behn-Krappa, A., Devriendt, K., Collen, D., Conway, S. J., and Carmeliet, P. (2003). VEGF: A modifier of the del22q11 (DiGeorge) syndrome? *Nat. Med.* **9**, 173–182.

Stalmans, I., Ng, Y. S., Rohan, R., Fruttlger, M., Bouche, A., Yuce, A., Fujisawa, H., Hermans, B., Shani, M., Jansen, S., Hicklin, D., Anderson, D. J., Gardiner, T., Hammes, H. P., Moons, L., Dewerchin, M., Collen, D., Carmeliet, P., and D'Amore, P. A. (2002). Arteriolar and venular patterning in retinas of mice selectively expressing VEGF isoforms. *J. Clin. Invest.* **109**, 327–336.

Steijlen, P. M., and van Steensel, M. A. (1999). Paradominant inheritance, a hypothesis explaining occasional familial occurrence of sporadic syndromes. *Am. J. Med. Genet.* **85**, 359–360.

Storkebaum, E., and Carmeliet, P. (2004). VEGF: A critical player in neurodegeneration. *J. Clin. Invest.* **113**, 14–18.

Takahashi, T., Nakamura, F., Jin, Z., Kalb, R. G., and Strittmatter, S. M. (1998). Semaphorins A and E act as antagonists of neuropilin-1 and agonists of neuropilin-2 receptors. *Nat. Neurosci.* **1**, 487–493.

Terman, B. I., Dougher-Vermazen, M., Carrion, M. E., Dimitrov, D., Armellino, D. C., Gospodarowicz, D., and Bohlen, P. (1992). Identification of the KDR tyrosine kinase as a receptor for vascular endothelial cell growth factor. *Biochem. Biophys. Res. Commun.* **187**, 1579–1586.

Testa, G., Zhang, Y., Vintersten, K., Benes, V., Pijnappel, W. W., Chambers, I., Smith, A. J., Smith, A. G., and Stewart, A. F. (2003). Engineering the mouse genome with bacterial artificial chromosomes to create multipurpose alleles. *Nat. Biotechnol.* **21**, 443–447.

Thisse, C., and Zon, L. I. (2002). Organogenesis—heart and blood formation from the zebrafish point of view. *Science* **295**, 457–462.

Towbin, J. A., and Belmont, J. (2000). Molecular determinants of left and right outflow tract obstruction. *Am. J. Med. Genet.* **97**, 297–303.

Valenzuela, D. M., Murphy, A. J., Frendewey, D., Gale, N. W., Economides, A. N., Auerbach, W., Poueymirou, W. T., Adams, N. C., Rojas, J., Yasenchak, J., Chernomorsky, R., Boucher, M., Elsasser, A. L., Esau, L., Zheng, J., Griffiths, J. A., Wang, X., Su, H., Xue, Y., Dominguez, M. G., Noguera, I., Torres, R., Macdonald, L. E., Stewart, A. F., DeChiara, T. M., and Yancopoulos, G. D. (2003). High-throughput engineering of the mouse genome coupled with high-resolution expression analysis. *Nat. Biotechnol.* **21**, 652–659.

Venter, J. C., Adams, M. D., Myers, E. W., Li, P. W., Mural, R. J., Sutton, G. G., Smith, H. O., Yandell, M., Evans, C. A., Holt, R. A., Gocayne, J. D., Amanatides, P., Ballew, R. M., Huson, D. H., Wortman, J. R., Zhang, Q., Kodira, C. D., Zheng, X. H., Chen, L., Skupski, M., Subramanian, G., Thomas, P. D., Zhang, J., Gabor Miklos, G. L., Nelson, C., Broder, S., Clark, A. G., Nadeau, J., McKusick, V. A., Zinder, N., Levine, A. J., Roberts, R. J., Simon, M., Slayman, C., Hunkapiller, M., Bolanos, R., Delcher, A., Dew, I., Fasulo, D., Flanigan, M., Florea, L., Halpern, A., Hannenhalli, S., Kravitz, S., Levy, S., Mobarry, C., Reinert, K., Abu-Threideh, J., Beasley, E., Biddick, K., Bonazzi, V., Brandon, R., Cargill, M., Chandramouliswaran, I., Charlab, R., Chaturvedi, K., Deng, Z., Di Francesco, V., Dunn, P., Eilbeck, K., Evangelista, C., Gabrielian, A. E., Gan, W., Ge, W., Gong, F., Gu, Z., Guan, P., Heiman, T. J., Higgins, M. E., Ji, r. R., Ke, Z., Ketchum, K. A., Lai, Z., Lei, Y., Li, Z., Li, J., Liang, Y., Lin, X., Lu, F., Merkulov, G. V., Milshina, N., Moore, H. M., Naik, A. K., Narayan, V. A., Neelam, B., Nusskern, D., Rusch, D. B., Salzberg, S., Shao, W., Shue, B., Sun, J., Wang, Z., Wang, A., Wang, X., Wang, J., Wei, M., Wides, R., Xiao, C., Yan, C., *et al.* (2001). The sequence of the human genome. *science* **291**, 1304–1351.

Vikkula, M., Boon, L. M., and Mulliken, J. B. (2001). Molecular genetics of vascular malformations. *Matrix Biol.* **20**, 327–335.

Wright, A. F., Carothers, A. D., and Pirastu, M. (1999). Population choice in mapping genes for complex diseases. *Nat. Genet.* **23,** 397–404.

Yagi, H., Furutani, Y., Hamada, H., Sasaki, T., Asakawa, S., Minoshima, S., Ichida, F., Joo, K., Kimura, M., Imamura, S., Kamatani, N., Momma, K., Takao, A., Nakazawa, M., Shimizu, N., and Matsuoka, R. (2003). Role of TBX1 in human del22q11.2 syndrome. *Lancet* **362,** 1366–1373.

Yamaguchi, T. P., Dumont, D. J., Conlon, R. A., Breitman, M. L., and Rossant, J. (1993). flk-1, an flt-related receptor tyrosine kinase is an early marker for endothelial cell precursors. *Development* **118,** 489–498.

Yang, Y., and Seed, B. (2003). Site-specific gene targeting in mouse embryonic stem cells with intact bacterial artificial chromosomes. *Nat. Biotechnol.* **21,** 447–451.

Yu, D., Ellis, H. M., Lee, E. C., Jenkins, N. A., Copeland, N. G., and Court, D. L. (2000). An efficient recombination system for chromosome engineering in *Escherichia coli. Proc. Natl. Acad. Sci. USA* **97,** 5978–5983.

Yuan, L., Moyon, D., Pardanaud, L., Breant, C., Karkkainen, M. J., Alitalo, K., and Eichmann, A. (2002). Abnormal lymphatic vessel development in neuropilin 2 mutant mice. *Development* **129,** 4797–4806.

Zambrowicz, B. P., Friedrich, G. A., Buxton, E. C., Lilleberg, S. L., Person, C., and Sands, A. T. (1998). Disruption and sequence identification of 2,000 genes in mouse embryonic stem cells. *Nature* **392,** 608–611.

Zhang, Y., Buchholz, F., Muyrers, J. P., and Stewart, A. F. (1998). A new logic for DNA engineering using recombination in *Escherichia coli. Nat Genet.* **20,** 123–128.

Zhong, T. P., Childs, S., Leu, J. P., and Fishman, M. C. (2001). Gridlock signalling pathway fashions the first embryonic artery. *Nature* **414,** 216–220.

8
Development of Coronary Vessels

Mark W. Majesky
Departments of Medicine and Genetics
Carolina Cardiovascular Biology Center
University of North Carolina
Chapel Hill, North Carolina 27599

  I. Introduction
 II. Formation of the Proepicardium
III. Formation of the Epicardium
 IV. Vasculogenesis in the Subepicardium
  V. Role of Growth Factors in Coronary Vasculogenesis
 VI. Epicardial to Mesenchymal Transformation (EMT)
VII. Epicardial EMT Leads to Changes in Gene Expression
VIII. Bidirectional Epicardial-Myocardial Signaling
 IX. Establishment of a Coronary Circulation: Making Connections to the Aorta
  X. Maturation of Coronary Vessels and Formation of a Tunica Media
 XI. Development of the Coronary Lymphatic Vessels
XII. Coronary Vessel Anomalies
XIII. Coronary Development and Disease
     Acknowledgments
     References

I. Introduction

Progenitors for the coronary vasculature are found in the proepicardium. Proepicardial cells arise independently of the heart itself and provide epicardial cells to the outer surface of the looped heart tube during cardiac development. Fate-mapping studies show that proepicardial cells are also forerunners for the endothelium and smooth muscle of the coronary vasculature, and the connective tissue cells that form the coronary adventitia and interstitial matrix of the myocardium (Gittenberger-de Groot et al., 1998; Mikawa and Fishman, 1992). All vertebrates, with the exception of primitive chordates, have an epicardium (Hirakow, 1985). The appearance of the epicardium in evolution and its derivative coronary vessels correlates with the transition of the heart from a primitive tubular structure with a thin-walled, epithelial-type myocardium, as found in pre- and early chordates, to a thick multilayered pumping organ found in most vertebrates, including fish, amphibia, birds, and mammals. Formation of the coronary vessels occurs through an epithelial-to-mesenchymal transformation of proepicardial cells, followed by vasculogenesis in the subepicardial layer, assembly and remodeling of a primitive coronary plexus, and recruitment of a smooth-muscle coating. This article will review our current understanding of the origins and development of the coronary vasculature, with an emphasis on formation of coronary smooth muscle. The interested reader is directed to several excellent recent reviews for areas not covered in-depth here (Manner *et al.*, 2001; Morabito *et al.*, 2002; Munoz-Chapuli *et al.*, 2002; Tomanek and Zheng, 2002; Wada *et al.*, 2003).

II. Formation of the Proepicardium

The first step in coronary development is formation of the proepicardium (PE), a transient structure that originates from pericardial serosa in the area of the sinoatrial junction (Fig. 1). Proepicardial cells are quite remarkable for the number of different cell types that they produce during heart development. The structure of the PE varies depending on the species in which it is described (Fransen and Lemanski, 1990; Hiruma and Hirakow, 1989; Komiyama *et al.*, 1987; Manner, 1992; Munoz-Chapuli *et al.*, 1996; Viragh *et al.*, 1993). It develops at the ventral midline as bilateral primordia on the pericardial surface of the septum transversum. In mouse embryos, these

Figure 1 The proepicardium. Scanning electron micrograph of the developing chick proepicardial villi at Hamburger–Hamilton (HH) stage 17.

8. Coronary Vessel Development

bilateral primordia continue to develop as such, whereas in avian embryos the left side regresses while the right side continues to form. Each villus is composed of a simple cuboidal mesothelium covering a proteoglycan- and hyaluronic-acid-rich extracellular matrix (ECM) core, with occasional clusters of stellate mesenchymal cells embedded within (Kalman et al., 1995; Kuhn and Liebherr, 1988; Munoz-Chapuli et al., 2002; Nahirney et al., 2003; Viragh and Challice, 1981). The number and density of mesenchymal cells within PE villi progressively increase as the PE continues to develop. Mitoses are frequently observed in the mesothelium (Viragh et al., 1993). Numerous microvilli project from the apical surface of PE cells into the coelomic cavity and toward the looped heart (Nahirney et al., 2003). In mouse embryos, the proepicardium appears around embryonic day 9.5 (E9.5). In the chick embryo, the PE appears at Hamburger–Hamilton (HH) stage 14 after about 48 h of incubation.

The first to propose that the epicardium originates from PE mesothelium was Kurkiewicz, in 1909 (Kurkiweicz, 1909; Manner et al., 2001). Prior to this, and continuing for many years afterward, it was common to see the term *epimyocardium*, an erroneous reference to the notion that the epicardium is a product of the myocardial layer. It was not until some 60 years later, however, when Manasek confirmed that the epicardium arises from sources outside of the myocardium, that the PE became a focus of detailed study (Manasek, 1969). Subsequent scanning electron microscopy (EM; Ho and Shimada, 1978; Shimada et al., 1981), and transmission EM studies (Hiruma and Hirakow, 1989; Komiyama et al., 1987; Viragh and Challice, 1981) identified the villus outgrowth of the PE as the definitive source of the epicardial layer (Poelmann et al., 1993). The next advance was the realization that proepicardial cells are progenitors for the coronary vasculature. This discovery came in the course of systematic lineage analysis studies of heart development. For example, using a replication-defective retrovirus expressing β-galactosidase, Mikawa and Fischman reported that when injections of the virus were made into embryonic chick hearts at HH stage 15 or earlier, the only blue-stained cells found at the time of hatching were small clusters of cardiac myocytes (Mikawa and Fishman, 1992). However, when injections were delayed until HH stage 17 or later, then elements of the coronary vasculature were also labeled. When low viral titers were used to tag individual progenitor cells, polyclones of either coronary endothelial cells or CoSMCs (but not both) were stained blue. These results argued that coronary vessels originate from already committed progenitors that enter the heart around stages 17–18. Later studies showed that retroviral injections into the proepicardium itself, or grafts of quail PE into chick embryo hosts, also produced labeled coronary vessels, suggesting that the source of coronary progenitors is the proepicardium itself (Mikawa and Gourdie, 1996). These findings raise two important questions that are active areas of investigation at the present time: (a) When do proepicardial cells become committed to coronary endothelial or SMC fates, and (b) what signals act on committed proepicardial cells to stimulate differentiation to mature vascular cell types?

The molecular signals that initiate formation of the PE from septum transversum mesenchyme are largely undefined. If we consider how questions of this type have been addressed for other developing systems, then we see that a combination of approaches will be needed. Molecular markers of the PE phenotype that are reliable and specific surrogates for PE cells are required. The currently used markers include Wilms' tumor-1 (WT1), α4-integrin, retinaldehyde dehydrogenase-2 (RALDH2), Tbx18, cytokeratin, and epicardin/capsulin/pod1 (Kraus et al., 2001; Moore et al., 1999; Sengbusch et al., 2002; Wada et al., 2003; Xavier-Neto et al., 2000). These genes are expressed in PE and epicardium but not in myocardium or endocardium. However, each of these markers is also expressed in additional, non-epicardial cell types, limiting their individual usefulness as proepicardial markers. On the other hand, specific enhancer elements that "read" a complex of transcription factors and coactivators that function to specify the proepicardial phenotype would be an important and necessary advance for the field. In that respect, it is interesting to note that an E-box/Ets-containing 270bp element in the serum response factor (SRF) promoter region conferred region-specific expression of a lacZ reporter gene to a subset of cells in the PE that appeared to mark cells destined to form coronary vessels (Nelson et al., 2004). Another requirement is the ability to explant culture the pericardial serosa and test its response to co-culture with "inducing" tissues, such as liver primordium or myocardium. Methods and approaches have been reported to support the feasibility of PE explant cocultures, and mitogenic signaling interactions between PE cells and myocardial cells have been identified (Chen et al., 2002b; Landerholm et al., 1999; Stuckmann et al., 2003). Finally, the ability to apply loss and gain of function approaches in genetically accessible model organisms will be essential to identify roles of specific gene products in formation and function of the PE *in vivo* (Begemann et al., 2002; Li et al., 2002; Sengbusch et al., 2002).

III. Formation of the Epicardium

Recognition of the PE as a principle source of epicardial cells raises the obvious question of how these epicardial precursors are transferred across the pericardial coelom to the surface of the heart. The answer to this question reveals one of the more remarkable and unexpected features of coronary vessel development. In avian and amphibian embryos, contact of the expanding proepicardial villi with the outer surface of the myocardium forms a connection that allows direct transfer of proepicardial cells to the developing heart (Manner, 1992; Vrancken Peeters *et al.*, 1995) (Fig. 2). In fact, Nahirney *et al.* (2003) identified a heparin sulfate- and fibronectin-rich extracellular matrix bridge between the PE and the myocardium of chick embryos that may guide proepicardial villi to their target site on the lesser curvature of the looped heart. In mammalian embryos, proepicardial cells reach the heart by release of cellular aggregates (or vesicles) from the PE into the pericardial cavity. The PE vesicles are then carried by fluid motion to the surface of the heart (Komiyama *et al.*, 1987; Perez-Pomares *et al.*, 1997). Attachment of these cellular aggregates might occur near the original PE, or at a distance from it (Sengbusch *et al.*, 2002). Release of PE vesicles also occurs to a limited extent in avian embryos. In either case, epicardial cells migrate as a continuous sheet of mesothelial cells from the initial points of attachment on the myocardium to cover the atrio-ventricular canal first and then to reach the inflow and outflow tracts (Moss *et al.*, 1998; Vrancken Peeters *et al.*, 1995; Xavier-Neto *et al.*, 2000).

Epicardial cell migration is remarkable not only for the extent of surface area covered, but also because epicardial cells maintain continuous cell–cell contacts during migration and do not loose the epithelial organization of their cytoskeleton despite rapid movements to cover the myocardium. Moreover, they accomplish this without converting into other cell types that

Figure 2 Formation of the epicardium. Scanning electron micrograph of the atrioventricular canal region of an HH19 chick embryo heart tube. Arrows indicate the extent of epicardial outgrowth.

proepicardial fate maps show they have the potential to become. Epicardial cells at the leading edge of the migrating sheet are in constant contact with the apical surface of the underlying myocardium, whereas the trailing cells acquire a matrix-filled space between their basal surface and the myocardium, with occasional cytoplasmic extensions that make contact with underlying myocardial cells (Nahirney *et al.*, 2003). Epicardial cells express E-cadherin at cell–cell contacts and connexin 43 at gap junctions (Li *et al.*, 2002; Wada *et al.*, 2001). Connexin 43 knockout mice die in the neonatal period from conotruncal heart malformations and outflow tract obstruction (Reaume *et al.*, 1995). In addition, they exhibit a variety of coronary artery patterning defects, including anomalies in the number or site of origin of the main coronary arteries (Li *et al.*, 2002). Epicardial cells also express bves, a membrane protein with cell–cell adhesive activity that is confined to the lateral membrane of epicardial cells (Reese and Bader, 1999). When epicardial cells are dissociated into single cells, and the distribution of bves is followed during restoration of cell–cell contacts, bves trafficks from the perinuclear region to the cell membrane (Wada *et al.*, 2001). Bves is expressed in the PE, in migrating epithelial epicardium, delaminated vasculogenic mesenchyme, and vascular smooth muscle cells (Reese and Bader, 1999). Moreover, antibodies to bves inhibit epithelial migration of vasculogenic cells from the proepicardium.

The vast majority of epicardial cells are derived from the proepicardium. However, ablation of the PE does not eliminate the entire epicardial layer. The remaining epicardial cells originate from pericardial mesothelium in the outflow tract region and continue to develop in the complete absence of a PE (Gittenberger-de Groot *et al.*, 2000; Manner, 1993; Perez-Pomares *et al.*, 2003). In this respect, it is interesting to note that, in addition to the PE, villuslike protrusions of the pericardial mesothelium have also been described at the junction of the pericardium with the aortic sac (Grant, 1926), (Manner, 1998) and elsewhere along the pericardial serosa (Nahirney *et al.*, 2003). Although pericardial-derived epicardial cells can partially compensate for ablation of the PE, epicardial coverage of the heart is incomplete and embryos with ablation of the PE die at midgestation due to cardiac failure (Gittenberger-de Groot *et al.*, 2000). It is intriguing to consider that the functional properties of pericardial-derived epicardium may differ from those of the PE-derived epicardium. For example, Perez-Pomares *et al.* (2003) report that the two types of epicardial cells exhibit pronounced differences in expression of RALDH2 and in the rate at which they undergo epithelial to mesenchymal transformation (EMT) when cultured on a collagen gel substrate in serum-containing medium. Whether and to what extent differences in epicardial origins might produce different types of CoSMCs has yet to be examined.

The epicardial layer forms closely apposed to the myocardium, suggesting that epicardial-myocardial cell–cell adhesion guides early epicardium formation. VCAM-1, a cell adhesion molecule of the immunoglobulin superfamily (Cybulsky and Gimbrone, 1991), is expressed by myocardial cells from the onset of heart development to the completion of epicardium formation (Kwee et al., 1995). Its principle ligand, the $\alpha_4\beta_1$ integrin (Chan et al., 1992), is expressed in the PE and in proepicardial-derived epicardial cells (Pinco et al., 2001; Yang et al., 1995). Inactivation of the VCAM-1 gene produces an embryonic lethal phenotype characterized by severe defects in cardiac development, as well as failure of the chorion to fuse with the allantois (Bjarnegard et al., 2004; Gurtner et al., 1995). In VCAM-1-deficient embryos, the epicardium begins to form normally around E9.5, but by E11.5 it is shed and disappears (Kwee et al., 1995). Loss of the epicardial layer results in bleeding into the pericardial space due to incomplete subepicardial vasculogenesis. In $\alpha 4$ integrin-deficient embryos, the epicardial layer fails to develop due to defects in two steps of epicardium formation (Yang et al., 1995). The first defect is a failure of the proepicardial villi to vesiculate (Sengbusch et al., 2002). The few proepicardial cell aggregates that do attach to the heart then exhibit a failure to migrate over the myocardium. In both VCAM-1 and $\alpha 4$ integrin-null embryos, the inability to form and maintain an intact epicardial layer leads to cardiac failure and death of the embryo (Gurtner et al., 1995; Kwee et al., 1995; Yang et al., 1995). In part, this failure of cardiac development can be attributed to a gain of invasive activity of $\alpha 4$ integrin-deficient epicardial cells (Dettman et al., 2003).

IV. Vasculogenesis in the Subepicardium

Shortly after migration of epicardial cells over the surface of the heart has begun, a subepicardial space appears that is composed of abundant extracellular matrix (ECM) and epicardium-derived cells embedded within. The subepicardial ECM is proteoglycan-rich (Kalman et al., 1995) and contains fibronectin (Tidball, 1992), vitronectin (Bouchey et al., 1996), laminin (Kim et al., 1999), tenascin X (Burch et al., 1995; Imanaka-Yoshida et al., 2003), collagens I, III, and IV (Rongish et al., 1996), fibrillin (Bouchey et al., 1996), fibulin-2 (Tsuda et al., 2001), and flectin (Tsuda et al., 1998). Both the epicardium and myocardium contribute to the subepicardial ECM. These ECM components exhibit dynamic patterns of expression and provide a local environment rich in angiogenic factors in which the initial steps of coronary vasculogenesis are carried out (Tomanek, 1996).

Coronary endothelial cells appear to arise from two sources. One source is from angioblasts that are formed elsewhere in the embryo and are carried to the heart by the proepicardial villi (Mikawa and Fishman, 1992; Poelmann *et al.*, 1993; Vrancken Peeters *et al.*, 1997a). This possibility is supported by the finding that retroviral markers injected into HH16 chick PEs showed that the ultimate fate of at least some proepicardial cells is already determined by the time PE contacts the heart (Mikawa and Fishman, 1992; Mikawa and Gourdie, 1996). Moreover, whole-mount immunostaining with antibody QH1, which recognizes angioblasts and endothelial and hematopoietic cells in quail embryos, identified individual cells within PEs that were in contact with the heart (Kattan *et al.*, 2004). The second source is from the epicardium itself. A variety of studies suggest that at least some epicardial cells are initially multipotential and that following EMT they are directed to various cell fates by instructive signals produced within the subepicardial layer or from the myocardium (Dettman *et al.*, 1998; Perez-Pomares *et al.*, 2002). The idea that some PE cells are committed to particular cell fates before they contact the heart, whereas others possess greater plasticity, is consistent with existing data and perhaps best explains the origins of coronary progenitors in heart development. It is important to point out that endocardial cells have a separate origin from that of coronary endothelium. Endocardial cells arise much earlier in embryogenesis from within the cardiogenic plate mesenchyme in response to inductive signals from foregut endoderm (Sugi and Markwald, 2003).

The first coronary endothelial cells are found in the fibronectin-rich subepicardial ECM and are frequently associated with CD45-positive hematopoietic cells in structures that resemble blood islands (Kattan *et al.*, 2004; Poelmann *et al.*, 1993; Rongish *et al.*, 1996). Subepicardial endothelial cells self-assemble into a capillary-like vessel network characterized by a laminin- and type IV collagen-rich basement membrane (Rongish *et al.*, 1996). The first lumenized vessels are found in subepicardial ECM within the atrioventricular canal (Vrancken Peeters *et al.*, 1997a). These early vessels are continuous with the sinus venosus and therefore, would be considered forerunners of the coronary venous system (Poelmann *et al.*, 1993; Vrancken Peeters *et al.*, 1997b). Coronary vasculogenesis occurs in a local environment that is rich in FGF-2 and vascular endothelial growth factors (VEGFs) (Tomanek *et al.*, 1999). The myocardium secretes FGF-1, FGF-2, FGF-7, VEGF-A, and angiopoietin-1 during epicardial cell EMT and coronary vasculogenesis (Cox and Poole, 2000; Morabito *et al.*, 2001; Pereira *et al.*, 1999; Tomanek *et al.*, 1999). Moreover, injection of either VEGF-A or FGF-2 *in ovo* stimulates coronary vasculogenesis, and blocking antibodies to either factor inhibit coronary vessel formation when tested in explanted embryonic hearts (Tomanek *et al.*, 1998; Yue and Tomanek, 2001). Early coronary vessel formation is driven by limited oxygen diffusion from the ventricular chambers, resulting in a gradient of hypoxia that reaches its maximum at the epicardial surface. The observation that levels of VEGF-A mRNA and protein were greatest in the epicardium and compact zone myocardium is consistent with hypoxia as an important stimulus for coronary vessel formation (Tomanek et al., 1999). In addition to hypoxia, metabolic and mechanical inputs also determine the timing and pattern of early coronary vessel formation (Zheng et al., 1999). As is characteristic of endothelium in other vascular beds that acquire specialized properties to optimize their function in different local environments, coronary endothelial cells are also heterogeneous. For example, location-specific differences in expression of constitutive endothelial nitric oxide synthase gene (eNOS) (Andries et al., 1998), and brain-derived neurotrophic factor (BDNF) (Donovan et al., 2000) in the coronary vasculature have been reported.

V. Role of Growth Factors in Coronary Vasculogenesis

Studies in gene-deficient mice have greatly advanced our understanding of the essential roles played by members of the VEGF gene family in coronary vessel formation. The mammalian VEGF family consists of five members, VEGF-A, -B, -C, -D, and placental growth factor (P1GF), which differ in receptor-binding specificities (for review, see Ferrara et al., 2003). Moreover, splice variants of VEGF-A are made, including $VEGF_{120}$, $VEGF_{164}$, and $VEGF_{188}$, that differ in tissue distribution of expression, interaction with cell surface co-receptors (neuropilins), and the ability to stimulate chemotaxis, survival, and mitogenic responses (Carmeliet et al., 1999; Robinson and Stringer, 2001). Mice expressing $VEGF_{120}$ only ($VEGF^{120/120}$) were generated by Cre/loxP-mediated recombination to remove exons 6 and 7 of the VEGF-A gene. $VEGF^{120/120}$ mice survive until birth, at which time some of the mice died while the remainder grew poorly for up to 2 weeks and then died of cardiac failure. Histologic analysis revealed that $VEGF^{120/120}$ hearts exhibited a failure of coronary vessels to keep pace with the postnatal growth of the myocardium (Carmeliet et al., 1999). Inability of the coronary vessel network to expand during the postnatal period in $VEGF^{120/120}$ mice is probably due to a combination of inadequate angiogenesis and defects in recruitment of circulating endothelial progenitor cells to the heart (Young et al., 2002). Moreover, smooth muscle investment of coronary arteries in $VEGF^{120/120}$ mice is also reduced. This may be due either to a reduction in blood flow or a lack of signaling via the particular VEGF isoforms that were deleted in these mice (e.g., $VEGF_{165}$). The finding that expression levels of the SMC recruitment factor PDGF-BB, and its receptor PDGF-receptor β, were also reduced is consistent with a role for this endothelial-derived paracrine factor in recruitment of CoSMCs (Hellstrom et al., 1999; Tallquist et al., 2003).

Although VEGF-B is highly expressed in developing myocardium, mice lacking VEGF-B are viable with no apparent abnormalities in coronary vasculature (Aase et al., 2001). Expression levels for VEGF-C were reduced in VEGF$^{120/120}$ hearts, although it is not known if this decrease contributed to the coronary defects in these mice. However, a recent study by Tomanek et al. (2002) suggests a role for VEGF-A, -B, and -C in coronary vessel development. In addition to assembly factors, newly formed coronary vessels also require survival factors for their maturation and maintenance. In part, survival factors are supplied by pericytes and mural cells that are recruited to form a stable vessel wall. At the same time, arrival of CoSMCs may signal endothelial cells to produce their own survival factors. One such factor may be brain-derived neurotrophic factor (BDNF). BDNF is made by coronary endothelial cells beginning about E17.5, and its receptor (trkB) is expressed a day or so later in both coronary endothelial cells and CoSMCs (Donovan et al., 2000). Mice deficient in BDNF exhibit early postnatal lethality with intramyocardial hemorrhage, reduced cardiac contractility, and endothelial cell apoptosis. Although widely expressed, BDNF deficiency results in vascular hemorrhage that is restricted to coronary vessels (Donovan et al., 2000).

Another case of myocardial hemorrhage during heart development is found in transgenic mice overexpressing angiopoietin-1 (Ang1) under control of the cardiac-specific α-myosin heavy chain promoter (Bellomo et al., 2000; Yu et al., 1996). Ninety percent of these mice die between E12.5 and E16.5 with thin ventricular walls, defective or missing epicardium, and a complete absence of coronary vessels (Ward and Dumont, 2004). Formation of the PE, epicardial colonization of the heart, and epicardial migration over the myocardium appeared normal. Maintenance of the epicardium and generation of a sub-epicardial vascular plexus were defective in Ang1 overexpressing mice. The myocardium normally expresses Ang1 during epicardium formation and coronary development. Tie2, a receptor for Ang1, is expressed by at least some epicardial cells, suggesting that direct paracrine signaling between myocytes and epicardium may be altered in these mice. It remains to be determined whether or not epicardial cell EMT is dependent upon proper levels of Ang1/Tie2 signaling or if the critical steps are more restricted to effects on coronary angioblasts per se.

VI. Epicardial to Mesenchymal Transformation (EMT)

Signals from the myocardium stimulate epicardial cells to undergo EMT and generate a population of epicardial-derived mesenchymal cells (EPDCs) that are progenitors for the coronary vessels (Dettman et al., 1998; Gittenberger-de Groot et al., 1998; Manner, 1999; Perez-Pomares et al., 1997; Viragh and

Challice, 1973). During EMT, epicardial cells acquire mesenchymal characteristics, including loss of cell–cell adhesion, reorganization of the actin cytoskeleton, degradation and remodeling of the ECM, and migration within the developing heart. Epicardial EMT is particularly active in the atrioventricular canal and interventricular sulcus, and it is a critical step in vertebrate heart development. Transplantation of quail PEs into chick embryos allows the movement and fate of epicardial cells to be closely monitored using antibodies to quail-specific nuclear antigens. These studies show that EPDCs are locally derived and that patches of quail epicardial cells produce matching patches of quail mesenchymal cells in the subepicardium of chick-quail chimeric embryos (Manner, 1999). A remarkable finding from these studies is that EPDCs are extremely invasive and migratory within the developing myocardium (Gittenberger-de Groot *et al.*, 1998; Manner, 1993; Manner, 1999; Perez-Pomares *et al.*, 1998). Gittenberger de-Groot *et al.* (2000) reported that EPDCs not only formed the coronary vessels but also gave rise to cells that produce the fibrous ECM skeleton of the heart and participate in valve and septal tissue development. Manner further showed that EPDCs also invade the subendocardial and endocardial layers (Manner, 1999). While it can be anticipated that EPDC migration within the heart depends on chemotactic gradients and reversible cell–matrix interactions, very little is currently known about the critical genes that control the movements and fate of EPDCs within the developing heart. In this regard, down-regulation of α4-integrin expression has been shown to enhance the invasive activity of epicardial cells (Dettman *et al.*, 2003).

Slug is a zinc finger transcription factor of the snail family that is highly expressed in the EPDCs of the subepicardium and atrioventricular canal region (Carmona *et al.*, 2000; Neito *et al.*, 1994). Slug is thought to play an important role in EMT by down regulating the expression of cell–cell adhesion molecules, including E-cadherin (Cano *et al.*, 2000), desmocolin, and desmoglein (Savagner *et al.*, 1997). The expression of slug is required for transformation of endocardium during valve tissue formation (Romano and Runyan, 1999) and for delamination of the cardiac neural crest (Sefton *et al.*, 1998). Other zinc finger transcription factors involved in epicardial cell EMT include Wilms' tumor-1 (WT1) and c-ets-1 (ets1). WT1 is expressed in the proepicardium, epicardial cells, and subepicardial mesenchymal cells, but not in differentiated coronary endothelial or CoSMCs (Carmona *et al.*, 2001; Moore *et al.*, 1999). Mice deficient in WT1 fail to develop tissues that depend on EMT, including kidney, gonads, spleen, and adrenal gland (Kreidberg *et al.*, 1993). WT1 −/− embryos die in midgestation due to heart failure resulting from disruption of the epicardial layer, lack of proliferative expanson of the compact myocardium, and absence of coronary vessel formation. WT1 is reported to be required for formation of retinoic acid-synthesizing EPDCs that migrate throughout the heart to provide important growth-promoting signals to the developing myocardium (Carmona et al., 2001). Macias et al. (1998) immunolocalized ets-1 protein to epicardium and subepicardial mesenchyme where it was coexpressed with the endothelial marker QH1. Consistent with the premise that c-ets-1 can transactivate a number of genes involved in the degradation of extracellular matrices and cell migration, Ets-1 immunostaining was also particularly strong in endocardial cells undergoing EMT during cushion tissue formation. More recently, Lie-Venema et al. (2003) used antisense sequences directed against Ets1 and Ets2 and showed that the process of epicardial cell EMT was inhibited, leading to defects in development of the coronary vascular system.

The critical, cardiomyocyte-derived signals that stimulate epicardial EMT have not yet been identified. One possibility is that one or more members of the BMP/TGF-β superfamily are involved. In very early stages of heart development, precardiac mesoderm is in close contact with endoderm that expresses BMPs 2, 4, and 7. Schultheiss et al. (1997) showed that explanted anterior mesoderm exposed to BMP-2 will differentiate into cardiac muscle even though it is not normally fated to do so in vivo. In later stages of heart development, expression of BMPs 2 and 4 become highly restricted to regions of the heart that exhibit ongoing EMT, including the outflow tract cushions, the atrioventricular canal, and the interventricular sulcus (Yamada et al., 1999). Allen et al. (2001) used a recombinant retrovirus expressing noggin, a secreted BMP antagonist, to inhibit the function of BMPs 2 and 4 in the developing chick heart. They found that BMP signaling is required for migration of neural crest cells into the developing outflow tract (OFT) and for development of the cushion tissue mesenchyme that will form the valves and septal tissues. Lineage tracing was not done in this study, so it is not clear whether exposure to noggin blocked epicardial EMT to an extent similar to that of endocardial EMT. In a study of the effects of various growth factors on epicardial cell EMT, Morabito et al. (2001) reported that recombinant FGFs 1, 2, and 7 stimulated, whereas TGF-βs 1 and 3 inhibited, epicardial cell EMT on collagen gel substrates. Moreover, neutralizing antibodies to FGF-1, -2, or -7 partially blocked epicardial EMT in response to cardiomyocyte conditioned medium. Using carboxyfluorescein to label surface epicardial cells, they developed an organ culture system to monitor EMT in intact HH-stage-28 chick hearts. These studies again found that FGF-2 stimulates, and TGF-βs 1 and 3 inhibit, epicardial cell EMT.

During development of cardiac valves, mesenchymal cells are generated via EMT from both endocardial and epicardial cells. Although one might predict that the same myocardial-derived signals operate to control endocardial as well as epicardial cell EMT, both gain and loss of function experiments suggest that TGF-β3 stimulates, rather than inhibits, endocardial cell EMT in developing chick hearts (Brown et al., 1999; Krishnan et al., 2004; Potts et al., 1991). Moreover, endocardial cell EMT stimulated by

TGF-β is mediated by the activin receptor-like kinase 2 (Alk-2) rather than the cannonical Alk5 receptor (Lai et al., 2000). The idea that different signaling molecules control EMT in the endocardium versus the epicardium may be consistent with genetic evidence that targeted mutations in genes that produce defects in valve and septal tissue formation have no effect on epicardial development, and vice versa (de la Pompa et al., 1998; Tevosian et al., 2000). Given the critical requirement for epicardial cell EMT in generation of coronary progenitor cells, identification of the myocardial factors that signal this essential step in coronary vasculogenesis is an important area of investigation (disussed later).

VII. Epicardial EMT Leads to Changes in Gene Expression

Epicardial cell EMT is characterized by dramatic reorganization of the actin cytoskeleton from subcortical bundles to elongated stress fibers (Landerholm et al., 1999) (Fig. 3). Cytoskeletal rearrangements are linked to changes in epicardial gene expression via nucleocytoplasmic shuttling of signal modifiers and transcriptional coactivators. Reorganization of actin filaments is a process orchestrated by members of the rhoGTPase family (Hall, 1998). Lu et al. (2001) reported that PDGF-BB stimulated proepicardial cells to undergo EMT ex vivo. Inhibition of rhoA-GTPase, or its downstream effector p160 rho kinase (rhoK), blocked epicardial cell EMT stimulated by PDGF-BB. When addition of rhoK inhibitor Y-27632 (RKI) was delayed until after epicardial cells completed EMT, then a large increase in cell death was observed. Likewise, when PEs were incubated with RKI, washed, and then grafted into host chick embryos to make chick-quail chimeras, according to methods described by Manner (1999), the number of EPDCs in the ventricular myocardium was dramatically reduced. Loss of EPDCs in the myocardium correlated with increased rates of apoptosis, suggesting that rhoK-dependent signaling was required for EPDC cell survival in the myocardial wall (Lu et al., 2001).

How is epicardial cell EMT coupled to changes in gene expression? One important pathway is shuttling of factors that relay information from cell surface adhesion molecules and the cortical actin cytoskeleton to the nucleus. A well-known example of this is β-catenin (Gottardi and Gumbiner, 2001). Loss of epicardial cell–cell adhesion leads to translocation of β-catenin from the cell surface to the nucleus and reorganization of cytoskeletal actin. Factors that stimulate CoSMC differentiation activate a rhoA-dependent pathway for actin reorganization and serum response factor (SRF)–dependent gene transcription (Lu et al., 2001). Maintenance of cell–cell contacts is therefore associated with inhibition of rhoA-dependent CoSMC gene expression. In this light, it is of considerable interest that

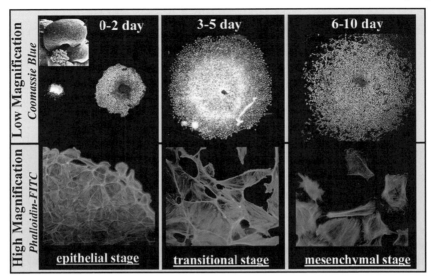

Figure 3 Epithelial to mesenchymal transformation (EMT) of explanted proepicardium. Top row: Coomassie blue stain to show overall cell shapes and colony morphology. Bottom row: Phalloidin-FITC staining to show the organization of the actin cytoskeleton. Inset, upper left: HH17 quail PE (arrow) extending toward the looped heart tube. Note that proepicardial cells initially have a mesothelial phenotype with extensive cell–cell contacts (Top left), and cytoskeletal actin organized as prominent subcortical bundles (Bottom left). Between 3 and 5 days after explant, proepicardial cells undergo EMT (Top center) with loss of cell–cell contacts, and rearrangement of cytoskeletal actin into stress fibers (Bottom center). By 6–10 days after explant, EMT is completed and the vast majority of cells now display a mesenchymal phentoype. (See Color Insert.)

Noren et al. (2003) report that cadherin engagement greatly reduces the levels of active rhoA in epithelial cells via activation of p190rhoGAP, a GTPase activating protein that maintains rhoA in the GDP-bound (inactive) state. Plating epithelial cells that express C-cadherin onto immobilized C-cadherin extracellular domain led to activation of p190rhoGAP (Noren et al., 2003). Moreover, p190rhoGAP activation could be correlated with overall levels of src-mediated tyrosine phosphorylation of p190rhoGAP. Therefore, cadherin-mediated cell–cell adhesion in epicardial cells may activate a src-dependent pathway that maintains p190rhoGAP in an active state, resulting in low rhoA activity, inhibition of stress fiber formation, and a block to CoSMC differentiation.

While loss of cell–cell adhesion is one consequence of EMT, the absence of functional adherens junctions also means that subcortical actin bundles are predisposed to reassemble into stress fiber filaments, leading to the release of actin-binding proteins and their redistribution within the cell. One such example is the cytoskeletal to nuclear translocation of the cysteine-rich LIM domain-containing proteins Crp1 and Crp2 (Chang *et al.*, 2003). In proepicardial cells, Crp2 is present at the cell surface, where it is colocalized with subcortical actin (Chang *et al.*, 2003). When explanted epicardial cells begin to undergo EMT, Crp2 translocates to the nucleus, where it is thought to organize SRF-dependent transcriptional coactivators via protein–protein binding interactions with its N- and C-terminal LIM domains. Once the cells have clearly adopted a mesenchymal phenotype, Crp2 leaves the nucleus and becomes reassociated with the actin cytoskeleton (Bogers *et al.*, 1989). A similar cytoplasmic to nuclear translocation has been reported for the myocardin-related SRF coactivator MRTF-A/Mal/MKL1 (Cen *et al.*, 2003; Miralles *et al.*, 2003; Wang *et al.*, 2002). Of particular interest is the observation that nuclear translocation of MRTF-A/Mal/MKL1 is dependent on activation of a rhoA-GTPase-dependent signaling pathway in NIH3T3 cells (Cen *et al.*, 2003; Miralles *et al.*, 2003).

VIII. Bidirectional Epicardial-Myocardial Signaling

Following attachment of the proepicardial cells to the surface of the heart, the epicardium and compact zone myocardium develop as a paired unit whose growth and differentiation are coordinated by the exchange of soluble signals. The compact zone is the region of the heart that shows the highest rate of cell proliferation during development (Jeter and Cameron, 1971; Pennisi *et al.*, 2003), and it lies adjacent to the epicardial layer. A common endpoint of surgical or genetic manipulations that disrupt formation of the epicardium is embryonic death due to cardiac failure. Morphologic examination of the hearts of these embryos shows lack of proliferative expansion of the compact myocardium, resulting in thin ventricular chambers that are unable to sustain cardiac output required to support embryonic growth and development. Thin ventricular myocardium associated with defects in formation of the epicardium or coronary vasculature has been reported for a number of targeted mutations in mice, including VCAM-1 (Kwee *et al.*, 1995), WT-1 (Moore *et al.*, 1999), Fog2 (Tevosian *et al.*, 2000), VEGF (Giordano *et al.*, 2001), and α4-integrin (Yang *et al.*, 1995).

Using viruses encoding a dominant-negative form of fibroblast growth factor receptor-1 (FGFR-1), Mikawa and colleagues showed that cardiac myocyte proliferation exhibits an early, FGFR1-dependent phase during the first week of chick development, followed by an FGFR1-independent phase that corresponds to the onset of dependence on the epicardium for mitogenic signaling (Mima *et al.*, 1995). Among the list of mouse mutants that exhibit cardiac failure accompanied by thin ventricular myocardium, those deficient in retinoid X receptor-alpha (RXRα) (Sucov et al., 1994), gp130 (Yoshida et al., 1996), and erythropoietin receptor (Wu et al., 1999) have been shown to exert their effects on the compact zone in a non–cell autonomous manner with respect to the myocardium (Chen et al., 1998; Tran and Sucov, 1998; Wu et al., 1999). Given the close spatial relationship of the epicardium to the compact zone myocardium, it is reasonable to consider that these genes may control the production of a cardiac myocyte-specific mitogenic activity by epicardial cells. Indeed, Chen et al. (2002b) identified a mitogenic activity for cardiac myocytes that was secreted by cultured epicardial cells. The functional activity of this epicardial-derived mitogen was not specific for cardiac myocytes, since it also stimulated proliferation of NIH3T3 cells. In addition, while fetal atrial and ventricular myocytes respond to epicardial cell–derived mitogenic activity, adult myocardial cells do not. The mitogenic activity is heat inactivated and ammonium sulfate precipitable, and its production by epicardial cells is inhibited by genestein, a tyrosine kinase inhibitor (Chen et al., 2002b). Likewise, Stuckmann et al. (2003) identified a mitogenic activity released from explanted epicardial cells that stimulates proliferation of embryonic cardiac myocytes. Release of this activity by epicardial cells required retinoid actid and erythropoietin signaling within the epicardium. The identity of the epicardial-derived mitogenic activity for myocardium is not known.

Mice deficient in Fog2 (Friend of GATA-2) also exhibit a thin ventricular myocardium while maintaining an intact epicardial layer (Svensson et al., 2000; Tevosian et al., 2000). Fog2−/− embryos die at midgestation with multiple cardiac defects, including dilated atria, common atrioventricular canal, and tetrology of Fallot, in addition to hypoplastic ventricular myocardium. Despite the fact that the epicardium forms normally and expresses wild-type levels of epicardial marker genes (WT1, capsulin, endoglin, and RALDH2; Tevosian et al., 2000), endothelial markers that normally appear in the subepicardial space, including PECAM-1, ICAM-2, and flk-1, are absent in Fog2−/− embryos. Known GATA factor target genes in the myocardium are all expressed at normal levels. Endocardium, endocardial-derived cushion mesenchyme, and large vessel endothelium were all present and appeared functionally normal in Fog2−/− mice. To test the requirement for Fog2 function in the epicardium, chimeric mice were made consisting of wild-type and Fog2−/− cells. In these chimeric embryos, Fog2−/− cells were frequently found in the subepicardial layer, indicating that Fog2 expression in epicardial cells is not required for formation of the proepicardium, the epicardial layer, or EMT. To test the requirement for Fog2 function in the myocardium, Fog2 expression was restored in cardiac myocytes by a transgene encoding Fog2 driven by the cardiac-specific α-myosin heavy-chain promoter in Fog2-null mice. Development of a normal coronary vasculature in these mice confirmed that Fog2 function in the myocardium is required to signal one or more critical steps for coronary vessel formation (Tevosian *et al.*, 2000). Additional studies using a knock in mutation of GATA4 that disrupts GATA4 binding to Fog2, but has no effect on GATA4 binding to DNA, showed clearly that Fog2 binding to GATA4 in the myocardium is required for normal coronary vessel development (Crispino *et al.*, 2001). The identity of myocardial-derived signals produced under the control of Fog2-GATA4-dependent transcription is not known. While VEGF-A is a logical candidate, expression levels of VEGF-A in the heart are normal in Fog2-deficient mice. Identification of genes whose expression is controlled by Fog2 in developing cardiac myocytes is an important objective to better understand the critical initial steps in coronary vessel development.

IX. Establishment of a Coronary Circulation: Making Connections to the Aorta

The means by which coronary arteries become connected to the systemic circulation is one of the most fascinating and unique aspects of coronary vessel development. In 1989, Bogers *et al.* (1989) reported that coronary arteries made contact with the aortic lumen not by outgrowth from the aorta, as had been generally accepted and taught for many years, but rather by ingrowth of endothelial strands from the peritruncal ring of preexisting capillary-like coronary microvessels. Waldo *et al.* (1990) independently reached the same conclusion in their examination of a larger number of embryos, where they concluded that the coronary orifice forms by a "controlled invasion of the aorta." Moreover, despite the presence of a dense collection of coronary endothelial cells surrounding the aortic root, there were usually never more than two stable connections made with the aortic lumen, and these were almost always found in the opposed aortic valve sinuses facing away from the pulmonary artery (Bogers *et al.*, 1989; Poelmann *et al.*, 1993; Vrancken Peeters *et al.*, 1997a; Waldo *et al.*, 1990; Waldo *et al.*, 1994). Even when penetration of endothelial strands into the noncoronary sinus was found, they were observed to ultimately fail to recruit coronary SMCs, regress, and disappear (Ando *et al.*, 2004; Hood and Rosenquist, 1992; Poelmann *et al.*, 1993; Vrancken Peeters *et al.*, 1997a). Many of these important aspects of coronary orifice formation still remain incompletely understood.

In the mouse, the smooth muscle coating that forms around the main coronary stems contains neural crest–derived SMCs for a short distance from the aorta (Jiang *et al.*, 2000). In the chick, there appears to be little or no contribution of cardiac neural crest that can be detected in chick-quail chimeric embryos (Gittenberger-de Groot et al., 1998; Hood and Rosenquist, 1992). Therefore, the SMCs in the coronary stems of chick embryos appears to be entirely proepicardium derived. The selection and growth of two coronary stems is enhanced by increased blood flow and pulsatile stretch, as well as by exposure to paracrine survival factors. One intriguing source of paracrine survival factors for the main coronary arteries is from parasympathetic ganglia. Waldo et al. (1994) observed that persisting coronary arteries were always associated with neural crest-derived parasympathetic ganglia near their origin. Moreover, the media of surviving coronary stems was frequently disrupted by clusters of parasympathetic nerves (Waldo et al., 1994). Although cardiac neural crest-derived parasympathetic ganglia do not induce capillary penetration directly, the frequent association of parasympathetic nerves with persisting coronary vessels suggests an important growth-promoting or survival effect on coronary vessel development (Bogers et al., 1993). This idea is supported by the observation that cardiac neural crest ablation results in defects in the number and position of persisting coronary stems (Hood and Rosenquist, 1992). Other survival factors are supplied by SMCs that are recruited to invest the coronary vessel walls and form the tunica media, including VEGF, angiopoietin 1, and FGF-1 (Benjamin et al., 1998; Carmeliet, 2000; Carmeliet et al., 1999; Fernandez et al., 2000).

Ingrowth of blind-ended capillary vessels from the peritruncal coronary ring into the aorta is accompanied by apoptosis of cells in the aortic media and eventually the aortic endothelium (Ando et al., 2004; Velkey and Bernanke, 2001). Moreover, high rates of cell proliferation are detected in close proximity to TUNEL-positive cells, suggesting that apoptosis is coupled with SMC proliferation (Velkey and Bernanke, 2001). Localized cell death can explain how coronary orifices are opened within a preformed aortic vessel wall, but the identity of the apoptotic signals involved has not yet been determined. It is important to note that the frequency of apoptotic cells in the aorta and the outflow tract myocardium is greatly reduced after ablation of the PE (Rothenberg et al., 2002). In fact, in their widespread migration throughout the heart, EPDCs appear to act instructively, via the production of secreted factors, to control the pool sizes of different cell populations that they encounter. Thus, in the compact layer, EPDCs are required for proliferative expansion of the myocardial cells that reside there. However, in the aortic root, the outflow tract myocardium, and valve cushion tissue, migrating EPDCs produce signals for apoptosis that direct remodeling of these structures to their more mature forms (Bernanke and Velkey, 2002; Manner, 1999; Rothenberg et al., 2002; Vrancken Peeters et al., 1997a; Wu et al., 1999).

X. Maturation of Coronary Vessels and Formation of a Tunica Media

Investment of the entire vascular system with smooth muscle requires the contribution of precursors from multiple embryonic origins (reviewed in Majesky, 2003). Evidence for a unique epicardial origin for coronary smooth muscle is now very strong. Lineage mapping studies show that when proepicardial cells are labeled with impermeant fluorescent dyes or infected with viral vectors carrying various reporter genes, the label ends up in the smooth muscle layers of coronary arteries (Dettman et al., 1998; Gittenberger-de Groot et al., 1998; Mikawa and Gourdie, 1996; Perez-Pomares et al., 2002; Vrancken Peeters et al., 1999). Similar studies that tag individual PE-derived epicardial cells provide evidence that EPDCs are the proximal progenitors for CoSMCs during heart development (Dettman et al., 1998). What little information exists about the walls of coronary veins suggests that they are made up of a combination of atrial myocytes (Vrancken Peeters et al., 1997b), EPDC-derived SMCs, and perivascular fibroblasts (Perez-Pomares et al., 2002).

The first appearance of smooth-muscle markers in the coronary vasculature appears around the main coronary artery stems at about E16.0 in the rat (Ratajska et al., 2001). Smooth-muscle α-actin (SMαA) was the first marker detected in the coronary wall on E16, followed by SM-myosin heavy chain on E17, the 1E12 antigen (a smooth muscle isoform of α-actinin, Hungerford et al., 1996) on E18, and finally smoothelin (van der Loop et al., 1996) in the early postnatal period (Ratajska et al., 2001). This timing, which starts after connection of the coronary plexus to the aorta has been established, suggests that CoSMC differentiation is dependent on blood flow through the coronary vessels. The markers first become expressed in the proximal coronary artery stems and then gradually progress in expression toward the apex of the heart. In coronary veins, the only marker that could be detected was SMαA. The gradual proximal-to-distal deployment of SMC marker expression is consistent with the investment of coronary artery smooth muscle by downstream migration of preformed CoSMCs from more proximal positions. A similar proximal-to-distal sequence of SMC differentiation was described for chick embryo CoSMCs by Hood and Rosenquist (1992). In the quail embryo, coronary stems make contact with the aortic lumen around E7.5 (Ando et al., 2004). Expression of SMαA was not detectable before that time but becomes evident around E8.0 (Hood and Rosenquist, 1992). This study also reported that SMC marker expression proceeds from the coronary stems in an orderly and continuous downstream sequence that continues until around E14 in the chick (Hood and Rosenquist, 1992).

A variety of studies have shown that vascular SMCs are formed by two pathways. One is by *de novo* recruitment from perivascular mesenchymal progenitors, and the other is by division and migration of pre-existing SMCs along capillary basement membranes. Coronary vessels acquire a smooth-muscle investment by a combination of both pathways (Hellstrom *et al.*, 1999). Existing evidence suggests that coronary stems at their origins from the aorta acquire SMCs from the aortic wall itself (Ando *et al.*, 2004; Jiang *et al.*, 2000). A short distance from the origin, the main coronary arteries appear to recruit SMCs *de novo* from epicardial-derived smooth-muscle progenitors in the local environment. The more distal segments of coronary arteries may become invested with SMCs by a combination of local recruitment and downstream migration of pre-existing SMCs along the endothelial basement membrane (Hellstrom *et al.*, 1999). The latter pathway could account for the long time course for coronary wall formation that extends over many days during the late fetal and early neonatal periods (Hood and Rosenquist, 1992; Ratajska *et al.*, 2001). One factor that is likely to be important for coronary wall formation is platelet-derived growth factor-B (PDGF-B; Hellstrom *et al.*, 1999; Lu *et al.*, 2001; Tallquist *et al.*, 2003). Hellstrom and coworkers studied the effects of gene deletion of PDGF-B and PDGF receptor-β on vascular smooth-muscle development (Hellstrom *et al.*, 1999). They reported that PDGF receptor-β-positive cells were preferentially found around developing vessels whose endothelium expressed higher levels of PDGF-B, rather than around mature, stable arteries. In the absence of PDGF-B/PDGF receptor-β signaling, coronary vessels had many fewer SMCs and pericytes than in wild-type hearts. PDGF-B is a strong chemoattractant for vascular SMCs (Grotendorst *et al.*, 1982). The results described by Hellstrom *et al.* (1999) are consistent with a gradual investment of coronary vessels with SMCs by downstream migration along endothelial basement membranes (Dettman *et al.*). Tallquist *et al.* (2003) used gene-targeting approaches to produce an allelic series of tyrosine to phenylalanine mutations in the intracellular domain of PDGF receptor-β. Their analysis showed a correlation between the extent of CoSMC investment, the amount of PDGF receptor-β expressed, and the number of signal transduction pathways that are activated by the receptor intracellular domain (Tallquist *et al.*, 2003). A critical source of PDGF-B for SMC recruitment is the endothelium (Bjarnegard *et al.*, 2004). In the late fetal and neonatal periods, rapid growth of the heart places increased metabolic demands for perfusion on the coronary vasculature. In the mouse, capillary density increases three- to four-fold during this time to match the overall rate of cardiac myocyte hypertrophy, and the number of coronary vessels that acquire an SMC covering increases at least 10-fold during the first 3 weeks after birth (Carmeliet *et al.*, 1999).

Effects of environmental cues and paracrine factors on stimulation of CoSMC differentiation are mediated by pathways that activate serum response factor (SRF)–dependent transcription (for review, see Miano, 2003; Owens, 1995; Parmacek, 2001). Using dominant-negative forms of SRF that are either incapable of binding to DNA or are missing the transactivation domain, Landerholm et al. (1999) showed that CoSMC differentiation from explanted proepicardial cells required transcriptionally active SRF. Although it is necessary, expression of SRF alone is not sufficient to drive CoSMC differentiation. Additional factors are required, and the expression or activation of these factors (including the SRF coactivators, described later) is closely correlated with EMT. Indeed, a particularly important feature of SMC differentiation in the coronary lineage is the dependence on EMT to produce a permissive state for transcriptional activation of SMC target genes (Chang et al., 2003; Lu et al., 2001).

CoSMC differentiation from isolated proepicardial cells requires rhoA GTPase–mediated cytoskeletal reorganization (Lu et al., 2001). Two types of SRF coactivators that reversibly associate with the actin cytoskeleton and shuttle to the nucleus in a rhoA GTPase–dependent manner have been described (Fig. 4). The myocardin family of coactivators was discovered by Wang et al. (2001) during an in silico screen for genes that are selectively expressed in early cardiac development. Myocardin contains conserved basic, polyglutamine-rich, and SAP domains, and its expression is restricted to cardiac and smooth muscle lineages in mouse embryos (Chen et al., 2002a; Du et al., 2003; Wang et al., 2001; Yoshida et al., 2003). Myocardin is capable of potent transactivation of CArG box-containing target genes in a manner that is strictly dependent on its association with SRF. Mice deficient for myocardin are embryonic lethal with defects in vascular SMC recruitment and/or differentiation (Li et al., 2003). The observation that some SRF-dependent genes are expressed in SMCs that have no detectable myocardin (Chen et al., 2002a) can be explained by the presence of one or more myocardin-related transcription factors (MRTFs). MRTF-A (also called Mal or MKL1) contains two conserved RPEL motifs that are not found in myocardin. These RPEL motifs help mediate rhoA GTPase–dependent shuttling of MRTF-A/MAL/MKL1 from the cytoplasm to the nucleus (Miralles et al., 2003). Dominant-negative MRTF-A/MAL/MKL1 specifically blocked SRF-dependent reporter-gene expression activated by serum or by rhoA-GTPase (Cen et al., 2003). A second group of SRF coactivators that exhibits nucleocytoplasmic shuttling in a rhoA GTPase–dependent manner is the family of cysteine-rich LIM only proteins (CSRPs; for review, see Weiskirchen and Gunther, 2003). Chang et al. (2003) reported that CSRP2 mediates formation of an SRF-GATA factor complex via binding interactions mediated by the LIM domains of CSRP2. Moreover,

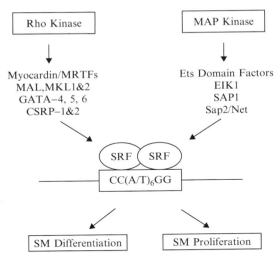

Figure 4 Coronary smooth muscle differentiation requires serum response factor (SRF)–mediated transcription. SRF is an evolutionarily conserved DNA binding protein that serves as a platform for protein–protein interactions. SRF coactivators are targets of upstream signaling pathways that respond to environmental cues for SMC growth (MAP kinase) or differentiation (Rho kinase). Critical SRF coactivators for CoSMC differentiation are the myocardin family of proteins and the cysteine-rich LIM domain containing proteins (CSRP1&2).

dominant-negative CSRP2 inhibited CoSMC differentiation from explanted proepicardial cells (Chang et al., 2003). Using a yeast two-hybrid interaction screen, CSRP2 was found to associate with a novel protein (called CRP2-binding protein) that has structural features of a histone acetyltransferase, a family of enzymes involved in chromatin remodeling (Weiskirchen and Gressner, 2000). Given the need to activate previously silent genes in epicardial cells, one can anticipate that SRF coactivators that respond to rhoA GTPase–mediated signaling, like myocardin, MRTF-A/MKL1/MAL, and CSRP2, will be found to play important roles in CoSMC formation during heart development (Fig. 4).

XI. Development of the Coronary Lymphatic Vessels

Blood and lymphatic vessels develop along parallel but independent pathways to form a complete and functional circulatory system. Lymphatic vessels collect protein-rich interstitial fluids and cells that have leaked out of the arterial circulation and return it to the venous circulation. The lymphatic system consists of an extensive series of capillaries, collecting vessels, and ducts. The larger lymphatic vessels are surrounded by a smooth-muscle layer that contracts automatically when the vessel becomes stretched with fluid. The lymphatic vessels also function as part of the immune system by transporting white blood cells within the lymphoid organs (thymus, spleen, lymph nodes, and tonsils) and bone marrow. Lymphatic drainage of the coronary arteries is via the adventitial lymphatic network and the subepicardial lymphatic plexus (Eliska et al., 1999; Sacchi et al., 1999). Prelymphatic channels are formed by collagen fibrils in the coronary media and function initially to transport interstitial fluid toward the adventitial and periadventitial lymphatic vessels. The smaller arterioles within the myocardium that branch off from penetrating coronary arteries have many more accompanying lymphatic channels than do the main epicardial coronary arteries. Coronary vessel contraction in the radial as well as the longitudinal direction likely plays an important role in propulsion of coronary interstitial fluid toward drainage vessels in the adventitial lymphatics.

One of the principle pathways controlling lymphatic vessel development is via the production of VEGF (for review, see Jussila and Alitalo, 2002). VEGF-C and VEGF-D are ligands for flt4/VEGF-R3, a receptor with intrinsic, ligand-activated tyrosine kinase activity whose expression is mostly limited to lymphatic endothelial cells. Overexpression of VEGF-C in the skin leads to excess production of lymphatic vessels in the dermis via activation of VEGF-R3 (Jeltsch et al., 1997). In the chorioallantoic membrane assay, VEGF-C acts as a highly specific lymphangiogenic factor, but other studies show that it can also stimulate angiogenesis via activation of VEGF-R2 (Cao et al., 1998). By contrast, VEGF-D is highly selective for VEGF-R3 and is a potent lymphangiogenic factor when overexpressed in the skin (Veikkola et al., 2001). Mouse embryos deficient in VEGF-R3 die around E9.5, with fluid accumulation in the pericardial cavity and cardiac failure secondary to defective large vessel formation (Dumont et al., 1998).

Interstitial myocardial edema occurs in a variety of clinical conditions and is usually thought of as producing decreased left ventricular function. In an experimental model of induced myocardial edema in dogs, left ventricular systolic function was not found to be altered, although end-diastolic pressure increased significantly from baseline by three hours after induction of lymphedema (Miyamoto et al., 1998). Thus, interstitial myocardial edema appears to have a direct causal association with diastolic stiffness while having no clear deleterious effect on systolic function in the left ventricle.

XII. Coronary Vessel Anomalies

Coronary anomalies are defined by deviations from what is generally accepted to be normal coronary anatomy. Given the variability in structure of coronary vessels commonly encountered in the population, the term *normal* should be defined as the interval between two standard deviations from the mean value. According to the classification scheme proposed by Angelini et al. (1999), coronary defects can be divided into either anomalies of origin and course or anomalies of intrinsic anatomy. In the former category, absence of a left main coronary trunk is defined as the condition in which both the LAD and circumflex artery originate directly from the center of the left sinus of Valsalva without having a common trunk. A Texas Heart Institute study concluded that absence of a left main coronary trunk occurs in 0.5–1.0% of the population (Angelini, 1995). The category of anomalous origins of coronary arteries includes ectopic origin of a coronary ostium at or near the non-coronary (right posterior) aortic sinus, ectopic origin arising within the ascending aorta, ectopic origin arising from the left ventricle, and ectopic origin from the pulmonary artery. These anomalous origins occur with a frequency of 0.1–1.2% (Angelini et al., 1999). The literature contains rare reports of coronary arteries arising from the aortic arch, innominate artery, internal mammary artery, subclavian artery, as well as descending thoracic aorta (Blake et al., 1964; Click et al., 1990). When only one aortic origin accounts for all of the coronary blood flow, the condition is usually called single coronary artery (Hillestad and Eie, 1971). This category includes a mixture of coronary origins whose common element is the presence of a single aortic ostium. The incidence of single coronary artery in the general population is approximately 0.025% (Shirani and Roberts, 1993).

Under the classification of anomalies of intrinsic coronary anatomy, the literature contains occasional reports of coronary arteries that are stenosed or atretic because of a membrane or fibrotic ridge that forms at or near the aortic orifice (Fortuin and Roberts, 1971; Harada et al., 1993). Coronary ectasia (dilation) can be divided into two types, a localized coronary segment with a disproportionately large diameter (primary ectasia) or a diffuse coronary dilatation in which the entire coronary artery is enlarged secondary to fistulus flow (secondary ectasia; Drexler et al., 1989; Seabra-Gomes et al., 1974). Intramural coronary artery is a condition where normally subepicardial coronaries prematurely penetrate the myocardial wall and interventricular septum. This condition is reported to occur in approximately 1% of the population (Reig et al., 1990). In rare cases, the right coronary artery, LAD, or left circumflex coronary artery pursues a subendocardial course after penetrating the myocardium, an example of a larger category of "coronary malpositions" (Kolodziej et al., 1994). For a more detailed discussion of these and other coronary anomalies and their pathophysiological significance, the reader is referred to several excellent general reviews on the subject (Angelini et al., 2002; Gittenberger-de Groot et al., 1997; Williams, 2000).

XIII. Coronary Development and Disease

Human fetal coronary arteries develop intimal masses at sites where atherosclerotic plaques form later in life (Fig. 4). Initially, these intimal masses are found in the left anterior descending (LAD) coronary artery beginning around 6 months of gestation and are eccentric in appearance (Ikari *et al.*, 1999; Velican and Velican, 1976). By 6 months after birth, the coronary intima has enlarged and become concentric and uniform. At the time of birth, 38% of human coronaries examined had histological evidence of intimal thickening, which rapidly increased to nearly 100% by 3–6 months of age (Ikari *et al.*, 1999). Intimal–medial ratios increase progressively in the left and right main coronary arteries from 6 months of gestation to birth, and in the left main coronary artery from birth to 2 years of age (Ikari *et al.*, 1999; Velican and Velican, 1979). Intimal masses consisted primarily of CoSMCs and associated extracellular matrix, with some studies reporting that macrophages could also be detected in these early intimal masses (Stary, 1985). However, other studies failed to confirm this finding (Ikari *et al.*, 1999). Human coronary vessels develop a complex architecture at branch points and bifurcations, including the presence of large pads or intimal cushions consisting of a very thick intima containing longitudinal muscle bundles (discussed later), some of which exhibit a spiral course at the level of branch origins and form muscle rings around the entrance orifices (Velican and Velican, 1979).

Shortly after birth, human coronary arteries develop a second layer of the media in which CoSMCs are organized into longitudinal muscle bundles in the subintimal space, rather than being arranged in the circumferential pattern seen in the outer media (Neufeld *et al.*, 1962; Velican and Velican, 1979). These longitudinal bundles comprise what is referred to as the musculoelastic layer (Ikari *et al.*, 1999) and are not present in similar-sized muscular arteries that develop in other organs such as renal, hepatic, bronchial, internal mammary, or basilar arteries. The striking and extensive development of this unique longitudinal muscle layer suggests that coronary arteries are subjected to greater mechanical forces in the longitudinal direction than major arteries in other organs. This possibility is consistent with the repeated deformation of main coronary arteries as a result of the cardiac cycle. Formation of this longitudinal muscle layer may also allow coronary arteries to contract lengthwise and thereby assist in propelling blood through the coronary circuit. Therefore, the early appearance of coronary intimal masses and the formation of a pronounced longitudinal smooth-muscle layer with the vessel wall are unique aspects of normal coronary development that may predispose these vessels to develop life-threatening coronary atherosclerosis and vasospasm later in life.

Acknowledgments

I thank the following individuals for their helpful input and constructive comments during the preparation of this review: Robert J. Schwartz, Robert Tomanek, Joey Barnett, John Schwarz, Joseph Miano, Victoria Bautch, Cam Patterson, Da-Zhi Wang, and Karen Hirschi. I would also like to thank the past and current members of my laboratory, including Xiu Rong Dong, Tom Landerholm, Jun Lu, and San-Pin Wu for their enthusiasm and constructive criticism of different aspects of this chapter. Funding was provided by the American Heart Association and the National Institutes & Health. The author is an established investigator of the American Heart Association.

References

Aase, K., von Euler, G., Li, X., Ponten, A., Thoren, P., Cao, R., Cao, Y., Olofsson, B., Gebre-Medhin, S., Pekny, M., Alitalo, K., Betsholtz, C., and Eriksson, U. (2001). Vascular endothelial growth factor-B-deficient mice display an atrial conduction defect. *Circulation* **104**, 358–364.

Allen, S., Bogardi, J., Barlow, A., Mir, S., Qayyum, S., Verbeek, F., Anderson, R., Francis-West, P., Brown, N., and Richardson, M. (2001). Misexpression of noggin leads to septal defects in the outflow tract of the chick heart. *Dev. Biol.* **235**, 98–109.

Ando, K., Nakajima, Y., Yamagishi, T., Yamamoto, S., and Nakamura, H. (2004). Development of proximal coronary arteries in quail embryonic heart. Multiple capillaries penetrating the aortic sinus fuse to form main coronary trunk. *Circ. Res.* **94**, 346–352.

Andries, L., Brutsaert, D., and Sys, S. (1998). Nonuniformity of endothelial constitutive nitric oxide synthase distribution in cardiac endothelium. *Circ. Res.* **82**, 195–203.

Angelini, P. (1995). Embryology and congenital heart disease. *Tex. Heart Inst. J.* **22**, 1–12.

Angelini, P., Velasco, J., and Flamm, S. (2002). Coronary anomalies: Incidence, pathophysiology and clinical relevance. *Circulation* **105**, 2449–2454.

Angelini, P., Villason, S., Chan, A., and Diez, J. (1999). Normal and anomalous coronary arteries in humans. *In* "Coronary Artery Anomalies" (P. Angelini, Ed.), pp. 27–79. Lippincott, Williams & Wilkins, Philadelphia.

Begemann, G., Gibert, Y., Meyer, A., and Ingham, P. (2002). Cloning of zebrafish T-box genes tbx15 and tbx18 and their expression during embryonic development. *Mech. Dev.* **114**, 137–141.

Bellomo, D., Headrick, J., Silins, G., Paterson, C., Thomas, P., Gartside, M., Mould, A., Cahill, M., Tonks, I., Grimmond, S., Twonson, S., Wells, C., Little, M., Cummings, M., Hayward, N., and Kay, G. (2000). Mice lacking the vascular endothelial growth factor-B gene (vegfb) have smaller hearts, dysfunctional coronary vasculature and impaired recovery from cardiac ischemia. *Circ. Res.* **86**, E29–E35.

Benjamin, L., Hemo, I., and Keshet, E. (1998). A plasticity window for blood vessel remodeling is defined by pericyte coverage of the preformed endothelial network and is regulated by PDGF-B and VEGF. *Development* **125**, 1591–1598.

Bernanke, D., and Velkey, J. (2002). Development of the coronary blood supply: Changing concepts and current ideas. *Anat. Rec. (New Anat.)* **269**, 198–208.

Bjarnegard, M., Enge, M., Norlin, J., Gustafsdottir, S., Fredriksson, S., Abramsson, A., Takemoto, M., Gustafsson, E., Fassler, R., and Betsholtz, C. (2004). Endothelium-specific ablation of PDGFB leads to pericyte loss and glomerular, cardiac and placental abnormalities. *Development* **131**, 1847–1857.

8. Coronary Vessel Development

Blake, H., Manion, W., and Mattingly, T. (1964). Coronary artery anomalies. *Circulation* **30**, 927-936.

Bogers, A., Bartelings, M., Bokenkamp, R., Stijnen, T., van Suylen, R., Poelmann, R., and Gittenberger-de Groot, A. (1993). Common arterial trunk, uncommon coronary arterial anatomy. *J. Thorac. Cardiovasc. Surg.* **106**, 1133-1137.

Bogers, A., Gittenberger-de Groot, A., Poelmann, R., Peault, B., and Huysmans, H. (1989). Development of the origin of the coronary arteries, a matter of ingrowth or outgrowth? *Anat. Embryol.* **180**, 437-441.

Bouchey, D., Drake, C., Wunsch, A., and Little, C. (1996). Distribution of connective tissue proteins during development and neovascularization of the epicardium. *Cardiovasc. Res.* **31**, E104-E115.

Brown, C., Boyer, A., Runyan, R., and Barnett, J. (1999). Requirement of type III TGF-beta receptor for endocardial cell transformation in the heart. *Science* **283**, 2080-2082.

Burch, G., Bedolli, M., McDonough, S., Rosenthal, S., and Bristow, J. (1995). Embryonic expression of tenascin-X suggests a role in limb, muscle and heart development. *Dev. Dynam.* **203**, 491-504.

Cano, A., Perez-Moreno, M., Rodrigo, I., Locascio, A., Blanco, M., del Barrio, M., Portillo, F., and Nieto, M. (2000). The transcription factor snail controls epithelial-mesenchymal transitions by repressing E-cadherin expression. *Nat. Cell. Biol.* **2**, 76-83.

Cao, Y., Linden, P., Farnebo, J., Cao, R., Eriksson, A., Kumar, V., Qi, J., Claesson-Welsh, L., and Alitalo, K. (1998). Vascular endothelial growth factor C induces angiogenesis *in vivo*. *Proc. Natl. Acad. Sci. USA* **95**, 14389-14394.

Carmeliet, P. (2000). Mechanisms of angiogenesis and arteriogenesis. *Nat. Med.* **6**, 389-395.

Carmeliet, P., Ng, Y.-S., Nuyens, D., Theilmeier, G., Brusselmans, K., Cornelissen, I., Ehler, E., Kakkar, V., Stalmans, I., Mattot, V., Perriard, J.-C., Dewerchin, M., Flameng, W., Nagy, A., Lupu, F., Moons, L., Collen, D., D'Amore, P., and Shima, D. (1999). Impaired myocardial angiogenesis and ischemic cardiomyopathy in mice lacking the vascular endothelial growth factor isoforms VEGF164 and VEGF188. *Nat. Med.* **5**, 495-502.

Carmona, R., Gonzalez-Iriarte, M., Macias, D., Perez-Pomares, J., Garcia-Garrido, L., and Munoz-Chapuli, R. (2000). Immunolocalization of the transcription factor Slug in the developing avian heart. *Anat. Embryol.* **201**, 103-109.

Carmona, R., Gonzalez-Iriarte, M., Perez-Pomares, J., and Munoz-Chapuli, R. (2001). Localization of the Wilm's tumour protein WT1 in avian embryos. *Cell Tissue Res.* **303**, 173-186.

Cen, B., Selvaraj, A., Burgess, R., Hitzler, J., Ma, Z., Morris, S., and Prywes, R. (2003). Megakaryoblastic leukemia 1, a potent transcriptional coactivator for serum response factor (SRF), is required for serum induction of SRF target genes. *Mol. Cell Biol.* **23**, 6597-6608.

Chan, B., Elices, M., Murphy, E., and Hemler, M. (1992). Adhesion to vascular cell adhesion molecule 1 and fibronectin. Comparison of alpha 4 beta 1 (VLA-4) and alpha 4 beta 7 on the human B cell line JY. *J. Biol. Chem.* **267**, 8366-8370.

Chang, D., Belaguli, N., Iyer, D., Roberts, W., Wu, S., Dong, X., Marx, J., Moore, M., Beckerle, M., Majesky, M., and Schwartz, R. (2003). Cysteine-rich LIM-only proteins CRP1 and CRP2 are potent smooth muscle differentiation cofactors. *Dev. Cell* **4**, 107-118.

Chen, J., Kitchen, C., Streb, J., and Miano, J. (2002a). Myocardin: A component of a molecular switch for smooth muscle differentiation. *J. Mol. Cell Cardiol.* **34**, 1345-1356.

Chen, J., Kubulak, S., and Chien, K. (1998). Ventricular muscle-restricted targeting of the RXRα gene reveals a non-cell-autonomous requirement in cardiac chamber morphogenesis. *Development* **125**, 1943-1949.

Chen, T., Chang, T., Kang, J., Choudhary, B., Makita, T., Tran, C., Burch, J., Eid, H., and Sucov, H. (2002b). Epicardial induction of fetal cardiomyocyte proliferation via a retinoic acid-inducible trophic factor. *Dev. Biol.* **250**, 198-207.

Click, R., Holmes, D., Vlietstra, R., Kosinski, A., and Kronmal, R. (1990). Anomalous coronary arteries: Location, degree of atherosclerosis and effect on survival—a report from the Coronary Artery Surgery Study. *J. Am. Coll. Cardiol.* **15**, 507–508.

Cox, C., and Poole, T. (2000). Angioblast differentiation is influenced by the local environment: FGF-2 induces angioblasts and patterns vessel formation in the quail embryo. *Dev. Dyn.* **218**, 371–382.

Crispino, J., Lodish, M., Thurberg, B., Litovsky, S., Collins, T., Molkentin, J., and Orkin, S. (2001). Proper coronary vascular development and heart morphogenesis depend on interaction of GATA4 with FOG cofactors. *Genes Dev.* **15**, 839–844.

Cybulsky, M., and Gimbrone, M. J. (1991). Endothelial expression of a mononuclear leukocyte adhesion molecule during atherogenesis. *Science* **251**, 788–791.

De la Pompa, J., Timmerman, L., Takimoto, H., Yoshida, H., Elia, A., Samper, E., Potter, J., Wakeham, A., L, M., Langille, B., Crabtree, G., and Mak, T. (1998). Role of the NF-ATc transcription factor in morphogenesis of cardiac valves and septum. *Nature* **392**, 182–186.

Dettman, R., Denetclaw, W., Ordahl, C., and Bristow, J. (1998). Common epicardial origin of coronary vascular smooth muscle, perivascular fibroblasts, and intermyocardial fibroblasts in the avian heart. *Dev. Biol.* **193**, 169–181.

Dettman, R., Paea, H., Morabito, C., and Bristow, J. (2003). Inhibition of α4-integrin stimulates epicardial–mesenchymal transformation and alters migration and cell fate of epicardially derived mesenchyme. *Dev. Biol.* **257**, 315–328.

Donovan, M., Lin, M., Weign, P., Ringstedt, T., Kraemer, R., Hahn, R., Wang, S., Ibanez, C., Rafii, S., and Hempstead, B. (2000). Brain-derived neurotrophic factor is an endothelial cell survival factor required for intramyocardial vessel stabalization. *Development* **127**, 4531–4540.

Drexler, H., Zeiher, A., Wollschlager, H., Meinertz, T., Just, H., and Bonzel, T. (1989). Flow-dependent coronary dilatation in humans. *Circulation* **80**, 466–474.

Du, K., Ip, H., Li, J., Chen, M., Dandre, F., Yu, W., Lu, M., Owens, G., and Parmacek, M. (2003). Myocardin is a critical serum response factor cofactor in the transcriptional program regulating smooth muscle cell differentiation. *Mol. Cell Biol.* **23**, 2425–2437.

Dumont, D., Jussila, L., Taipale, J., Lymboussaki, A., Mustonen, T., Pajusola, K., Breitman, M., and Alitalo, K. (1998). Cardiovascular failure in mouse embryos deficient in VEGF receptor-3. *Science* **282**, 946–949.

Eliska, O., Eliskova, M., and Miller, A. (1999). The morphology of the lymphatics of the coronary arteries in the dog. *Lymphology* **32**, 45–57.

Fernandez, B., Buehler, A., Wolfram, S., Kostin, S., Espanion, G., Franz, W., Niemann, H., Doevendans, P., Schaper, W., and Zimmermann, R. (2000). Transgenic myocardial overexpression of fibroblast growth factor-1 increases coronary artery density and branching. *Circ. Res.* **87**, 207–213.

Ferrara, N., Gerber, H., and LeCouter, J. (2003). The biology of VEGF and its receptors. *Nat. Med.* **9**, 669–676.

Fortuin, N., and Roberts, W. (1971). Congenital atresia of the left main coronary artery. *Am. J. Med.* **50**, 385–389.

Fransen, M., and Lemanski, L. (1990). Epicardial development in the axolotl, Ambystoma mexicanum. *Anat. Rec.* **226**, 228–236.

Giordano, F., Gerber, H., Williams, S., VanBruggen, N., Bunting, S., Ruiz-Lozano, P., Gu, Y., Nath, A., Huang, Y., Hickey, R., Dalton, N., Peterson, K., Ross, J. J., Chien, K., and Ferrara, N. (2001). A cardiac myocyte vascular endothelial growth factor paracrine pathway is required to maintain cardiac function. *Proc. Natl. Acad. Sci. USA* **98**, 5780–5785.

Gittenberger-de Groot, A., Polmann, R., and Bartelings, M. (1997). Embryology of congenital heart disease. *In* "Atlas of Heart Diseases: Congenital Heart Disease" (E. Braunwald, Ed.), pp. 3.1–3.10. Current Medicine, Philadelphia.

Gittenberger-de Groot, A., Vrancken-Peeters, M., Bergwerff, M., Mentink, M., and Poelmann, R. (2000). Epicardial outgrowth inhibition leads to compensatory mesothelial outflow tract collar and abnormal cardiac septation and coronary formation. *Circ. Res.* **87,** 969–971.

Gittenberger-de Groot, A., Vrancken Peeters, M., Mentink, M., Gourdie, R., and Poelmann, R. (1998). Epicardium-derived cells contribute a novel population to the myocardial wall and the atrioventricular cushions. *Circ. Res.* **82,** 1043–1052.

Gottardi, C., and Gumbiner, B. (2001). Adhesion signaling: How beta-catenin interacts with its partners. *Curr. Biol.* **11,** R792–794.

Grant, R. (1926). Development of the cardiac coronary vessels in the rabbit. *Heart* **13,** 261–271.

Grotendorst, G., Chang, T., Seppa, H., Kleinman, H., and Martin, G. (1982). Platelet-derived growth factor is a chemoattractant for vascular smooth muscle cells. *J. Cell Physiol.* **113,** 261–266.

Gurtner, G., Davis, V., Li, H., McCoy, M., Sharpe, A., and Cybulsky, M. (1995). Targeted disruption of the murine VCAM1 gene: Essential role of VCAM-1 in chorioallantoic fusion and placentation. *Genes Dev.* **9,** 1–14.

Hall, A. (1998). Rho GTPases and the cytoskeleton. *Science* **279,** 509–514.

Harada, K., Ito, K., and Suzuki, Y. (1993). Congenital atresia of the left coronary ostium. *Eur. J. Pediatr.* **152,** 539–540.

Hellstrom, M., Kalen, M., Lindahl, P., Abramsson, A., and Betsholtz, C. (1999). Role of PDGF-B and PDGFR-β in recruitment of vascular smooth muscle cells and pericytes during embryonic blood vessel formation in the mouse. *Development* **126,** 3047–3055.

Hillestad, L., and Eie, H. (1971). Single coronary artery. *Acta Med. Scand.* **189,** 409–413.

Hirakow, R. (1985). The vertebrate heart in phylogenetic relation to the prechordates. In "Vertebrate morphology" (H. Duncker and G. Fleischer, Eds.), pp. 367–369. Fischer, Stuttgart, Germany.

Hiruma, T., and Hirakow, R. (1989). Epicardial formation in embryonic chick heart: Computer-aided reconstruction, scanning and transmission electron microscopic studies. *Am. J. Anat.* **184,** 129–138.

Ho, E., and Shimada, Y. (1978). Formation of the epicardium studied with the scanning electron microscope. *Dev. Biol.* **66,** 579–585.

Hood, L., and Rosenquist, T. (1992). Coronary artery development in the chick: Origin and deployment of smooth muscle cells, and the effects of neural crest ablation. *Anat. Rec.* **234,** 291–300.

Hungerford, J., Owens, G., Aargraves, W., and Little, C. (1996). Development of the aortic vessel wall as defined by vascular smooth muscle and extracellular markers. *Dev. Biol.* **178,** 375–392.

Ikari, Y., McManus, B., Kenyon, J., and Schwartz, S. (1999). Neonatal intima formation in the human coronary artery. *Arterioscler. Thromb. Vasc. Biol.* **19,** 2036–2040.

Imanaka-Yoshida, K., Matsumoto, K., Hara, M., Sakakura, T., and Yoshida, T. (2003). The dynamic expression of tenascin-C and tenascin-X during early heart development in the mouse. *Differentiation* **71,** 291–298.

Jeltsch, M., Kaipainen, A., Joukov, V., Meng, X., Lakso, M., Rauvala, H., Swartz, M., Fukumura, D., Jain, R., and Alitalo, K. (1997). Hyperplasia of lymphatic vessels in VEGF-C transgenic mice. *Science* **276,** 1423–1425.

Jeter, J., and Cameron, I. (1971). Cell proliferation patterns during cytodifferentiation in embryonic chick tissues: Liver, heart, and erythrocytes. *J. Embryol. Exp. Morphol.* **25,** 405–422.

Jiang, X., Rowitch, D. H., Soriano, P., McMahon, A. P., and Sucov, H. M. (2000). Fate of the mammalian cardiac neural crest. *Development* **127,** 1607–1616.

Jussila, L., and Alitalo, K. (2002). Vascular growth factors and lymphangiogenesis. *Physiol. Rev.* **82,** 673–700.

Kalman, F., Viragh, S., and Modis, L. (1995). Cell surface glycoconjugates and the extracellular matrix of the developing mouse embryo epicardium. *Anat. Embryol. (Berl.)* **191**, 451–464.

Kattan, J., Dettman, R., and Bristow, J. (2004). Formation and remodeling of the coronary vascular bed in the embryonic avian heart. *Dev. Dyn.* **230**, 34–43.

Kim, H., Yoon, C., Kim, H., and Rah, B. (1999). Expression of extracellular matrix components fibronectin and laminin in the human fetal heart. *Cell Struct. Funct.* **24**, 19–26.

Kolodziej, A., Lobo, F., and Walley, V. (1994). Intra-arterial course of the right coronary artery and its branches. *Can. J. Cardiol.* **10**, 263–267.

Komiyama, M., Ito, K., and Shimada, Y. (1987). Origin and development of the epicardium in the mouse embryo. *Anat. Embryol.* **176**, 183–189.

Kraus, F., Haenig, B., and Kispert, A. (2001). Cloning and expression analysis of the mouse T-box gene *Tbx18*. *Mech. Dev.* **100**, 83–86.

Kreidberg, J., Sariola, H., Loring, J., Maeda, M., Pelletier, J., Housman, D., and Jaenisch, R. (1993). WT-1 is required for early kidney development. *Cell* **74**, 679–691.

Krishnan, S., Deora, A., Annes, J., Osoria, J., Rifkin, D., and Hajjar, K. (2004). Annexin II-mediated plasmin generation activates TGF-β3 during epithelial-mesenchymal transformation in the developing avian heart. *Dev. Biol.* **265**, 140–154.

Kuhn, H., and Liebherr, G. (1988). The early development of the epicardium in Tupaia belangeri. *Anat. Embryol. (Berl.)* **177**, 225–234.

Kurkiweicz, T. (1909). O histogenezie miesnia sercowego zwierzat kregowych. *Bull. l'Acad. Sci. Cracovie.* 148–191.

Kwee, L., Baldwin, H., Shen, H., Stewart, C., Buck, C., Buck, C., and Labow, M. (1995). Defective development of the embryonic and extraembryonic circulatory systems in vascular cell adhesion molecule (VCAM-1) deficient mice. *Development* **121**, 489–503.

Lai, Y., Beason, K., Brames, G., Desgrosellier, J., Cleggett, M., Shaw, M., Brown, C., and Barnett, J. (2000). Activin receptor-like kinase 2 can mediate atrioventricular cushion transformation. *Dev. Biol.* **222**, 1–11.

Landerholm, T., Dong, X.-R., Lu, J., Belaguli, N., Schwartz, R., and Majesky, M. (1999). A role for serum response factor in coronary smooth muscle differentiation from proepicardial cells. *Development* **126**, 2053–2062.

Li, S., Wang, D., Wang, Z., Richardson, J., and Olson, E. (2003). The serum response factor coactivator myocardin is required for vascular smooth muscle development. *Proc. Natl. Acad. Sci. USA* **100**, 9366–9370.

Li, W., Waldo, K., Kinask, K., Chen, T., Wessels, A., Parmacek, M., Kirby, M., and Lo, C. (2002). An essential role for connexin43 gap junctions in mouse coronary artery development. *Development* **129**, 2031–2042.

Lie-Venema, H., Gittenberger-de Groot, A., van Empel, L., Boot, M., Kerkdijk, H., de Kant, E., and DeRuiter, M. (2003). Ets-1 and Ets-2 transcription factors are essential for normal coronary and myocardial development in chicken embryos. *Circ. Res.* **92**, 749–756.

Lu, J., Landerholm, T., Wei, J., Dong, X., Wu, S., Liu, X., Nagata, K., Inagaki, M., and Majesky, M. (2001). Coronary smooth muscle differentiation from proepicardial cells requires rhoA-mediated actin reorganization and p160 rho-kinase activity. *Dev. Biol.* **240**, 404–418.

Macias, D., Perez-Pomares, J., Garcia-Garrido, L., Carmona, R., and Munoz-Chapuli, R. (1998). Immunoreactivity of the ets-1 transcription factor correlates with areas of epithelial-mesenchymal transition in the developing heart. *Anat. Embryol.* **198**, 307–315.

Majesky, M. (2003). Vascular smooth muscle diversity: Insights from developmental biology. *Curr. Athero. Repts.* **5**, 208–213.

Manasek, F. (1969). Embryonic development of the heart II. Formation of the epicardium. *J. Embryol. Exp. Morph.* **22**, 333–348.

Manner, J. (1992). The development of pericardial villi in the chick embryo. *Anat. Embryol.* **186**, 379–385.

8. Coronary Vessel Development

Manner, J. (1993). Experimental study on the formation of the epicardium in chick embryos. *Anat. Embryol.* **187,** 281–289.

Manner, J. (1998). The origin and course of coronary vessels: Embryological considerations. *Cardiol. Young* **8,** 534–535.

Manner, J. (1999). Does the subepicardial mesenchyme contribute myocardioblasts to the myocardium of the chick embryo? A quail-chick chimera study tracing the fate of the epicardial primordium. *Anat. Rec.* **255,** 212–226.

Manner, J., Perez-Pomares, J., Macias, D., and Munoz-Chapuli, R. (2001). The origin, formation and developmental significance of the epicardium: A review. *Cells Tissues Organs* **169,** 89–103.

Miano, J. (2003). Serum response factor: Toggling between disparate programs of gene expression. *J. Mol. Cell Cardiol.* **35,** 577–593.

Mikawa, T., and Fishman, D. (1992). Retroviral analysis of cardiac morphogenesis: Discontinuous formation of coronary vessels. *Proc. Natl. Acad. Sci. USA* **89,** 9504–9508.

Mikawa, T., and Gourdie, R. (1996). Pericardial mesoderm generates a population of coronary smooth muscle cells migrating into the heart along with ingrowth of the epicardial organ. *Dev. Biol.* **174,** 221–232.

Mima, T., Ueno, H., Fischman, D., Williams, L., and Mikawa, T. (1995). Fibroblast growth factor receptor is required for *in vivo* cardiac myocyte proliferation at early embryonic stages of heart development. *Proc. Natl. Acad. Sci. USA* **92,** 467–471.

Miralles, F., Posern, G., Zaromytidou, A., and Treisman, R. (2003). Actin dynamics control SRF activity by regulation of its coactivator MAL. *Cell* **113,** 329–342.

Miyamoto, M., McClure, D., Schertel, E., Andrews, P., Jones, G., Pratt, J., Ross, P., and Myerowitz, P. (1998). Effects of hypoproteinemia-induced myocardial edema on left ventricular function. *Am. J. Physiol.* **274,** H937–H944.

Moore, A., McInnes, L., Kreidberg, J., Hastie, N., and Schedl, A. (1999). YAC complementation shows a requirement for WT1 in the development of epicardium, adrenal gland, and throughout nephrogenesis. *Development* **126,** 1845–1857.

Morabito, C., Dettman, R., Kattan, J., Collier, J., and Bristow, J. (2001). Positive and negative regulation of epicardial-mesenchymal transformation during avian heart development. *Dev. Biol.* **234,** 204–215.

Morabito, C., Kattan, J., and Bristow, J. (2002). Mechanisms of embryonic coronary artery development. *Curr. Opin. Cardiol.* **17,** 235–241.

Moss, J., Xavier-Neto, J., Shapiro, M., Nayeem, S., McCaffery, P., Drager, U., and Rosenthal, N. (1998). Dynamic patterns of retinoic acid synthesis and response in the developing mammalian heart. *Dev. Biol.* **199,** 55–71.

Munoz-Chapuli, R., Gonzalez-Iriate, M., Carmona, R., Atencia, D., Macias, D., and Perez-Pomares, J. (2002). Cellular precursors of the coronary arteries. *Tex. Heart Inst. J.* **29,** 243–249.

Munoz-Chapuli, R., Macias, D., Ramos, C., Gallego, A., and Andres, V. (1996). Development of the subepicardial mesenchyme and the early cardiac vessels in the dogfish *(Scyliorhinus canicula)*. *J. Exp. Zool.* **275,** 95–111.

Nahirney, P., Mikawa, T., and Fischman, D. (2003). Evidence for an extracellular matrix bridge guiding proepicardial cell migration to the myocardium of chick embryos. *Dev. Dyn.* **227,** 511–523.

Neito, M., Sargent, M., Wilkinson, D., and Cooke, J. (1994). Control of cell behaviour by slug, a zinc finger gene. *Science* **264,** 835–839.

Nelson, T., Duncan, S., and Misra, R. (2004). Conserved enhancer in the serum response factor promoter controls expression during early coronary vasculogenesis. *Circ. Res.* **94,** 1059–1066.

Neufeld, H., Wagnevoort, C., and Edwards, J. (1962). Coronary arteries in fetuses, infants, juveniles and young adults. *Lab. Invest.* **11,** 837–844.

Noren, N., Arthur, W., and Burridge, K. (2003). Cadherin engagement inhibits RhoA via p190RhoGAP. *J. Biol. Chem.* **278**, 13615–13618.

Owens, G. (1995). Regulation of differentiation of vascular smooth muscle cells. *Physiol. Rev.* **75**, 487–517.

Parmacek, M. (2001). Transcriptional programs regulating vascular smooth muscle cell development and differentiation. *Curr. Top. Dev. Biol.* **51**, 69–89.

Pennisi, D., Ballard, V., and Mikawa, T. (2003). Epicardium is required for the full rate of myocyte proliferation and levels of expression of myocyte mitogenic factors FGF2 and its receptor, FGFR-1, but not for transmural myocardial patterning in the embryonic chick heart. *Dev. Dyn.* **228**, 161–172.

Pereira, F., Qiu, Y., Zhou, G., Tsai, M.-J., and Tsai, S. (1999). The nuclear orphan receptor COUP-TFII is required for angiogenesis and heart development. *Genes Dev.* **13**, 1037–1049.

Perez-Pomares, J., Carmona, R., Gonzalez-Iriarte, M., Atencia, G., Wessels, A., and Munoz-Chapuli, R. (2002). Origin of coronary endothelial cells from epicardial mesothelium in avian embryos. *Int. J. Dev. Biol.* **46**, 1005–1013.

Perez-Pomares, J., Macias, D., Garcia-Garrido, M., and Munoz-Chapuli, R. (1997). Contribution of the primitive epicardium to the subepicardial mesenchyme in hamster and chick embryos. *Dev. Dyn.* **210**, 96–105.

Perez-Pomares, J., Macias, D., Garcia-Garrido, L., and Munoz-Chapuli, R. (1998). The origin of the subepicardial mesenchyme in the avian embryo: An immunohistochemical and quail-chick chimera study. *Dev. Biol.* **200**, 57–68.

Perez-Pomares, J., Phelps, A., Sedmerova, M., and Wessels, A. (2003). Epicardial-like cells on the distal arterial end of the cardiac outflow tract do not derive from the proepicardium but are derivatives of the cephalic pericardium. *Dev. Dyn.* **227**, 56–68.

Pinco, K., Liu, S., and Yang, J. (2001). α4 integrin is expressed in a subset of cranial neural crest cells and in epicardial progenitor cells during early mouse development. *Mech. Dev.* **100**, 99–103.

Poelmann, R., Gittenberger-de Groot, A., Mentink, M., Bokenkamp, R., and Hogers, B. (1993). Development of the cardiac coronary vascular endothelium, studied with antiendothelial antibodies, in chicken-quail chimeras. *Circ. Res.* **73**, 559–568.

Potts, J., Dagle, J., Walder, J., Weeks, D., and Runyan, R. (1991). Epithelial-mesenchymal transformation of embryonic cardiac endothelial cells is inhibited by a modified antisense oligonucleotide to transforming growth factor β3. *Proc. Natl. Acad. Sci. USA* **88**, 1516–1520.

Ratajska, A., Zarska, M., Quensel, C., and Krämer, J. (2001). Differentiation of the smooth muscle cell phenotypes during embryonic development of coronary vessels in the rat. *Histochem. Cell Biol.* **116**, 79–87.

Reaume, A., de Sousa, P., Kulkarni, S., Langille, B., Zhu, D., Davies, T., Juneja, S., Kidder, G., and Rossant, J. (1995). Cardiac malformation in neonatal mice lacking connexin43. *Science* **267**, 1831–1834.

Reese, D., and Bader, D. (1999). Cloning and expression of hbves, a novel and highly conserved mRNA expressed in the developing and adult heart and skeletal muscle in the human. *Mamm. Genome* **10**, 913–915.

Reig, J., Ruiz de Miguel, C., and Moragas, A. (1990). Morphometric analysis of myocardial bridges in children with ventricular hypertrophy. *Pediatr. Cardiol.* **11**, 186–190.

Robinson, C., and Stringer, S. (2001). The splice variants of vascular endothelial growth factor (VEGF) and their receptors. *J. Cell Sci.* **114**, 853–865.

Romano, L., and Runyan, R. (1999). Slug is a mediator of epithelial-mesenchymal cell transformation in the developing chicken heart. *Dev. Biol.* **212**, 243–254.

8. Coronary Vessel Development

Rongish, B., Hinchman, G., Doty, M., Baldwin, H., and Tomanek, R. (1996). Relationship of the extracellular matrix to coronary neovascularization during development. *J. Mol. Cell. Cardiol.* **28**, 2203–2215.

Rothenberg, F., Hitomi, M., Fisher, S., and Watanabe, M. (2002). Initiation of apoptosis in the developing avian outflow tract myocardium. *Dev. Dyn.* **223**, 469–482.

Sacchi, G., Weber, E., Agliano, M., Cavina, N., and Comparini, L. (1999). Lymphatic vessels of the human heart: Precollectors and collecting vessels. A morpho-structural study. *J. Submicrosc. Cytol. Pathol.* **31**, 515–525.

Savagner, P., Yamada, K., and Thiery, J. (1997). The zinc finger protein slug causes desmosome dissociation, an initial and necessary step for growth factor-induced epithelial-mesenchymal transition. *J. Cell Biol.* **137**, 1403–1419.

Schultheiss, T., Burch, J., and Lassar, A. (1997). A role for bone morphogenetic proteins in the induction of cardiac myogenesis. *Genes Dev.* **11**, 451–462.

Seabra-Gomes, R., Somerville, J., Ross, D., Emanuel, R., Parker, D., and Wong, M. (1974). Congenital coronary artery aneurysms. *Br. Heart J.* **36**, 329–335.

Sefton, M., Sanchez, S., and Neito, M. (1998). Conserved and divergent roles for members of the snail family of transcription factors in the chick and mouse embryo. *Development* **125**, 3111–3121.

Sengbusch, J., He, W., Pinco, K., and Yang, J. (2002). Dual functions of $\alpha 4\beta 1$ integrin in epicardial development: Initial migration and long-term attachment. *J. Cell Biol.* **157**, 873–882.

Shimada, Y., Ho, E., and Toyota, N. (1981). Epicardial covering over myocardial wall in the chicken embryo as seen with the scanning electron microscope. *Scan. Elect. Micros.* **11**, 275–280.

Shirani, J., and Roberts, W. (1993). Solitary coronary ostium in the aorta in the absence of other major congenital cardiovascular anomalies. *J. Am. Coll. Cardiol.* **21**, 137–143.

Stary, H. (1985). Macrophage foam cells in the coronary artery intima of human infants. *Ann. NY Acad. Sci.* **454**, 5–8.

Stuckmann, I., Evans, S., and Lassar, A. (2003). Erythropoietin and retinoic acid, secreted from the epicardium, are required for cardiac myocyte proliferation. *Dev. Biol.* **255**, 334–349.

Sucov, H., Dyson, E., Gumeringer, C., Price, J., Chien, K., and Evans, R. (1994). RXRα mutant mice establish a genetic basis for vitamin A signaling in heart morphogenesis. *Genes Dev.* **8**, 1007–1018.

Sugi, Y., and Markwald, R. (2003). Endodermal growth factors promote endocardial precursor cell formation from precardiac mesoderm. *Dev. Biol.* **263**, 35–49.

Svensson, E., Huggins, G., Lin, H., Clendenin, C., Jiang, F., Tufts, R., Dardik, F., and Leiden, J. (2000). A syndrome of tricuspid atresia in mice with a targeted mutation of the gene encoding Fog-2. *Nature Genet.* **25**, 353–356.

Tallquist, M., French, W., and Soriano, P. (2003). Additive effects of PDGF receptor β signaling pathways in vascular smooth muscle cell development. *PLoS Biol.* **1**, 288–299.

Tevosian, S., Deconinck, A., Tanaka, M., Schinke, M., Litvosky, S., Izumo, S., Fujiwara, Y., and Orkin, S. (2000). FOG-2, a cofactor for GATA transcription factors, is essential for heart morphogenesis and development of coronary vessels from epicardium. *Cell* **101**, 729–739.

Tidball, J. (1992). Distribution of collagens and fibronectin in the subepicardium during avian cardiac development. *Anat. Embryol.* **185**, 155–162.

Tomanek, R. (1996). Formation of the coronary vasculature: A brief review. *Cardiovasc. Res.* **31**, E46–E51.

Tomanek, R., Holifield, J., Reiter, R., Sandra, A., and Lin, J. (2002). Role of VEGF family members and receptors in coronary vessel formation. *Dev. Dyn.* **225**, 233–240.

Tomanek, R., Lotun, K., Clark, E., Suvarna, P., and Hu, N. (1998). VEGF and bFGF stimulate myocardial vascularization in embryonic chick. *Am. J. Physiol.* **274**, H1620–H1626.

Tomanek, R., Ratajska, A., Kitten, G., Yue, X., and Sandra, A. (1999). Vascular endothelial growth factor expression coincides with coronary vasculogenesis and angiogenesis. *Dev. Dyn.* **215**, 54–61.

Tomanek, R., and Zheng, W. (2002). Role of growth factors in coronary morphogenesis. *Texas Heart Inst. J.* **29**, 250–254.

Tran, C., and Sucov, H. (1998). The RXRα gene functions in a non-cell-autonomous manner during mouse cardiac morphogenesis. *Development* **125**, 1951–1956.

Tsuda, T., Majumder, K., and Linask, K. (1998). Differential expression of flectin in the extracellular matrix and left–right asymmetry in mouse embryonic heart in mouse embryonic heart during looping stages. *Dev. Genet.* **23**, 203–214.

Tsuda, T., Wang, H., Timpl, R., and Chu, M. (2001). Fibulin-2 expression marks transformed mesenchymal cells in developing cardiac valves, aortic arch vessels, and coronary vessels. *Dev. Dyn.* **222**, 89–100.

Van der Loop, F., Shaart, G., Timmer, E., Ramaekers, F., and van Eys, G. (1996). Smoothelin, a novel cytoskeletal protein specific for smooth muscle cells. *J. Cell Biol.* **134**, 401–411.

Veikkola, T., Jussila, L., Makinen, T., Karpanen, T., Jeltsch, M., Petrova, T., Kubo, H., Thurston, G., McDonald, D., Achen, M., Stacker, S., and Alitalo, K. (2001). Signalling via vascular endothelial growth factor receptor-3 is sufficient for lymphangiogenesis in transgenic mice. *EMBO J.* **20**, 1223–1231.

Velican, D., and Velican, C. (1976). Intimal thickening in developing coronary arteries and tis relevance to atherosclerotic involvement. *Atherosclerosis* **23**, 345–355.

Velican, C., and Velican, D. (1979). Some particular aspects of the microarchitecture of human coronary arteries. *Atherosclerosis* **33**, 191–200.

Velkey, J., and Bernanke, D. (2001). Apoptosis during coronary artery orifice development in the chick embryo. *Anat. Rec.* **262**, 310–317.

Viragh, S., and Challice, C. (1973). Origin and differentiation of cardiac muscle cells in the mouse. *J. Ultrastruct. Res.* **42**, 1–24.

Viragh, S., and Challice, C. (1981). The origin of the epicardium and the embryonic myocardial circulation in the mouse. *Anat. Rec.* **201**, 157–168.

Viragh, S., Gittenberger-de, G. A., Poelmann, R., and Kalman, F. (1993). Early development of quail heart epicardium and associated vascular and glandular structures. *Anat. Embryol. (Berl.)* **188**, 381–393.

Vrancken Peeters, M., Gittenberger-de, G. A., Mentink, M., Hungerford, J., Little, C., and Poelmann, R. (1997a). The development of the coronary vessels and their differentiation into arteries and veins in the embryonic quail heart. *Dev. Dyn.* **208**, 338–348.

Vrancken Peeters, M., Gittenberger-de Groot, A., Mentink, M., Hungerford, J., Little, C., and Poelmann, R. (1997b). Differences in development of coronary arteries and veins. *Cardiovasc. Res.* **36**, 101–110.

Vrancken Peeters, M., Gittenberger-de Groot, A., Mentink, M., and Poelmann, R. (1999). Smooth muscle cells and fibroblasts of the coronary arteries derive from epithelial-mesenchymal transformation of the epicardium. *Anat. Embryol.* **199**, 367–378.

Vrancken Peeters, M., Mentink, M., Poelmann, R., and Gittenberger-de, G. A. (1995). Cytokeratins as a marker for epicardial formation in the quail embryo. *Anat. Embryol. (Berl.)* **191**, 503–508.

Wada, A., Reese, D., and Bader, D. (2001). Bves: Prototype of a new class of cell adhesion molecules expressed during coronary artery development. *Development* **128**, 2085–2093.

Wada, A., Willet, S., and Bader, D. (2003). Coronary vessel development: A unique form of vasculogenesis. *Atheroscler. Thromb. Vasc. Biol.* **23**, 2138–2145.

Waldo, K., Kumiski, D., and Kirby, M. (1994). Association of the cardiac neural crest with development of the coronary arteries in the chick embryo. *Anat. Rec.* **239**, 315–331.

Waldo, K., Willner, W., and Kirby, M. (1990). Origin of the proximal coronary artery stems and a review of ventricular vascularization in the chick embryo. *Am. J. Anat.* **188**, 109–120.

Wang, D., Chang, P., Wang, Z., Sutherland, L., Richardson, J., Small, W., Krieg, P., and Olson, E. (2001). Activation of cardiac gene expression by myocardin, a transcriptional cofactor for serum response factor. *Cell* **105**, 851–862.

Wang, D., Li, S., Hockemeyer, D., Sutherland, L., Wang, Z., Schratt, G., Richardson, J., Nordheim, A., and Olson, E. (2002). Potentiation of serum response factor activity by a family of myocardin-related transcription factors. *Proc. Natl. Acad. Sci. USA* **99**, 14855–14860.

Ward, N., and Dumont, D. (2004). Angiopoietin 1 expression levels in the myocardium direct coronary vessel development. *Dev. Dyn.* **229**, 500–509.

Weiskirchen, R., and Gressner, A. (2000). The cysteine- and glycine-rich LIM domain protein CRP2 specifically interacts with a novel human protein (CRP2BP). *Biochem. Biophys. Res. Commun.* **274**, 655–683.

Weiskirchen, R., and Gunther, K. (2003). The CRP/MLP/TLP family of LIM domain proteins: Acting by connecting. *Bioessays* **25**, 152–162.

Williams, R. (2000). "The Athlete and Heart Disease: Diagnosis, Evaluation and Management." Lippincott, Williams & Wilkins, Philadelphia.

Wu, H., Lee, S., Gao, J., Liu, X., and Iruela-Arispe, M. (1999). Inactivation of erythropoietin leads to defects in cardiac morphogenesis. *Development* **126**, 3597–3605.

Xavier-Neto, J., Shapiro, M., Houghton, L., and Rosenthal, N. (2000). Sequential programs of retinoic acid synthesis in the myocardial and epicardial layers of the developing avian heart. *Dev. Biol.* **219**, 129–141.

Yamada, M., Szendro, P., Prokscha, A., Schwartz, R., and Eichele, G. (1999). Evidence for a role of SMAD6 in chick cardiac development. *Dev. Biol.* **215**, 48–61.

Yang, J., Rayburn, H., and Hynes, R. (1995). Cell adhesion events mediated by $\alpha 4$ integrins are essential in placental and cardiac development. *Development* **121**, 549–560.

Yoshida, K., Taga, T., Saito, M., Suematsu, S., Kumanogoh, A., Tanaka, T., Fujiwara, H., Hirata, M., Yamagami, T., Nakahata, T., Hirabayashi, T., Yoneda, Y., Tanaka, K., Wang, W., Mori, C., Shiota, K., Yoshida, N., and Kishimoto, T. (1996). Targeted disruption of gp130, a common signal transducer for the interleukin 6 family of cytokines, leads to myocardial and hematological disorders. *Proc. Natl. Acad. Sci. USA* **93**, 407–411.

Yoshida, T., Sinha, S., Dandre, F., Wamhoff, B., Hoofnagle, M., Kremer, B., Wang, D., Olson, E., and Owens, G. (2003). Myocardin is a key regulator of CArG-dependent transcription of multiple smooth muscle marker genes. *Circ. Res.* **92**, 856–864.

Young, P., Hofling, A., and Sands, M. (2002). VEGF increases engraftment of bone marrow-derived endothelial progenitor cells (EPCs) into vasculature of newborn murine recipients. *Proc. Natl. Acad. Sci. USA* **99**, 11951–11956.

Yu, Z., Redfern, C., and Fishman, G. (1996). Conditional transgene expression in the heart. *Circ. Res.* **79**, 691–697.

Yue, X., and Tomanek, R. (2001). Effects of VEGF(165) and VEGF(121) on vasculogenesis and angiogenesis in cultured embryonic quail hearts. *Am. J. Physiol. Heart. Circ. Physiol.* **280**, H2240–2247.

Zheng, W., Brown, M., Brock, T., Bjercke, R., and Tomanek, R. (1999). Bradycardia-induced coronary angiogenesis is dependent on vascular endothelial growth factor. *Circ. Res.* **85**, 192–198.

9

Identifying Early Vascular Genes Through Gene Trapping in Mouse Embryonic Stem Cells

Frank Kuhnert*,† and Heidi Stuhlmann*
*Department of Cell Biology
Division of Vascular Biology
The Scripps Research Institute
La Jolla, California 92037
†Department of Medicine
Division of Hematology
Stanford University Medical Center
Stanford, California 94305

I. Introduction
II. Gene Discovery Through Mutagenesis Screens: Chemical Mutagenesis Versus Gene Trap Mutagenesis
III. Entrapment Vectors
IV. Gene Trap Screens
    A. Sequence Information
    B. Mutagenesis
    C. Expression Screens
V. ES Cell *In Vitro* Differentiation as a Model System to Identify Trapped Genes Expressed in Vascular Lineages
VI. Designs of Cardiovascular Entrapment Screens
VII. Conclusions and Future Directions
    References

I. Introduction

The formation of a circulatory system consisting of the heart, a network of blood and lymphatic vessels, and blood cells is crucial for the development of the vertebrate embryo. The extensively branched lattice of blood vessels carries nutrients, gases, and waste products to and from distant tissues, and it may also provide developmental signals for the formation of endoderm-derived organs like liver and pancreas. The lymphatic system regulates tissue fluid balance by draining excessive fluid and macromolecules into the venous system and also trafficks immune cells, thus complementing the blood vascular system. The development of the blood and lymphatic vasculature is controlled by several signaling pathways, including vascular endothelial growth factor/Flk-1, Flt-1, VEGFR-3; angiopoietin/Tie; ephrin/Eph; transforming growth factor-β and its receptor; and Notch and its ligands (Rossant and Howard, 2002; Yancopoulos *et al.*, 2000). Many of the genes involved in these pathways are expressed during early stages of vascular development in midgestation mouse embryos. While the importance of these factors for the formation of the vascular networks has been demonstrated by gain- and loss-of-function studies, many genes that act downstream or in parallel to the known pathways still remain unidentified. Thus, the discovery of novel vascular genes and the examination of their function will greatly advance our understanding of the molecular processes underlying vascular development. In this review, we will focus on one particular approach, the gene trap screen in mouse embryonic stem (ES) cells, for discovering genes that are specifically expressed during vascular development.

II. Gene Discovery Through Mutagenesis Screens: Chemical Mutagenesis Versus Gene Trap Mutagenesis

Two major types of forward genetic screens have been developed that hold great promise for large-scale discovery of developmental genes and their functional annotation in mice: chemical mutagenesis and gene trap mutagenesis. A thorough comparison of these two screens can be found in two excellent recent reviews (Jackson, 2001; Stanford *et al.*, 2001). The first type of mutagenesis screen uses N-ethyl-N-nitrosurea (ENU), a powerful DNA alkylating agent that efficiently generates single-gene mutations in stem cells of spermatogonia in mice. ENU induces primarily point mutations and, infrequently, small deletions, thus generating a large spectrum of mutations, including loss- and gain-of-function, hypomorphic, and antimorphic alleles. Large numbers of mutant mice can easily be generated and screened for a variety of dominant and recessive mutant phenotypes. However, identification of the affected gene requires positional cloning, and therefore recovery of the mutated gene can be cumbersome. Several large-scale chemical mutagenesis screens are underway (Hrabe de Angelis *et al.*, 2000; Kile *et al.*, 2003; Nolan *et al.*, 2000). One of these screens utilizes an inversion-containing balancer chromosome to simplify the isolation of mutations in a given interval of a selected chromosome (Kile *et al.*, 2003). To this date, more than 1000 new mutations, including defects affecting cardiovascular development, have been identified.

A second powerful approach for identifying developmental genes is to generate insertional mutations in the germ line of mice. The strategy is based on the random insertion of foreign DNA elements into coding or regulatory regions of the genome, thereby frequently disrupting or altering the expression of surrounding genes. The insertion serves to "tag" the endogenous genes, allowing for subsequent cloning. This approach has been used successfully in *Drosophila, C. elegans,* and in plants, and, more recently, in vertebrate systems (for review, see Stanford et al., 2001). In mice, the availability of pluripotent ES cells, together with the development of sophisticated entrapment vectors (discussed later), has made it possible to efficiently introduce insertional mutations into ES cells and to screen for candidate genes in ES cells and embryos derived from the ES cells. Thus, this type of mutagenesis screen, referred to as gene trap, generates random insertional mutations *in vitro*, and it also allows for selection and characterization of the targeted endogenous gene before introducing the mutation into the germ line. Of importance, ES cells can also differentiate *in vitro*, either spontaneously by removal of the feeder layer of mouse embryonic fibroblasts (MEFs) and leukemia inhibitory factor (LIF), or directed by removal of differentiation inhibitors and the addition of specific growth factors and cytokines. The principal method for differentiation *in vitro* employs cell aggregation in suspension and formation of 3-dimensional embryoid bodies. The process of embryoid body formation closely mimics the schedule of cell lineage appearance during early postimplantation development, however, without proper axial organization or body plan (for review, see Smith, 2001). Remarkably, with the exception of trophoblast cells, there appears to be no limitation to the differentiation potential of ES cells *in vitro*. Thus, progenitor cells from all three germ layers and their differentiated cell derivatives can be readily detected in mixed cultures (see Table I). Selective culture conditions can be used to enrich for a given cell population, such as the hematopoietic and endothelial lineages (Hirashima et al., 2004; Wiles and Keller, 1991). Alternatively, lineage-committed progenitors or differentiated cell populations can be purified based on specific marker gene expression (Nishikawa et al., 1998). Thus, differentiation of ES cells provides an opportunity to preselect *in vitro* for entrapment vector insertions into developmentally regulated or lineage-restricted genes (discussed later).

III. Entrapment Vectors

The essential feature of entrapment vectors is a promoterless reporter gene, usually the *E. coli lacZ* gene expressing β-galactosidase (β-gal), or a human alkaline phosphatase (AP) gene, whose expression becomes dependent on transcription initiated from cis-acting regulatory regions upon insertion into a cellular gene. Therefore, candidate genes can be identified based on the pattern of reporter gene expression rather than a mutant phenotype. To provide a selection for random insertions, the vectors also contain a neomycin resistance (neo^R) gene under an autonomous promoter. Alternatively, a β-galactosidase-neo^R fusion (β-geo) is used in entrapment vectors, allowing only selection of clones with insertions into genes that are transcriptionally

Table I Differentiated Cell Types from Mouse ES Cells *In Vitro*[a]

Cell type	Reference
Yolk sac endoderm	Doetschman et al., 1985
Yolk sac mesoderm	Bautch et al., 1996; Doetschman et al., 1985
Primitive hematopoietic	Choi et al., 1998; Doetschman et al., 1985; Nakano et al., 1996
Definitive hematopoietic	Eto et al., 2002; Nakano et al., 1996; Nishikawa et al., 1998; Wiles and Keller, 1991
Lymphoid precursor	Potocnik et al., 1994
Mast cell	Tsai et al., 2000
Dendritic cell	Fairchild et al., 2000; Senju et al., 2003
Endothelial cell	Bautch et al., 1996; Risau et al., 1988; Vittet et al., 1996; Wang et al., 1992; Yamashita et al., 2000
Cardiomyocyte	Doetschman et al., 1985; Maltsev et al., 1993
Striated muscle	Rohwedel et al., 1994
Smooth muscle	Yamashita et al., 2000
Adipocyte	Dani et al., 1997
Osteoblast	Buttery et al., 2001
Chondrocyte	Kramer et al., 2000
Keratinocyte	Bagutti et al., 1996; Yamashita et al., 2000
Neuron, astrocyte, oligodendrocyte	Bain et al., 1995; Brustle et al., 1997; Fraichard et al., 1995; Liu et al., 2000; Strübing et al., 1995

[a]Adapted from (Smith, 2001).

active in ES cells (Fig. 1). The entrapment constructs are introduced into cells either as plasmids or as retroviral vectors.

Three prototypes of entrapment vectors have been developed for gene trap screens: enhancer-, promoter-, and gene trap vectors (Friedrich and Soriano, 1991; Gossler et al., 1989; Korn et al., 1992; Skarnes et al., 1992; von Melchner et al., 1992; for review, see Gossler and Zachgo, 1993; Stanford et al., 2001). Enhancer trap vectors contain a basal promoter and require integration near a cis-acting cellular enhancer for expression of the reporter gene. Promoter trap vectors contain only the coding sequence of the reporter gene. Its expression is dependent on insertion into an exon and formation of an in-frame fusion transcript with upstream exon sequences derived from the targeted gene. Gene trap vectors contain a splice acceptor site positioned 5′ of the promoterless reporter gene, resulting in a spliced fusion transcript from an upstream exon. Since this vector facilitates selection of trapped genes upon its insertion into introns, the frequency with which genes are targeted is about 50 times higher when compared to promoter traps.

9. Identifying Early Vascular Genes

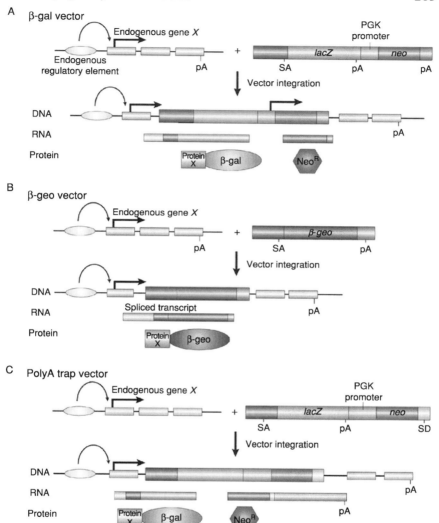

Figure 1 Schematic structure of gene trap β-gal, β-geo, and polyA trap vectors. See text for details (adapted from Stanford et al., 2001). (See Color Insert.)

As a result, β-gal or β-geo gene trap vectors (Figs. 1A and 1B) are the most commonly used entrapment vectors for gene trap screens. However, gene trap integration into introns sometimes leads to alternatively spliced transcripts, resulting in low expression of the wild-type gene and the generation of hypomorphic alleles.

Recently, several variations of the basic vectors have emerged that lead to increased trapping efficiencies or allow detection and recovery of specific classes of proteins. Several groups have developed polyadenylation (polyA)-trap vectors in which the neo^R selectable marker is under control of a constitutive promoter but lacks a polyA signal (Salminen et al., 1998; Zambrowicz et al., 1998). A splice donor site is provided in the vector, and therefore expression of the neo^R marker requires a correctly spliced transcript onto a downstream exon of the trapped gene (Fig. 1C). Modified vectors that contain Cre recombination sites allow for postinsertional modification of the trapped gene or facilitate the identification of transiently expressed genes (Araki et al., 1999; Hardouin and Nagy, 2000; Thorey et al., 1998). Specialized vectors have been developed to identify secreted and transmembrane proteins that are expressed in ES cells. These "secretory trap" vectors contain a transmembrane domain upstream of the β-geo reporter and capture genes that encode N-terminal signal sequences (Skarnes et al., 1995). Finally, a novel gene trap vector has been generated that transduces an EGFP-neo fusion reporter gene and allows monitoring expression of trapped genes in live cells and embryos (Chen and Chen, 2004).

IV. Gene Trap Screens

Following entrapment vector transduction and selection, large numbers of ES cell clones with vector insertions can be rapidly isolated, expanded, and frozen as reference libraries. By using vectors that contain a neo^R selectable marker under an autonomous promoter (Fig. 1A), clones are isolated regardless of reporter gene expression at the trapped locus. Consequently, the trapping frequency is relatively low (1–4%) when compared to vectors that contain the β-geo fusion marker (Fig. 1B; trapping efficiency $\sim$30%). However, in the latter case, only insertions into genes that are expressed in undifferentiated ES can be selected. Trapped endogenous genes can be readily cloned by RCR-based methods, such as 5'RACE or inverse PCR, or by cloning of genomic DNA flanking the vector insertions. Individual ES cell clones can be passaged through the germ line of mice by blastocyst injection or aggregation with morula-stage embryos. *In vitro* cultures of differentiating embryoid bodies and embryos derived from the ES cells can be monitored for *lacZ*, AP, or EGFP reporter gene expression.

Consequently, three categories of information can be obtained from gene trap screens: (1) sequence of the trapped endogenous gene, (2) phenotype of mutation, and (3) expression of the endogenous gene. Information has been primarily accumulated from several high-throughput screens that are presently performed by four academic and one biotech consortia (see Table II); (Hansen et al., 2003; Hicks et al., 1997; Mitchell et al., 2001; To et al., 2004;

Table II Large-Scale Gene Trap Screens in Mouse ES Cells

Consortium	Screen	Entrapment vector(s)	Reference	Website
The Centre for Modeling Human Disease (CMHD) Gene Trap Resource, Canada	Expression screen	β-gal, β-geo gene trap vectors, various new vectors	To et al., 2004	http://www.cmhd.ca
BayGenomics Consortium, USA	Phenotypic screen	Promoter, gene trap, and secretory trap vectors	Mitchell et al., 2001	http://baygenomics.ucsf.edu
German Gene Trap Consortium (GGTC) of the German Human Genome Project, Germany	Sequence tag screen	β-geo promoter and gene trap vectors	Hansen et al., 2003; Wiles et al., 2000	http://tikus.gsf.de
Mammalian Functional Genomics Centre, Canada	Sequence tag screen	Promoter trap vectors	Hicks et al., 1997	http://www.escells.ca
Omnibank, Lexicon Genetics, USA	Sequence tag screen	polyA trap vectors	Zambrowicz et al., 1998, 2003	http://www.lexicon-genetics.com/omnibank

Wiles et al., 2000; Zambrowicz et al., 1998). Online searchable databases from each of these consortia are available to the scientific community with information on the types of entrapment vectors used, phenotypes and expression patterns, and sequence information of the entrapped genes. Mutated ES cell clones with vector insertions are provided to researchers. In addition, a number of smaller or specialized screens have been carried out by many different laboratories (Chen et al., 2004; Friedrich and Soriano, 1991; Forrester et al., 1996; Gossler et al., 1989; Korn et al., 1992; Mainguy et al., 2000; Skarnes et al., 1992; von Melchner et al., 1992; Voss et al., 1998; Wurst, 1995; Xiong et al., 1998).

A. Sequence Information

Entrapment vector insertions are dispersed throughout the genome and can be found on all chromosomes except the Y chromosome. Insertions occur more frequently in chromosomes with a high density of genes (Hansen et al., 2003; Wiles et al., 2000). Whereas the majority of the insertions appear to be random, hot spots for preferred insertions exist, and they appear to be specific for particular types of entrapment vectors (Hansen et al., 2003; Mitchell et al., 2001; Wiles et al., 2000; Zambrowicz et al., 1998). The trapped genes identified thus far represent genes encoding all known classes of proteins. However, since the screens have not yet reached saturation, it is unknown if every gene is amenable to entrapment vector insertion. Finally, a preference for insertion of retroviral gene trap vectors into the 5'UTR and first half of the coding region has been found (Chen et al., 2004; Hansen et al., 2003).

B. Mutagenesis

Evidence accumulated from several screens shows that about 90% of ES cell clones with vector insertions contribute to the germ line, and between 30–60% of these cause recessive lethal phenotypes. As discussed, alternative splicing, or splicing onto the 5'UTR, can occur with gene trap vectors, and this can result in hypomorphic mutations or lack of a mutant phenotype. Prescreening for insertions with efficient splicing to coding sequences can help to increase the frequency of mutagenic insertions.

C. Expression Screens

Valuable information about the *in vivo* expression profile of trapped genes can be obtained by monitoring reporter gene expression in embryos. In most cases, reporter expression appears to faithfully recapitulate expression of the trapped cellular gene. However, it is possible that the site of vector insertion modulates the level and range of reporter expression. Of note, two large-scale screens focus on identifying genes that are expressed in hematopoietic and cardiovascular lineages (Mitchell *et al.*, 2001; To *et al.*, 2004). We therefore anticipate the discovery of novel vascular genes by combining expression screening and sequence data analysis.

V. ES Cell *In Vitro* Differentiation as a Model System to Identify Trapped Genes Expressed in Vascular Lineages

Expression screens by gene trapping in ES cells identify genes solely based on the pattern of reporter gene expression rather than a mutant phenotype or a sequence tag. However, the time-consuming and expensive nature of *in vivo* analysis limits its practicality for large-scale entrapment screens. Thus, after generating a library of randomly trapped ES cell clones, a robust, preferably high-throughput assay has to be in place to identify genes that are expressed in a certain cell type or tissue before analyzing expression and, potentially, a mutant phenotype in mice generated from the trapped ES cell line. In this context, the *in vitro* differentiation system of ES cells has proven to be a powerful instrument for prescreening trapped genes that are expressed in a particular cell lineage. Under appropriate culture conditions, ES cells can be induced to differentiate *in vitro*, recapitulating many aspects of early lineage commitment and organogenesis during embryonic development and giving rise to derivatives of all three germ layers (Table I) (for current protocols, see *Meth. Enzym.*, 2003, Vol. 365). By all accounts, differential reporter gene expression during ES cell *in vitro* differentiation is a faithful predictor for its expression *in vivo*.

Embryonic vascular development is characterized by two distinct phases, vasculogenesis and angiogenesis. Vasculogenesis involves the differentiation of endothelial progenitor cells from the lateral plate mesoderm, their maturation into endothelial cells, and coalescence into primitive blood vessels to form a primary vascular plexus. Angiogenesis describes the subsequent growth, remodeling, and maturation processes that ultimately give rise to the formation of the organized blood vasculature. The formation of blood islands in the extraembryonic yolk sac marks the onset of vasculogenesis as well as hematopoiesis in the developing mouse embryo (Risau, 1997). The mammalian yolk sac consists of an endodermal and a mesodermal layer and is continuous with the splanchnopleure of the embryo proper. Blood islands develop from aggregates of mesodermal cells at approximately 7 days post-coitum (dpc) of mouse development. They consist of an inner layer of primitive hematopoietic cells and a peripheral population of angioblasts. These angioblasts differentiate into endothelial cells, form a lumen, migrate, and interconnect to form a primary vascular plexus (Risau and Flamme, 1995). Intraembryonic vasculogenesis is initiated in the cranial region of 7.5 dpc embryos with the emergence of endocardial progenitor cells. Concomittantly, the first aortic primordia become discernable (Drake and Fleming, 2000). The larger vessels of the embryo, as well as the primary vascular plexus in the lung, pancreas, spleen, and heart are also formed via vasculogenesis in the embryo (Wilting and Christ, 1996).

ES cell *in vitro* differentiation protocols have been established that recapitulate the endothelial differentiation program and maturation steps observed during vasculogenesis and angiogenesis in the embryo. This methodology has also been crucial for defining progenitor populations of the endothelial and hematopoietic lineages that have been notoriously difficult to assess *in vivo*. These progenitor cells include the bipotential hemangioblast and the common endothelial precursor. The classic model of ES cell *in vitro* differentiation is the formation of embryoid bodies (EB) in suspension culture, either in low-adhesion tissue culture dishes or as hanging drops, in the absence of differentiation inhibitory factors (Doetschman *et al.*, 1985; Robertson, 1987). Under these culture conditions, ES cells robustly and reproducibly form aggregates of differentiating cells. Within 2–4 days in suspension culture, an outer layer of endodermal-like cells is observed, followed by the development of an ectodermal columnar epithelium and the appearance of mesodermal cell types. The formation of vascular structures reminiscent of blood islands and cordlike channels is easily detectable in these EBs (Doetschman *et al.*, 1985; Risau *et al.*, 1988; Wang *et al.*, 1992). Furthermore, by monitoring the expression kinetics of the vascular markers *Flk-1*, *Tie1*, *Tie2*, VE-cadherin, and MECA-32, it was demonstrated that endothelial cell differentiation in the ES/EB system recapitulates the developmental program and gene expression profiles of the endothelial differentiation in the embryo (Vittet *et al.*, 1996). Using this EB system in combination with colocalization of reporter gene expression and that of the endothelial lineage marker *Flk-1* (Fig. 2) the vascular endothelial transcription factor *Vezf1* has been identified in a retroviral entrapment screen (Xiong *et al.*, 1999).

In a modification of the EB suspension culture system, EBs are replated after 4–7 days onto tissue culture dishes to promote the outgrowth of differentiated cells (Bautch *et al.*, 1996). This flat culture method results in the formation of blood islands and vascular structures that are more accessible to observation and can be morphologically identified. This method considerably facilitated the identification of novel vascular genes through *in vitro* prescreening (Stanford *et al.*, 1998). These cultures also contain aggregates of beating cardiomyocytes and contracting smooth-muscle cell bundles, thus allowing to screen for ES cell clones in which trapped genes are expressed in all cardiovascular lineages (Chen and Chen, 2004; Wobus *et al.*,

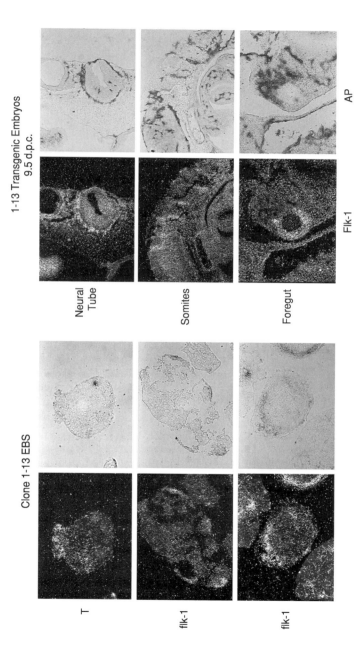

Figure 2 Colocalization of *flk-1* and AP reporter gene expression in embryoid bodies and transgenic mice derived from ES cell clone 1–13. *In Situ* hybridization on sectioned, AP stained clone 1–13 EBs (Left panel) using *Brachyury T* and *flk-1* riboprobes; each section is shown as a dark-field (Left) and a bright-field image. Transverse sections through E9.5 transgenic embryos derived from clone 1–13 (Right panel); each section is shown as a dark-field (Left) and bright-field image (adapted from Leahy *et al.*, 1999). (See Color Insert.)

1991). The ES/EB system can be further exploited by analyzing reporter gene expression in EBs grown in a three-dimensional collagen network. In fact, this culture system represents an *in vitro* approximation of the angiogenic sprouting and maturation process (Feraud *et al.*, 2003) and therefore enables the preselection of trapped genes that are expressed during later stages of vascular development.

Several ES cell *in vitro* differentiation protocols have been devised that specifically recapitulate the earliest stages of vasculogenesis, that is, the differentiation of the mesodermal precursors of the endothelial cell lineage. First, the culture of cells isolated from 3-day-old EBs in semisolid methylcellulose resulted in the formation of blast cell colonies (Kennedy *et al.*, 1997). When exposed to the appropriate cytokines, individual blast cells derived from the colonies possess the potential to give rise to both the endothelial and the hematopoietic cell lineage (Choi *et al.*, 1998) and as such represent the *in vitro* equivalent of the bipotential hemangioblast, long hypothesized to be the precursor of the hematopoietic and endothelial lineage. Second, similar hemangiogenic progenitors were identified via cell sorting as a Flk-1^+, VE-cadherin$^+$, CD45$^-$ cell population in a two-dimensional ES cell differentiation protocol (Nishikawa *et al.*, 1998). In this protocol, ES cells are cultured as a monolayer on collagen type IV–coated dishes. Presumptive lateral plate mesoderm Flk$^+$ cells are isolated by flow cytometry and cultured on OP9 stromal cells to induce endothelial and hematopoietic cell differentiation. More recently, this protocol has led to the isolation of a new type of blood vessel precursor, designated the common vascular precursor, from which both endothelial and PDGF-BB-dependent smooth-muscle cells develop (Yamashita *et al.*, 2000). The observation that OP9 stromal cells efficiently support endothelial differentiation has led to a third experimental culture protocol in which ES are differentiated by growth on OP9 cells. In this scenario, robust endothelial differentiation can be observed after 5 days of culture (Hirashima *et al.*, 2004). Because of its ease of use, this protocol holds great promise for a large-scale, high-throughput screen for trapped genes that are expressed during early vascular development.

VI. Designs of Cardiovascular Entrapment Screens

In this section, we will discuss four different experimental designs of entrapment screens for cardiovascular genes (summarized in Table III). The screens carried out in the Stuhlmann and Bernstein lab represent the first examples of gene trap approaches specifically aimed at the identification of vascular genes, whereas the Rossant and Chen lab protocols introduce important modifications of the original designs geared toward high-throughput screening.

Table III Prescreen for Cardiovascular Genes

Screen	Stuhlmann lab	Bernstein lab	Rossant lab	Chen lab
Entrapment vector design	Retroviral, IRES-AP (alkaline phosphatase) reporter, independent neo selection cassette	Plasmid-based, SA-lacZ and SA-βgeo reporter, independent neo selection cassette for SA-lacZ	Plasmid-based, IRESlacZ reporter, independent neo selection cassette, polyA trap	Plasmid-based, SA-Eno (EGFP-neo fusion gene) reporter
ES cell differentiation protocol	Suspension culture for up to 10 days	Suspension culture for 3 days followed by attachment culture for up to 16 days	Culture on OP9 feeder cells for 5 days	Hanging drop culture for 4 days followed by attachment culture for up to 20 days
Principle of identification of vascular genes	Colocalization with endothelial marker Flk-1	Morphological criteria	Morphological criteria	Morphological criteria
Generation of sequence tag	inverse PCR, supF complementation	5' RACE	5' and 3' RACE	5' RACE
Differentially expressing clones/number of ES cell clones screened	41/2400	79/1288	5/864	53/118
Vascular genes identified	Vezf1, Egfl7	several novel EST	Endoglin, ASPP1, Hes1	Hig1, Stag1
References	Fitch et al., 2004; Stuhlmann, 2003; Xiong et al., 1998, 1999	Stanford et al., 1998	Hirashima et al., 2004	Chen and Z, 2004

In the Stuhlmann screen, the bifunctional retroviral entrapment vector PT-IRES-AP with alkaline phosphatase (AP) is used as a reporter gene (Xiong et al., 1998). The human placental AP gene was chosen as a reporter because it was found to display significantly less variability in its expression in ES cells and EBs as compared to the *lacZ* gene. To facilitate the selection of infected ES cells independent of reporter expression, the neo^R gene was transcribed from an internal phosphoglycerokinase (PGK-1) promoter. An internal ribosomal entry site (IRES) element was inserted upstream of the AP gene. Integration of PT-IRES-AP in a transcribed gene leads to cap-independent translation of an authentic AP protein from a fusion transcript, independent of the reading frame. As an additional feature, PT-IRES-AP contains the bacterial suppressor tRNA gene to allow cloning of proviral and flanking genomic host sequences. Upon infection and G418 selection, ES cell clones were differentiated into EBs by suspension culture for up to 10 days. To identify clones that contain entrapment vector insertions into developmentally regulated genes, the expression pattern of the AP reporter gene was analyzed during the course of *in vitro* differentiation and compared to the expression profiles of a panel of germ layer and cell lineage markers. This allowed to correlate differential reporter gene expression patterns during EB formation with distinct phases of embryogenesis, that is, pregastrulation, gastrulation, and early organogenesis (Leahy et al., 1999; Stuhlmann, 2003). Using this screening methodology, 41 out of 2400 ES cell clones analyzed (1.7%) displayed differential reporter gene expression during EB formation. Furthermore, to identify clones that show reporter gene expression in the vascular lineage, colocalization studies of AP reporter expression with the RNA *in situ* hybridization signal for the endothelial marker *Flk-1* were performed in EBs. This screening approach, as a proof of principle, has lead to the identification of the endothelial transcription factor *Vezf1* (Xiong et al., 1999) and the secreted endothelial protein *Egfl7* (Fitch et al., 2004).

Two different plasmid-based entrapment vectors, PT1 and GT1.8geo, were used in the Stanford and Bernstein lab screen (Stanford et al., 1998). PT1 contains a splice acceptor site (SA) immediately upstream of a promoter-less *lacZ* reporter gene and a neo^R gene driven by a PGK-1 promoter, while GT1.8geo carries the splice acceptor site upstream of a β-geo fusion gene. Integration of either vector leads to the generation of a spliced fusion transcript between the endogenous gene and *lacZ*. However, GT1.8geo function is dependent on integration into a gene that is transcriptionally active in undifferentiated ES cells, whereas PT1 also allows trapping of genes that are not expressed in ES cells. Selected ES cell clones were differentiated in EB suspension culture for 3 days, followed by an attachment culture for up to 16 days. Blood islands and vascular structures were easily detectable in these flat cultures, and identification of vascular reporter gene expression was based on morphological criteria. For PT1, 32 out of 779 ES cells clones analyzed were found to display differential expression of the *lacZ* reporter during differentiation. Of those, 3% showed *lacZ* signal specifically in endothelial cells, while another 3% was expressed both in hematopoietic and endothelial cells. Using the GT1.8geo vector, among 353 *lacZ* expressing clones identified, 4% displayed reporter gene expression that was confined to the endothelial lineage, and another 19% displayed expression in both the hematopoietic and endothelial compartment. DNA sequence tags were generated for 11 of the identified clones, 8 of which corresponded to novel genes or EST. Vascular expression of several of these novel genes was confirmed by *in vivo* reporter gene expression analysis in transgenic mice (Stanford *et al.*, 1998).

The design of the Rossant lab screen incorporates two significant modifications (Hirashima *et al.*, 2004). First, a new generation polyA entrapment vector is used (Salminen *et al.*, 1998) in which the neo^R expression cassette lacks a polyA signal, but instead contains a splice donor (compare Fig. 1C). A spliced polyA signal from an endogenous gene is therefore required for the expression of the selection marker, ensuring that only integrations into genes and not intergenic regions are selected for, while at the same time not requiring expression on the targeted gene in undifferentiated ES cells. Second, the ES cell differentiation protocol has been significantly simplified, rendering it suitable for large-scale, high-throughput screens. Differentiation into the endothelial lineage is accomplished by coculture of small numbers of ES cells with OP9 feeder cells. In this protocol, PECAM-1^+/Flk-1^+ cells, indicative of endothelial differentiation, were easily recognized after 5 days of culture as spindle-shaped cells spreading out at the periphery of each colony. In a proof-of-principle expression screen, 5 out of 866 G418-resistant ES cell clones showed increased *lacZ* reporter expression during endothelial differentiation. Three of the entrapped genes are known genes, *Endoglin, ASPP1*, and *Hes1*. Endoglin is a TGF-β type III receptor known to be strongly expressed in endothelial cells, while endothelial expression of the p53-binding protein ASPP1 was confirmed in transgenic embryos derived from the trapped ES cell clone. The basic helix-loop-helix transcription factor Hes-1 is a downstream target of Notch signaling and has been implicated in endothelial morphogenesis (Hirashima *et al.*, 2004).

In the entrapment screen reported by Chen and Chen, a gene trap vector was developed that uses enhanced green fluorescent protein (EGFP) as a noninvasive reporter to monitor the expression of trapped genes throughout the entire culture period of a given *in vitro* differentiation protocol (Chen and Chen, 2004). The EGFP gene was fused in frame to the neo^R gene to create a reporter/selection marker hybrid, Eno (*EGFP-neo*). Eno was confirmed to possess both GFP and neo^R activities comparable to its paternal proteins. Furthermore, it was verified that Eno fluorescence was sufficient to identify cardiomyocytes during *in vitro* differentiation of ES cells by expressing Eno under the control of the α-cardiac myosin heavy-chain promoter. Applying G418 selection in the differentiation culture was found to enrich for reporter-expressing cells with intensified EGFP signals. To induce cardiovascular differentiation, ES cell clones were grown in hanging drops cultures for 4 days, followed by up to 20 days of attachment cultures. This protocol resulted in the formation of cells of all cardiovascular lineages: aggregates of beating cardiomyocytes, contracting smooth-muscle bundles, and vessels and blood islands comprising vascular endothelial and hematopoietic cells. Using this design for a gene trap screen, ES cell clones with Eno reporter expression in cardiovascular lineages could be detected. However, it was necessary to apply G418 selection during differentiation to generate detectable GFP signals in many clones, a procedure that rendered 30% of differentiation cultures uninformative with respect to reporter gene expression. Eighteen out of 118 clones analyzed showed detectable GFP expression in cardiomyocytes, whereas in 35 clones, green fluorescence was detected in blood islands or blood vessel–like structures. Cardiac expression *in vivo* was confirmed for 3 of the 18 clones in E11.5 transgenic embryos. Some of the identified sequence tags, for example, hypoxia inducible gene 1 (*hig1*), confirm cardiac expression patterns. However, at least two of the seven genes identified appear to be ubiquitously expressed. This may reflect a present limitation of the Eno gene trap screen that selects for genes expressed in undifferentiated ES cells, while often mosaic patterns of reporter expression are observed during *in vitro* differentiation (Chen and Chen, 2004).

VII. Conclusions and Future Directions

The first example of a gene trap mutagenesis screen in mouse ES cells was reported in 1989 (Gossler *et al.*, 1989). Many different screen designs have been developed since then, and the technology continues to evolve. As the result of these efforts, several public and private large-scale gene traps with the goal of saturation mutagenesis have been established. At the same time, several smaller, investigator-driven specialized screens are under way that remain an important source of innovation.

Gene trapping has proven to be a powerful means to identify early vascular genes. This is due to the fact that ES cell differentiation *in vitro* faithfully recapitulates early vascular development *in vivo*, thus providing an excellent screening tool for genes expressed in vascular lineages. As illustrated in this review, as a proof-of-principal, several important endothelial genes have been identified through gene trapping in several labs, including ours. In parallel, significant modifications in the screen design are being implemented and are aimed at increasing the efficiency of the process and to allow for high-throughput screening. With respect to entrapment vector design, accumulating data indicate that including a polyA trap selection cassette and an unbiased reporter gene (i.e., one that does not rely on expression of the trapped gene in undifferentiated ES cells) will increase the percentage of trapped genes with a restricted expression pattern during *in vitro* differentiation. Regarding the screening process itself, the simplified yet robust differentiation protocol described by Hirashima *et al.* (2004) constitutes a major improvement both in terms of time and ease of use, and it enables screening at a much higher throughput rate. In addition, the utilization of a noninvasive reporter (i.e., EGFP) might further facilitate the screening process. These advances in gene trapping, in combination with other molecular genetic technologies such as microarrays and site-specific recombination, provide the individual investigator with powerful tools for gene discovery and vascular development.

References

Araki, K., Imaizumi, T., Sekimoto, T., Yoshinobu, K., Yoshimuta, J., Akizuki, M., Miura, K., Araki, M., and Yamamura, K. (1999). Exchangeable gene trap using the Cre/mutated lox system. *Cell. Mol. Biol.* **45,** 737–750.

Bagutti, C., Wobus, A. M., Fassler, R., and Watt, F. M. (1996). Differentiation of embryonal stem cells into keratinocytes: Comparison of wild-type and beta 1 integrin-deficient cells. *Dev. Biol.* **179,** 184–196.

Bain, G., Kitchens, D., Yao, M., Huettner, J. E., and Gottlieb, D. I. (1995). Embryonic stem cells express neuronal properties *in vitro*. *Dev. Biol.* **168,** 342–357.

Bautch, V. L., Stanford, W. L., Rapoport, R., Russel, S., Byrum, R. S., and Futch, T. A. (1996). Blood island formation in attached cultures of murine embryonic stem cells. *Dev. Dyn.* **205,** 1–12.

Brustle, O., Spiro, A. C., Karram, K., Choudhary, K., Okabe, S., and McKay, R. D. (1997). *In Vitro* generated neural precursors participate in mammalian brain development. *Proc. Natl. Acad. Sci. USA* **94,** 14809–14814.

Buttery, L. D., Bourne, S., Xynos, J. D., Wood, H., Hughes, F. J., Hughes, S. P., Episkopu, V., and Polak, F. M. (2001). Differentiation of osteoblasts and *in vitro* bone formation from murine embryonic stem cells. *Tissue Eng.* **7,** 89–99.

Chen, W. V., Delrow, J., Corrin, P. D., Frazier, J. P., and Soriano, P. (2004). Identification and validation of PDGF transcriptional targets by microarray-coupled gene-trap mutagenesis. *Nat. Genet.* **36,** 304–312.

Chen, W. V., and Chen., Z. (2004). Differentiation trapping screen in live culture for genes expressed in cardiovascular lineages. *Dev. Dyn.* **229,** 319–327.

Choi, K., Kennedy, M., Kazarov, A., Papadimitriou, J. C., and Keller, G. (1998). A common precursor for hematopoietic and endothelial cells. *Development* **125,** 725–732.

Dani, C., Smith, A. G., Dessolin, S., Leroy, P., Stacini, L., Villageois, P., Darimont, C., and Ailhaud, G. (1997). Differentiation of embryonic stem cells into adipocytes *in vitro*. *J. Cell Sci.* **110,** 1279–1285.

Doetschman, T. C., Eistetter, H., Katz, M., Schmidt, W., and Kemler, R. (1985). The *in vitro* development of blastocyst-derived embryonic stem cell lines: Formation of visceral yolk sac, blood islands and myocardium. *J. Embryol. Exp. Morph.* **87,** 27–45.

Drake, C., and Fleming, P. (2000). Vasculogenesis in the day 6.5 to 9.5 mouse embryo. *Hemostasis Thrombosis Vascular Biol.* **95**, 1671–1679.

Eto, K., Murphy, R., Kerrigan, S. W., Bertoni, A., Stuhlmann, H., Nakano, T., Leavitt, A. D., and Shattil, S. J. (2002). Megakaryocytes derived from embryonic stem cells implicate CalDAG-GEFI in integrin signaling. *Proc. Natl. Acad. Sci. USA* **99**, 12819–12824.

Fairchild, P. J., Brook, F. A., Gardner, R. L., Graca, L., and Strong, V. (2000). Directed differentiation of dendritic cells from mouse embryonic stem cells. *Curr. Biol.* **10**, 1515–1518.

Feraud, O., Prandini, M. H., and Vittet, D. (2003). Vasculogenesis and angiogenesis from *in vitro* differentiation of mouse embryonic stem cells. *Methods Enzymol.* **365**, 214–228.

Fitch, M. J., Campagnolo, L., Kuhnert, F., and Stuhlmann, H. (2004). Egfl7, a novel epidermal growth factor-domain gene expressed in endothelial cells. *Dev. Dyn.* **230**, 316–324.

Forrester, L., Nagy, A., Mehran, S., Watt, A., Stevenson, L., Bernstein, A., Joyner, A., and Wurst, W. (1996). An induction gene trap screen in embryonic stem cells: Identification of genes that respond to retinoic acid *in vitro*. *Proc. Natl. Acad. Sci. USA* **93**, 1677–1682.

Fraichard, A., Chassande, O., Bilbaut, G., Dehay, C., Savatier, P., and Samarut, J. (1995). In Vitro differentiation of embryonic stem cells into glial cells and functional neurons. *J. Cell Sci.* **108**, 3181–3188.

Friedrich, G., and Soriano, P. (1991). Promoter traps in embryonic stem cells: A genetic screen to identify and mutate developmental genes in mice. *Genes Dev.* **5**, 1513–1523.

Gossler, A., Joyner, A. L., Rossant, J., and Skarnes, W. C. (1989). Mouse embryonic stem cells and reporter constructs to detect developmentally regulated genes. *Science* **244**, 463–465.

Gossler, A., and Zachgo, J. (1993). Gene and enhancer trap screens in ES cell chimeras. *In* "Gene targeting: A practical approach" (A. L. Joyner, Ed.), pp. 181–227. Oxford University Press, New York.

Hansen, J., Floss, T., Van Sloun, P., Fuchtbauer, E. M., Vauti, F., Arnold, H. H., Schnutgen, F., Wurst, W., von Melchner, H., and Ruiz, P. (2003). A large-scale, gene-driven mutagenesis approach for the functional analysis of the mouse genome. *Proc. Natl. Acad. Sci. USA* **100**, 9918–9922.

Hardouin, N., and Nagy, A. (2000). Gene-trap-based target site for cre-mediated transgenic insertion. *Genesis* **26**, 245–252.

Hicks, G. G., Shi, E. G., Li, X. M., Li, C. H., Pawlak, M., and Ruley, H. E. (1997). Functional genomics in mice by tagged sequence mutagenesis. *Nat. Genet.* **16**, 338–344.

Hirashima, M., Bernstein, A., Stanford, W. L., and Rossant, J. (2004). Gene trap expression screening to identify endothelial-specific genes. *Blood* **104**, 711–718.

Hrabe de Angelis, M. H., Flaswinkel, H., Fuchs, H., Rathkolb, B., Soewarto, D., Marschall, S., Heffner, S., Pargent, W., Wuensch, K., Jung, M., Reis, A., Richter, T., Alessandrini, F., Jakob, T., Fuchs, E., Kolb, H., Kremmer, E., Schaeble, K., Rollinski, B., Roscher, A., Peters, C., Meitinger, T., Strom, T., Steckler, T., Holsboer, F., Klopstock, T., Gekeler, F., Schindewolf, C., Jung, T., Avraham, K., Behrendt, H., Ring, J., Zimmer, A., Schughart, K., Pfeffer, K., Wolf, E., and Balling, R. (2000). Genome-wide, large-scale production of mutant mice by ENU mutagenesis. *Nat. Gent.* **25**, 444–447.

Jackson, I. J. (2001). Mouse mutagenesis on target. *Nat. Genet.* **28**, 198–200.

Kennedy, M., Firpo, M., Choi, K., Wall, C., Robertson, S., Kabrun, N., and Keller, G. (1997). A common precursor for primitive erythropoiesis and definitive haematopoiesis. *Nature* **386**, 488–493.

Kile, B. T., Hentges, K. E., Clark, A. T., Nakamura, H., Salinger, A. P., Liu, B., Box, N., Stockton, D. W., Johnson, R. L., Behringer, R. R., Bradley, A., and Justice, M. J. (2003). Functional genetic analysis of mouse chromosome 11. *Nature* **425**, 81–86.

Korn, R., Schoor, M., Neuhaus, H., Henseling, U., Soininen, R., Zachgo, J., and Gossler, A. (1992). Enhancer trap integrations in mouse embryonic stem cells give rise to staining

9. Identifying Early Vascular Genes

patterns in chimaeric embryos with a high frequency and detect endogenous genes. *Mech. Dev.* **39,** 95–109.

Kramer, J., Hegert, C., Guan, K., Wobus, A. M., Muller, P. K., and Rohwedel, J. (2000). Embryonic stem cell-derived chondrogenic differentiation *in vitro*: Activation by BMP-2 and BMP-4. *Mech. Dev.* **92,** 193–205.

Leahy, A., Xiong, J.-W., Kuhnert, F., and Stuhlmann, H. (1999). Use of developmental marker genes to define temporal and spatial patterns of differentiation during embryoid body formation. *J. Exp. Zool.* **284,** 67–81.

Liu, S., Qu, Y., Stewart, T. J., Howard, M. J., and Chakrabortty, S. (2000). Embryonic stem cells differentiate into oligodendrocytes and myelinate in culture and after spinal cord transplantation. *Proc. Natl. Acad. Sci. USA* **97,** 6126–6131.

Mainguy, G., Montesinos, M. L., Lesaffre, B., Zevnik, B., Karasawa, M., Kothary, R., Wurst, W., Prochiantz, A., and Volovitch, M. (2000). An induction gene trap for identifying a homeoprotein-regulated locus. *Nat. Biotechnol.* **18,** 746–749.

Maltsev, V. A., Rohwedel, J., Hescheler, J., and Wobus, A. M. (1993). Embryonic stem cells differentiate *in vitro* into cardiomyocytes representing sinusnodal, atrial and ventricular cell types. *Mech. Dev.* **44,** 41–50.

Mitchell, K. J., Pinson, K. I., Kelly, O. G., Brennan, J., Zupicich, J., Scherz, P., Leighton, P. A., Goodrich, L. V., Lu, X., Avery, B. J., Tate, P., Dill, K., Pangilinan, E., Wakenight, P., Tessier-Lavigne, M., and Skarnes, W. C. (2001). Functional analysis of secreted and transmembrane proteins critical to mouse development. *Nat. Genet.* **28,** 241–249.

Nakano, T., Kodama, H., and Honjo, T. (1996). *In Vitro* development of primitive and definitive erythrocytes from different precursors. *Science* **272,** 722–724.

Nishikawa, S. I., Nishikawa, S., Hirashima, M., Matsuyoshi, N., and Kodama, H. (1998). Progressive lineage analysis by cell sorting and culture identifies FLK1+VE-cadherin+ cells at a diverging point of endothelial and hemopoietic lineages. *Development* **125,** 1747–1757.

Nolan, P. M., Peters, J., Strivens, M., Rogers, D., Hagan, J., Spurr, N., Gray, I. C., Vizor, L., Brooker, D., Whitehill, E., Washbourne, R., Hough, T., Greenaway, S., Hewitt, M., Liu, X., McCormack, S., Pickford, K., Selley, R., Wells, C., Tymowska-Lalanne, Z., Roby, P., Glenister, P., Thornton, C., Thaung, C., Stevenson, J. A., Arkell, R., Mburu, P., Hardisty, R., Kiernan, A., Erven, A., Steel, K. P., Voegeling, S., Guenet, J. L., Nickols, C., Sadri, R., Nasse, M., Isaacs, A., Davies, K., Browne, M., Fisher, E. M., Martin, J., Rastan, S., Brown, S. D., and Hunter, J. (2000). A systematic, genome-wide, phenotype-driven mutagenesis programme for gene function studies in the mouse. *Nat. Genet.* **25,** 440–443.

Potocnik, A. J., Nielsen, P. J., and Eichmann, K. (1994). *In Vitro* generation of lymphoid precursors from embryonic stem cells. *EMBO J.* **13,** 5274–5283.

Risau, W. (1997). Mechanisms of angiogenesis. *Nature* **386,** 671–674.

Risau, W., and Flamme, I. (1995). Vasculogenesis. *Annu. Rev. Cell. Dev. Biol.* **11,** 73–91.

Risau, W., Sariola, H., Zerwes, H.-G., Sasse, J., Ekblom, P., Kemler, R., and Doetschman, T. (1988). Vasculogenesis and angiogenesis in embryonic-stem-cell-derived embryoid bodies. *Development* **102,** 471–478.

Robertson, E. J. (1987). Embryo-derived stem cell lines. *In* "Teratocarcinomas and embryonic stem cells: A practical approach" (E. J. Robertson, Ed.), pp. 71–112. IRL Press, Oxford, England.

Rohwedel, J., Maltsev, V., Bober, E., Arnold, H.-H., Hescheler, J., and Wobus, A. M. (1994). Muscle cell differentiation of embryonic stem cells reflects myogenesis *in vivo*: Developmentally regulated expression of myogenic determination genes and functional expression of ionic currents. *Dev. Biol.* **154,** 87–101.

Rossant, J., and Howard, L. (2002). Signaling pathways in vascular development. *Annu. Rev. Cell. Dev. Biol.* **18,** 541–773.

Salminen, M., Meyer, B. I., and Gruss, P. (1998). Efficient poly A trap approach allows the capture of genes specifically active in differentiated embryonic stem cells and in mouse embryos. *Dev. Dyn.* **212,** 326–333.

Senju, S., Hirata, S., Matsuyoshi, H., Masuda, M., Uemura, Y., Araki, K., Yamamura, K. I., and Nishimura, Y. (2003). Generation and genetic modification of dendritic cells derived from mouse embryonic stem cells. *Blood* **101,** 3501–3508.

Skarnes, W., Auerbach, B. A., and Joyner, A. L. (1992). A gene trap approach in mouse embryonic stem cells: The *lacZ* reporter is activated by splicing, reflects endogenous gene expression, and is mutagenic in mice. *Genes & Development* **6,** 903–918.

Skarnes, W., Moss, J., Hurtley, S., and Beddington, R. (1995). Capturing genes encoding membrane and secreted proteins important for mouse development. *Proc. Natl. Acad. Sci. USA* **92,** 6592–6596.

Smith, A. G. (2001). Embryo-derived stem cells: Of mice and men. *Annu. Rev. Cell Dev. Biol.* **17,** 435–462.

Stanford, W. L., Caruana, G., Vallis, K. A., Inamdar, M., Hidaka, M., Bautch, V. L., and Bernstein, A. (1998). Expression trapping: Identification of novel genes expressed in hematopoietic and endothelial lineages by gene trapping in ES cells. *Blood* **92,** 4622–4631.

Stanford, W. L., Cohn, J. B., and Cordes, S. P. (2001). Gene-trap mutagenesis: Past, present and beyond. *Nat. Rev. Genet.* **2,** 756–768.

Strübing, C., Ahnert-Hilger, G., Shan, J., Wiedemann, B., Hescheler, J., and Wobus, A. M. (1995). Differentiation of pluripotent embryonic stem cells into the neuronal lineage *in vitro* gives rise to mature inhibitory and excitatory neurons. *Mech. Dev.* **53,** 275–287.

Stuhlmann, H. (2003). Gene trap vector screen for developmental genes in differentiating ES cells. *In* "Methods in Enzymology" (P. M. Wassarman and G. M. Keller, Eds.), Vol. 365 pp. 386–406. Elsevier Academic Press, San Diego, CA.

Thorey, I. S., Muth, K., Russ, A. P., Otte, J., Reffelmann, A., and von Melchner, H. (1998). Selective disruption of genes transiently induced in differentiating mouse embryonic stem cells by using gene trap mutagenesis and site-specific recombination. *Mol. Cell. Biol.* **18,** 3081–3088.

To, C., Epp, T., Reid, T., Lan, Q., Yu, M., Li, C. Y., Ohishi, M., Hant, P., Tsao, N., Casallo, G., Rossant, J., Osborne, L. R., and Stanford, W. L. (2004). The Centre for Modeling Human Disease Gene Trap resource. *Nucleic Acids Res.* **32,** D557–D559.

Tsai, M., Wedemeyer, J., Ganiatsas, S., Tam, S. Y., Zon, L. I., and Galli, S. J. (2000). *In Vivo* immunological function of mast cells derived from embryonic stem cells: An approach for the rapid analysis of even embryonic lethal mutations in adult mice *in vivo*. *Proc. Natl. Acad. Sci. USA* **97,** 9186–9190.

Vittet, D., Prandini, M. H., Berthier, R., Schweitzer, A., Martin-Sisteron, H., Uzan, G., and Dejana, E. (1996). Embryonic stem cells differentiate *in vitro* to endothelial cells through successive maturation steps. *Blood* **88,** 3424–3421.

von Melchner, H., DeGregori, J. V., Rayburn, H., Reddy, S., Friedel, C., and Ruley, H. E. (1992). Selective disruption of genes expressed in totipotent embryonal stem cells. *Genes & Development* **6,** 919–927.

Voss, A. K., Thomas, T., and Gruss, P. (1998). Efficiency assessment of the gene trap approach. *Dev. Dyn.* **212,** 171–180.

Wang, R., Clark, R., and Bautch, V. L. (1992). Embryonic stem cell-derived cystic embryoid bodies form vascular channels: An *in vitro* model of blood vessel development. *Development* **114,** 303–316.

Wiles, M. V., and Keller, G. (1991). Multiple hematopoietic lineages develop from embryonic stem cells (ES) cells in culture. *Development* **111,** 259–267.

Wiles, M. V., Vauti, F., Otte, J., Fuchtbauer, E. M., Ruiz, P., Fuchtbauer, A., Arnold, H. H., Lehrach, H., Metz, T., von Melchner, H., and Wurst, W. (2000). Establishment of a genetrap

9. Identifying Early Vascular Genes

sequence tag library to generate mutant mice from embryonic stem cells. *Nat. Genet.* **24,** 13–14.
Wilting, J., and Christ, B. (1996). Embryonic angiogenesis: A review. *Naturwissenschaften* **83,** 158–163.
Wobus, A. M., Wallukat, G., and Hescheler, J. (1991). Pluripotent mouse embryonic stem cells are able to differentiate into cardiomyocytes expressing chronotropic responses to adrenergic and cholinergic agents and Ca2+ channel blockers. *Differentiation* **48,** 173–182.
Wurst, W., Rossant, J., Priedeaux, V., Kownacka, M., Joyner, A., Hill, D., Guillemot, F., Gasca, S., Cado, D., Auerbach, A., and Ang, S. (1995). A large scale gene trap screen for insertional mutants in developmentally regulated genes in mice. *Genetics* **139,** 889–899.
Xiong, J.-W., Battaglino, R., Leahy, A., and Stuhlmann, H. (1998). Large-scale screening for developmental genes in ES cells and embryoid bodies using retroviral entrapment vectors. *Dev. Dyn.* **212,** 181–197.
Xiong, J.-W., Leahy, A., Lee, H.-H., and Stuhlmann, H. (1999). *Vezf1*: A Zn finger transcription factor restricted to endothelial cells and their precursors. *Dev. Biol.* **206,** 123–141.
Yamashita, J., Itoh, H., Hirashima, M., Ogawa, M., Nishikawa, S., Yurugi, T., Naito, M., and Nakao, K. (2000). Flk1-positive cells derived from embryonic stem cells serve as vascular progenitors [In Process Citation]. *Nature* **408,** 92–96.
Yancopoulos, G. D., Davis, S., Gale, N. W., Rudge, J. S., Wiegand, S. J., and Holash, J. (2000). Vascular-specific growth factors and blood vessel formation. *Nature* **407,** 242–248.
Zambrowicz, B. P., Abuin, A., Ramirez-Solis, R., Richter, L. J., Piggott, J., BeltrandelRio, H., Buxton, E. C., Edwards, J., Finch, R. A., Friddle, C. J., Gupta, A., Hansen, G., Hu, Y., Huang, W., Jaing, C., Key, B. W. J., Kipp, P., Kohlhauff, B., Ma, Z. Q., Markesich, D., Payne, R., Potter, D. G., Qian, N., Shaw, J., Schrick, Shi, Z. Z., Sparks, M. J., Van Sligtenhorst, I., Vogel, P., Walke, W., Xu, N., Zhu, Q., Person, C., and Sands, A. T. (2003). Wnk1 kinase deficiency lowers blood pressure in mice: A gene-trap screen to identify potential targets for therapeutic intervention. *Proc. Natl. Acad. Sci. USA* **100,** 14109–14114.
Zambrowicz, B. P., Friedrich, G. A., Buxton, E. C., Lilleberg, S. L., Person, C., and Sands, A. T. (1998). Disruption and sequence identification of 2,000 genes in mouse embryonic stem cells. *Nature* **392,** 608–611.

Index

A

AAA. *See* Aortic arch artery
ACE. *See* Adrenal-cortex derived microvascular endothelial cells
Actin
  in cytoskeleton, 237, 238
  markers of, 243
  in smooth muscle, 179, 243
Activin receptor-like kinase (Alk), 237
ADCL. *See* Autosomal dominant cutis laxa
Adhesion, cellular, 231, 237. *See also* Vascular cell adhesion molecules
  cadherins in, 235, 238
  desmocolin and, 235
  desmoglein and, 235
  EMILIN/Multimerin family and, 168
  MAdCAM and, 23
Adipocytes, in blood vessel development, 10–11
Adrenal gland, in blood vessel development, 18
Adrenal-cortex derived microvascular endothelial cells (ACE), 107
Affymetrix microarrays, 156–57
Alagille syndrome, 18, 92
Angioblasts, 10, 42, 56–57, 58, 232, 269
  in patterning, 68, 69, 70, 72, 129, 143, 193
Angiogenesis, 2–4, 8, 11, 12, 28, 57, 89, 96, 106, 129
  angioblasts in, 10, 42, 56–57, 58, 193, 269
  Bv8 and, 9
  EG-VEGF in, 9
  limitations in adult, 130
  myodullin and, 9
  organ-specific molecules in, 9
  thrombospondins in, 176
Angiopoietins, 64, 129, 234
  from vascular smooth muscle, 89
  vessel wall formation and, 155, 232
Angiostatins, 99

Aorta development, 19, 21, 141, 190–92. *See also* Vascular matrix/aortic development
  angiogenesis in, 89
  coronary circulation and, 241–42
  deficiencies of, 160, 161, 163
  patterning in, 65, 67, 68–71, 73–75, 92, 138, 143, 171, 179
  tenascins in, 178
Aortic arch artery (AAA), 211
  defects of, 213–14
Apoptosis, 242
ARNT. *See* Aryl hydrocarbon nuclear translocator
Arrest chemokines, 23
Artemin (ARTN), 13, 111
  tyrosine kinase receptors and, 13, 111–12
ARTN. *See* Artemin
Aryl hydrocarbon nuclear translocator (ARNT), gene expression and, 38–39, 41, 43–44
ASD. *See* Atrial septal defect
ASPP1, p-53 binding protein, 275
Astrocytes, in nerve development, 12, 110
  LIF and, 12
Atrial septal defect (ASD), 194, 196–97
Atrioventricular (AV) field, 204
Atrioventricular septal defect (AVSD), 196
Autacoids, 24–25
Autosomal dominant cutis laxa (ADCL), 163
AV. *See* Atrioventricular field
Avian model, 24, 25, 57, 228–29
  grafting studies in, 100
  patterning in, 57, 58, 60, 72, 74, 75, 76, 105
AVSD. *See* Atrioventricular septal defect
Axial vein, in zebrafish, 67, 69, 70, 71, 91

B

Bacterial artificial chromosomes (BAC), 199, 200
Bardet-Biedl syndrome, 198

283

Basement membrane (BM), 2, 4–5, 10
  components of, 8, 169–70
  in kidney, 14
Basic fibroblast growth factor (b-FGF), 45
  in quail, 60
BDNF. See Brain-derived neurotrophic factor
Bethlem myopathy, 162
b-FGF. See Basic fibroblast growth factor
Biglycan, 174
Bilateral primordia, in mouse, 226
Blood brain barrier, endothelium and, 5, 6
Blood islands, 129, 274–75
Blood vessel patterning, embryonic midline, 55–56, 58
  animal models used in, 58–61
  axial structures in, 67–75
  development of, 56–58
  differences in features of, 61
  early studies of, 56
  endoderm in, 68, 69–70
  ephrins/eph signaling pathways in, 65–66, 76
  mutant deficiencies in, 70–72
  neural tube in, 68, 73, 74–75
  Notch signaling pathways in, 64–65, 76, 106, 138
  notochord in, 68, 70–74, 75
  semaphorin/neuropilin/plexin pathways in, 55, 66–67, 76, 105
  Tie/Tek signaling pathways in, 55, 60, 64, 76, 129
  universal features of, 61
  VEGF pathways in, 61–63, 66, 67, 68, 70, 71, 72, 74–75
  vessel formation in, 57–58
Blood vessel signals, 1–2
  adult tissues and, 21–26
  basement membrane and, 8
  cell-cell interactions and, 3–4, 10
  development/differentiation and, 9–21
  experimental considerations of, 9
  fenestrae, 5–6
  heterogeneity of ECs in, 8–9
  molecular pathways for, 10
  primary vascular plexus and, 2–3, 16, 21, 42, 55, 75, 109, 129
  reciprocal interaction in, 3–4, 27
  secretion, 6–7
  vascular development and, 2–3, 109
  VEGF and, 3
  VSMC recruitment, 10
Blood vessels
  functions of, 2
  lumen and, 2
BM. See Basement membrane
BMEC. See Bone marrow endothelial cells
BMP. See Bone morphogenetic proteins
Bone cells, in blood vessel development, 11–12
Bone marrow endothelial cells (BMEC), 23, 24, 57
Bone morphogenetic proteins (BMP), 13, 14
  in NCC migration, 236
Brachyury, mouse model and, 71, 271
Brain-derived neurotrophic factor (BDNF), 24, 25, 111, 233
  defects of, 234
Bv8, in angiogenesis, 9

C

Cadherins, in cell adhesion, 235, 238
Cancer, HIFs, hypoxia, vascular development in, 45–47
Cardinal veins, 89, 92, 133, 143
Cardiomycetes, in heart development, 190, 270, 276
Caveolae, 5, 6
  in secretion, 6–7
CD34 receptor, 129
Cell signaling, 5
  juxtacrine, 4, 8, 21, 22, 92, 100, 106
  paracrine, 4, 7, 8, 11, 109, 247
CHD. See Congenital heart defects
Chemokines, 21, 22
  arrest, 23
  keratinocyte-derived, 23
Chondrogenesis, 11
  defects in, 44–45
Cloche mutants, 12, 14, 26
Collagen, 8
  fibrillar, 158–62
  functions of, 159–60
  integrins and, 160
  mutant defects in, 160–62
  proteoglycans and, 174
  structure/expression of, 158–62
  vessel wall formation and, 155–56, 158–62
Congenital heart defects (CHD), 194–98
Connexin, 110
Coronary smooth muscle cells (CoSMC), 227, 230, 233–35, 237, 238, 243–46, 249, 270

Index

Coronary vessel development, 225–26
  anomalies in, 247–48
  bidirectional signaling in, 239–41
  coronary circulation in, 241–42
  defects in, 247–49
  disease in, 249
  EMT gene expression in, 237–39
  epicardial/mesenchymal transformation in, 234–37
  epicardium formation in, 228–31
  lymphatic vessels and, 246–47
  maturation and, 243–46
  proepicardium formation in, 226–28
  subepicardium vasculogenesis in, 231–33
  vasculogenesis/growth factors in, 233–34
CoSMC. See Coronary smooth muscle cells
Cre-lox, in signaling, 61
CSRP. See Cysteine-rich LIM only proteins
Cysteine-rich LIM only proteins (CSRP), 245–46
Cytokines, 38, 88, 89, 199. See also Vascular endothelial growth factor
  interleukins and, 17
  VEGF and, 42
Cytoskeleton
  actin in, 237, 238
  CoSMC differentiation in, 245
  markers of, 154, 179

D

Decorin, 174
Desmocolin, in cell adhesion, 235
Diapedesis, 21, 22
Differentiation
  blood vessel cells and, 9–21
  CoSMC cells and, 245
  ES cells and, 270, 271, 272, 274
  organogenesis and, 87
  vascular systems and, 109
Di-George syndrome, 190, 196, 214–15
  Tbx1 mutation in, 209, 215
  VEGF birth defects, haplotypes in, 209–15
  VEGF isoforms in, 209–11, 212
Double outlet right ventricle (DORV), 196
Down's syndrome (trisomy), 195, 196
Drosophila, 38, 59, 91, 262
  axonal defects in, 103
  neuronal pathfinding in, 106
  sprouty gene in, 99

E

EB. See Embroid bodies
EC. See Endothelial cells
ECM. See Extracellular matrix
EDS. See Ehlers-Danlos syndrome
EE. See Endocardial epithelial cells
EG-VEGF. See Endocrine gland-vascular endothelial growth factor
Ehlers-Danlos syndrome (EDS), 160
Elastic lamella, extracellular matrix and, 156, 157, 160
Elastins
  tropoelastin monomer and, 162, 165
  from vascular smooth muscle cells, 89–90
  vessel wall formation and, 155–56, 164
Embroid bodies (EB), in ES cell differentiation, 270, 271, 272, 274
Embryogenesis, 3–4, 21, 87, 232, 274
  diversification in, 8–9
  HIF role in, 42–44
Embryonic stem cells (ES), 43, 62, 199, 200, 263, 264
  embroid bodies and, 270
  in gene trapping, 263
EMILIN/Multimerin family, 168–69
  cell adhesion and, 168
EMT. See Epithelial-mesenchymal transformation
Endocardial epithelial cells (EE), 24–25
Endocardium, 3, 228
Endocrine gland-vascular endothelial growth factor (EG-VEGF), 9
Endocytosis, 6–7
Endoderm, 3
  blood vessel patterning and, 68, 69–70
Endoglin, 275
Endostatins, 99
Endothelial cells (EC), 2, 56–57, 58, 60, 128, 138–40
  from adrenal cortex, 107
  from blood vessels, 24–25, 207
  from bone marrow, 23–24, 57
  cell biology of, 4–8
  diversification of, 8
  from endocardium, 24–25, 190
  functions of, 4–5, 23
  heterogeneity of, 8–9
  HUVECS, 72
  markers of, 240
  in secretion, 6–7
  sinusoid, in liver, 16

Endothelium
  in brain, 5, 6
  caveolae and, 5, 6
  in embryo, 7
  fenestrae in, 5–6
  in heart development, 225
  homeostasis regulation by, 2
  in liver, 5, 6
  in lung, 5, 6, 8
  in molecular transport, 128
  pancreatic islets and, 5, 6
  in smooth muscle, 5
Eno gene, in gene trapping, limitation of, 276
Entactin/nidogen, 8
  basement membrane and, 171–72
Entrapment vectors, in gene trapping, 263–66
  mechanisms of, 264–65
  polyadenylation trap vectors and, 266
  retroviral trap vectors and, 274
EPDC. *See* Epicardial-derived mesenchymal cells
Ephrins/eph, in signaling, 55, 65–66, 76, 102, 129, 207
  arteriovenous generation and, 91, 92, 93, 99, 100, 107, 110
  reverse signaling with, 103–4
Epicardial-derived mesenchymal cells (EPDC), 234–35, 237, 242
  CoSMCs and, 243
Epicardium, 228–31
  cell migration in, 229–30
  EMT in, 234
  marker genes of, 240
  in signaling, 239–41
Epidermal growth factor, 165
Epithelial-mesenchymal transformation (EMT), 204–5, 226, 230, 234–37, 238
  gene expression in, 237–39
Erythropoietin (EPO), 38, 43
  receptor of, 240
ES. *See* Embryonic stem cells
Exocytosis, 7
Extracellular matrix (ECM), 2, 8, 11, 208, 227
  elastic lamella in, 156, 157, 160
  gene expression in, 154–55
  proteoglycans and, 156–57, 226
  in subepicardium, 231
  vascular smooth muscle role in, 155
  from VSMC, 155

F

F-actin, semaphorins and, 142
Fate mapping, 193, 225
Fenestrae, in blood vessels, 5–6
FGF. *See* Fibroblast growth factors
Fibrillin/microfibrils, 165–66
  integrin-binding sequence in, 165
Fibroblast growth factors (FGF), 45, 60, 89, 129, 232, 236
  inhibitors of, 90
  Spry regulator of, 90
Fibroblasts, from muscle, 155
Fibronectin, 229
  patterning and, 111, 169
Fibulins, 166–68
  ECM interaction with, 166
  mutant defects in, 168
Filopodia, 205, 206
Flk-1 receptor, 17, 20, 42–43, 62–63, 104–5, 110, 137, 271, 272
  VGEF and, 112–13, 132, 138–39, 202
Flt-1 receptor, 42, 43, 62–63, 104–5
  VEGF and, 137–38, 202
Fog2 mutant, 239–41
Frog model, 228–29
  eph gene in, 106–7
  ephrin-lacking mutants in, 97
  patterning in, 55, 58, 59, 76

G

Gastrulation, 72, 274
  HIF role in, 42
Gene trap screens, 266–69
  types of information obtained by, 266, 267, 268–69
Gene trapping
  entrapment screen design in, 272–76
  entrapment vectors in, 263–66
  ES cell model in, 269–72
  mutagenesis screens, chemical and, 262
  screens in, 199, 266–69
  vector-induced mutagenesis and, 199
Genetics, role of VEGF, 189–90
  abnormal remodeling of PAAs in, 207–14
  animal models in, 198–201
  in arterial EC specification, 207
  cardiovascular development and, 201–3
  chromosome anomalies in, 195–96
  comparison of animal models in, 198–201

Index

congenital heart defects in, 194–98
conserved body plan in, 190–93
early heart development, 190–93
expression patterns and, 203–4
gene variations of DGS in, 214–15
malformation syndromes in, 196–97
multifactorial CHDs in, 197
septation and, 204–5, 206
transcription factors in, 196–97
in vascular/neuronal development, 205–6
GFP. *See* Green fluorescent protein
Glomulin gene, 198
Glucagon, 19
GM-CSF. *See* Granulocyte/macrophage colony stimulating factor
Granulocyte/macrophage colony stimulating factor (GM-CSF), 23, 24
Green fluorescent protein (GFP), 59–60, 134–35
enhanced form of, 275

H

Hamburger-Hamilton (HH), developmental stage, 227, 229
Heart cells, in blood vessel development, 14
adults and, 24–25
Hemangioblasts, in blood vessel development, 129, 270
Hemangiogenic progenitors, 272
Hemangiomas, 195
Hematopoietic cells, in blood vessel development, 12, 129, 269
adult tissue and, 23–24
stem cells and, 23–24
Hepatocyte growth factor (HGF), 25
Hepatocytes, 16
necrosis of, 25
HGF. *See* Hepatocyte growth factor
HH. *See* Hamburger-Hamilton
HIF. *See* Hypoxia inducible factors
HIFs, hypoxia, vascular development.
 See also Hypoxia inducible factors
cancer and, 45–47
disease and, 44–47
mechanisms of, 40–41
mutations of, 43–44
oxygen homeostasis and, 37–38, 42–44
physical structure, 38–39
physiological limits and, 38

regulation of, 39–41
target genes of, 38–39
Higher vocal center (HVC), in songbirds, 24, 25, 111
HNF mutant mice, 18
HRE. *See* Hypoxia response element
Human umbilical vein endothelial cells (HUVEC), 72
HVC. *See* Higher vocal center
Hyperoxia, 37
Hypochord, in blood vessel patterning, 68–69, 71
Hypoxia, ATP production and, 38
Hypoxia inducible factors (HIF). *See also* HIFs, hypoxia, vascular development
disease of, 44–47
family structure of, 38–39
mechanisms of, 40–41
mutations of, 43–44
target genes of, 38–39, 40
tumorigenesis and, 46–47
Hypoxia response element (HRE), 39

I

ICAM. *See* Intercellular adhesion molecules
Identifying early vascular genes, 261–62
chemical v. gene trap mutagenesis in, 262–63
entrapment screen design in, 272–76
entrapment vectors in, 263–66
ES cell gene trap model in, 269–72
gene trap screens in, 266–69
IEL. *See* Internal elastic lamina
IHBD. *See* Intrahepatic biliary ducts
IL-1. *See* Interleukin-1
IL-6. *See* Interleukin-6
IL-8. *See* Interleukin-8
In situ hybridization (ISH)
in study of zebrafish, 134, 136
thrombospondin study by, 175
Insulin, 19, 20
Integrins, 23, 231
collagen interaction with, 160
fibrillin interaction with, 165
netrins in signaling and, 108
in quail, 60
in signaling, 239
Intercellular adhesion molecules (ICAM), 23

Interleukin-1 (IL-1), 23
Interleukin-6 (IL-6), 26
Interleukin-8 (IL-8), 7, 22, 45
Internal elastic lamina (IEL), 155–56
Intersegmental vessels (ISV), 130, 133, 137, 138, 140
Intrahepatic biliary ducts (IHBD), 17
ISH. *See In situ* hybridization
ISV. *See* Intersegmental vessels

J

Jagged-1, Notch ligand, 106
Juxtacrine cell signaling, 4, 8, 21, 22, 92
  eph proteins and, 100, 106

K

Keratinocyte-derived chemokines (KC), 23
Kidney cells, in blood vessel development, 14–15
Kidney glomerular cells, 6, 8, 14

L

lacZ promoter-less reporter gene, 263, 266, 274–75
LAD coronary artery, 248
  significance in arterial development, 249
Laminins, 8, 169, 170–71
Leukemia inhibitory factor (LIF), 263
Leukocytes, 6, 7
  cell recruitment and, 21, 22
  function of, 21–23
LIF. *See* Leukemia inhibitory factor
Lineage mapping studies, 243
Liver
  in blood vessel development, 15–18
  endothelium of, 5, 6, 16
  parenchymal cells in, 25
  primordium in, 16, 26
  septum transversum in, 16, 17
  VEGF cells in, 25
Liver cells, in blood vessel development, adults and, 25–26
Lumican, 174
Lymphatic system development, 93, 95, 246–47
  VEGF and, 247
Lysosomes, 6

M

MAdCAM. *See* Mucosal addressin cellular adhesion molecules
MAGP. *See* Microfibril-associated glycoproteins
Marfan syndrome, 166
MEF. *See* Mouse embryonic fibroblasts
Mesenchyme, 16, 17, 18, 19, 73, 106. *See also* Epithelial-mesenchymal transformation (EMT)
  stellate, 227
  vessel development and, 155, 157, 232, 234–37
Mesoderm, 3, 269
  somitic, 20
Mesothelium, 227
Metalloproteinases, 7
Microangiography, 132–34, 140
Microarrays, 156, 157, 158, 179
Microfibril-associated glycoproteins (MAGP), 165
Microvascular epithelial cells (MVE), 24–25
Mitogens, 87, 98, 100, 113, 233
MO. *See* Morpholino phosphorodiamidate
Morpholino antisense knockdown, in study of zebrafish model, 135, 136, 141–42, 202, 206, 207, 213
Morpholino phosphorodiamidate (MO), 135
Mouse embryonic fibroblasts (MEF), 263
Mouse model, 14, 15, 21, 26, 62, 109, 110, 165, 189
  bilateral primordia in, 226
  Brachyury probe, 71, 271
  differentiation in, 263
  ephrin-lacking mutants in, 97
  experimental tools for, 199
  Fog2 mutant, 239, 240, 241
  gridlock mutant, 91
  human model homology with, 199
  knockin, 105
  knockout, 45, 67, 106, 108, 136, 138, 141, 169, 206, 208
  Notch gene in, 18, 137–38
  patterning in, 60–61, 62
  robo receptors in, 107–8
  TBX1 deficient gene in, 205, 206, 209, 211, 213, 215
  transgenic, 20, 64, 137, 209, 234, 275
  VEGF deficient gene in, 202–3, 207–11
  VEGF gene overexpression in, 202–3
  VEGFR2 and, 20

Index

MRTF. *See* Myocardin-related transcription factors
Mucosal addressin cellular adhesion molecules (MAdCAM), 23
Mural cells, in blood vessel development, 10
Mutagenesis, zebrafish and, 131–32
Mutagenesis screens, chemical, 262
  balancer chromosome in, 262
Mutagenesis screens, gene trap, 199, 262–63
  altered expression in, 262–63
  DNA insertion in, 262–63
MVE. *See* Microvascular epithelial cells
Myocardin, 245
Myocardin-related transcription factors (MRTF), 245, 246
Myocardium, 225, 228
  defects of, 239, 240, 241
  in signaling, 239–41
Myodullin, in angiogenesis, 9
Myosin, from smooth muscle, 179

N

NCC. *See* Neural crest cells
Netrins, in signaling, 108
  integrin receptor for, 108
  neogenin receptor of, 102, 108
  patterning of, 111
  UNC-5 receptor of, 102
Neural crest cells (NCC)
  caudal hindbrain and, 192
  development of, 155, 191–93
  PA development and, 209, 210
  transforming growth factor-β and, 13, 236–37
Neural stem cells (NSC), 24
Neural tubes, in blood vessel patterning, 68, 73, 74–75
Neuregulin, in heart tissue development, 14, 112
Neuroblastoma cells, 106
Neuron cells, in blood vessel development, 12–14
Neuropilins (NRP), in signaling, 62, 66–67, 93, 104, 105, 110, 127, 129
  defects in, 66, 211
  extracellular domains of, 139–40
  neuronal development and, 205–6
  overexpression of, 67
  receptor of, 67

  vascular development and, 140, 141, 205–6
  VGEF and, 112–13, 139–42
NFAT transcription factors, 107
Nidogen, 8
Nitric oxide synthase (e-NOS), 7, 233
Notch gene, 18, 55
  overexpression of, 106
  stroke/dementia and, 64–65, 92
Notch pathway
  in cell fate determination, 105
  in signaling, 64–65, 72, 76, 99, 136, 138
  in vascular development, 91, 92
  in VEGF regulation, 137–38
Notch proteins, 64–65, 91
Notochord, 20, 21
  angioblasts and, 193
  aorta assembly, role in, 73–74
  blood vessel patterning and, 67, 68, 70–74, 75, 193
  VEGF and, 91, 207
NRP. *See* Neuropilins (NRP), in signaling
NSC. *See* Neural stem cells
Null embryos, 17, 21, 43, 45, 46, 65, 174
  integrin, 231
  mutations of, 106, 161, 167, 168, 178–79

O

OI. *See* Osteogenesis imperfecta
Organogenesis, 21, 274
Osteogenesis, 11, 28
Osteogenesis imperfecta (OI), 160
Out-of-bound mutants, in zebrafish, 97

P

PA. *See* Pharyngeal arches
PAA. *See* Pharyngeal arch artery system
Pancreas cells, in blood vessel development, 19–21
Pancreatic islets, 5, 6, 8
Paracrine cell signaling, 4, 7, 8, 11, 109
  from parasympathetic ganglia, 247
Passive diffusion, HIF role in, 42
Pax-6, endocrine progenitor marker, 20
PCR. *See* Polymerase chain reaction
PDGF. *See* Platelet derived growth factor
PE. *See* Proepicardium
PE molecular markers, 228

PECAM, in vasculogenesis, 205, 275
PEDF. *See* Pigment epithelium-derived factor
Pericardial serosa, 226
Pericytes, 3, 42, 234, 244
Perineural vascular plexus (PNVP), 73
  neural tube patterning and, 74–75
Pharyngeal arch artery system (PAA), 191–92, 208
Pharyngeal arch, development, neural crest cells and, 209, 210
Pharyngeal arches (PA), 191, 192
Pigment epithelium-derived factor (PEDF), 15–16
Placental growth factor, 233
Platelet derived growth factor (PDGF), 10, 39, 173, 233, 237, 244
  dependent smooth muscle cells and, 272
Plexin, in signaling, 66, 67, 105
Polyadenylation trap vectors, in gene trapping, 266, 277
Polymerase chain reaction (PCR), 21, 164
Primary vascular plexus, 2–3, 16, 21, 42, 55, 75, 109, 129
Proepicardium (PE), 225, 226–28
  coronary vasculature from, 226, 227
  epicardial cell origins in, 230
  vesicles in, 229
Prostate cells, in blood vessel development, 15–16
Proteoglycans
  basement membrane and, 169
  collagen fibrillogenesis and, 174
  expression patterns of, 174
  extracellular matrix and, 156–57, 226
  hyaluronan interaction with, 172–74
  leucine rich family of, 174–75
  patterning affected by, 173–74, 208
PubMed, search engine, 3
pVHL. *See* von Hippel-Lindau tumor suppressor protein

Q

QH1, endothelial marker, 236

R

RACE technology, 199
RALDH-2. *See* Retinaldehyde dehydrogenase-2
Reactive oxygen species (ROS), 37

Reciprocal interaction, in cell signaling, 3–4, 14, 27
  adrenals and, 24
  in liver, 18
  in songbirds, 24, 25
Respiratory distress syndrome, 44
Retinaldehyde dehydrogenase-2 (RALDH-2), PE cell marker, 228, 240
Retinoid X receptors (RXR), 240
Roundabout (robo) receptors, in signaling, 103, 107–8, 112

S

Schwann cells, 110, 112
Secreted protein acidic, rich in cysteine (SPARC), 178–79
  mutant defects from, 178–79
Selectins, 6, 22–23
Semaphorins, in signaling, 55, 76, 102, 104–5
  f-actin and, 142
  inhibitory effect of, 112–13, 142, 205–6
  neuropilin binding with, 66, 67, 142–43
  overexpression with, 143
  plexin interaction with, 66, 67, 105
  tumors and, 142
Septation
  heart development and, 190, 192, 204–5
  VEGF in, 204–5, 206
Septum transversum, 226
  in liver, 16, 17
Serum response factor (SRF), 228, 237, 239, 245, 246
SIV. *See* Subintestinal vein
Slits, in signaling, 102
  roundabout (robo) receptors of, 103, 108
SLRP. *See* Small leucine rich proteoglycans
Small leucine rich proteoglycans (SLRP), 174–75
  collagen fibrillogenesis by, 174
Smooth muscle, 42, 136, 179, 243, 272. *See also* Vascular smooth muscle cells
  endothelium and, 5
  in lymphatic system, 246–47
  vascular, 88, 225, 276
Smooth muscle cells (SMC), 228, 241, 244, 246
  defects in, 242
Songbird model, 24, 25
  higher vocal center, 24, 25, 111

Index 291

Sonic hedgehog (Shh) mutant, zebrafish model and, 69, 91, 132, 137–38, 207
SPARC. *See* Secreted protein acidic, rich in cysteine
SPARC/osteonectin, 178
SRF. *See* Serum response factor
Stem cells, 28
  embroid bodies and, 270
  embryonic, 43, 62, 199, 200, 263, 264
  hematopoietic, 23–24
  neural, 24
Stenosis
  aortic, 163, 248
  pulmonary artery, 210
  pulmonary valve, 196
Stroke/dementia (CADASIL), notch gene in, 64–65, 92
Subependymal zone/lateral ventricle (SVZ), 24
Subepicardial layer, extracellular matrix of, 226
Subepicardium, 231–33
  vasculogenesis in, 231–33
Subintestinal vein (SIV), 139
Superior cervical ganglion (SCG), 13
Supravalvular aortic stenosis (SVAS), 163, 248
SVAS. *See* Supravalvular aortic stenosis
SVZ. *See* Subependymal zone/lateral ventricle

T

T cells, 23, 136
T-box transcription factor (TBX1), 205, 206, 211, 213, 215
Tenascins, cell proliferation and, 176–78
Testosterone, 15, 24
  VEGF regulation and, 24, 111
Tetralogy of Fallot (ToF), 196, 197, 210, 214
TGF-β. *See* Transforming growth factor-β
Thrombospondins (TSP), 175–76
  TGF-β interaction with, 176
Tie/tek
  in coronary wall formation, 234
  in signaling, 55, 60, 64, 76, 129
  in vessel wall formation, 155
TNF. *See* Tumor necrosis factor
ToF. *See* Tetralogy of Fallot
Transcription factor NF-AT, in signaling, 205
Transcytosis, 5, 6–7, 22

Transforming growth factor-β (TGF-β), 10, 89
  in cancer, 45
  in heart tissue development, 14
  in neural crest development, 13, 236–37
Transgenesis, high throughput automated, 199
TSP. *See* Thrombospondins
Tumor necrosis factor (TNF), 23
  in cancer, 45
Tumors, 136, 142
Tunica adventitia, 155
Tunica intima, 155
Tunica media, 155, 243

U

Ullrich's disease, 162

V

Vascular cell adhesion molecules (VCAM), 23
  functions of collagen and, 159–60
Vascular endothelial growth factor (VEGF), 3, 27–28, 39, 42–43, 45–47, 55. *See also* Genetics role of VEGF; VEGF, in blood vessel development
  in adipose tissue development, 11
  in arteriogenesis, 99
  in bone development, 11
  chemoattractant properties of, 109
  in endocrine gland cells, 9
  as global endothelial cell mitogen, 114
  in heart development, 14, 232, 233–34
  in hematopoietic cells, 24
  isoforms of, 207–14, 233–34
  in kidney development, 15
  in liver cells, 25
  in lymphatic system, 247
  mitogenic properties of, 136
  in pancreas development, 20, 21
  in patterning, 68, 70, 71, 72, 74–75, 94, 95, 98–99, 110, 136, 139
  in prostate development, 15
  in quail, 60
  receptors for, 42–43, 129, 138–39
  in signaling, 61–63, 65, 66–67, 72, 76, 89, 104–5, 239, 241
  in songbirds, 25
  testosterone effect on, 24, 111
  in tumor development, 136
  in zebrafish, 127–44

Vascular matrix/aortic development
  basement membrane in, 169–70
  collagen mutations in, 160–62
  cytoskeletal markers of, 154, 179
  elastic fiber in, 162–66
  EMILIN/Multimerins in, 168–69
  evolution of, 153–55
  extracellular matrix, role in, 156–58
  fibrillin/microfibrils, 165–66
  fibronectins in, 169
  fibulins in, 166–68
  formation/structure of, 155–56
  functions of collagen in, 159–60
  gene expression in, 154
  hemodynamics of, 156
  matricellular proteins in, 175–79
  matrix pattern in, 157, 158
  morphology/histology of, 154–55, 157, 158, 159, 164, 173, 177
  mutant defects in, 163
  proteoglycans in, 172–75
Vascular patterning, primary vessel networks and, 55–77, 156–58
Vascular smooth muscle cells (VSMC), 3, 8, 13, 14, 155, 160
  angiopoietins from, 89
  elastins from, 89–90
  extracellular matrix from, 155
  migration of, 173–74
  mural cells from, 9, 10
Vasculogenesis, 2–3, 42–44, 57, 89, 106, 129, 193, 269
  in adults, 130
  coronary growth factors and, 233–34
  PECAM in, 205
  in subepicardium, 231–33
VCAM. See Vascular cell adhesion molecules
VEGF. See Vascular endothelial growth factor
VEGF, in blood vessel development, 127–28, 136–38
  EC markers and, 128
  isoforms and, 207–14
  mutant deficiencies and, 136–38, 207–14
  neuropilins and, 139–42, 206
  vascular growth factors in zebrafish and, 128–29
  VEGF receptors and, 138–39
  in zebrafish, 129–30
  zebrafish model advantages and, 130–32

zebrafish model analysis methods and, 132–36
Ventricular septal defect (VSD), 194, 196–97, 214
Versican, 173–74
von Hippel-Lindau tumor suppressor protein (pVHL), 40, 41
  tumor suppressor qualities of, 45–46
von Willebrand factor (vWF), 6–7
VSD. See Ventricular septal defect
VSMC. See Vascular smooth muscle cells
vWF. See von Willebrand factor

W

Weibel-Palade body (WPB), 6–7
  vWF from, 6–7
Western blotting, 135
William's syndrome, 163
Wilms' tumor-1 (WT1), 228, 235, 239
Wiring/vascular circuitry
  angiogenesis and, 89
  attractive v. repulsive cues in, 100
  defective mutations and, 92–93, 95
  diseases and, 92, 95
  factors involved in, 89–90
  gene-targeting in, 89
  genetic evidence for vascular guidance in, 96–98
  guidance molecules in, 101, 102, 103–4
  independent v. interdependent guidance, 112–13
  local cues/arterial-venous identity and, 95–96
  lymphatic system development and, 93, 95
  neural guidance pathways and, 100–4, 101, 113
  neural/vessel guidance colinearity in, 112–14
  signaling pathways in, 88–89, 104–12
  test methods for, 89–90
  vasculogenesis and, 89
  VEGF as guidance cue in, 98–99
  VEGF overexpression in, 96
  vessel network formation and, 90–95
WPB. See Weibel-Palade body
WT1. See Wilms' tumor-1

X

Xenograft model, 46–47
Xenopus. See Frog model

Index

Z

Zebrafish information network, 132
Zebrafish international resource center, 132
Zebrafish model, 12, 14, 26, 68–69, 76, 189
  advantages of, 130–32, 143–44, 200, 201
  alkaline phosphatase staining, in study of, 133, 134
  floating head mutant, 70–72, 193
  gridlock mutant, 90–91, 132
  high throughput assays, in study of, 136, 200–1
  in situ hybridization, in study of, 133, 134
  jekyll mutant, 132
  microangiography, in study of, 132–34, 140
  MO treated genes in, 141–42, 201, 202
  morpholino knockdown, in study of, 135, 136, 141–42, 202, 206, 207, 213
  mosaic embryos in, 71
  mutagens used with, 131–32
  no tail mutant, 70–72, 193
  NRP genes in, 141–42
  one-eyed pinhead mutant, 132
  out-of-bound mutants, 97
  patterning in, 55, 59–60, 62, 65, 67
  robo receptors in, 107–8
  semaphorin overexpression in, 143
  silent heart mutant, 96
  sonic hedgehog (Shh) mutant, 69, 91, 132, 137–38, 207
  transgenics, in study of, 134–35
  van gogh mutant, 209
  vascular endothelial growth factor in, 127–44
Zona glomerulosa, 18

Contents of Previous Volumes

Volume 47

1. **Early Events of Somitogenesis in Higher Vertebrates: Allocation of Precursor Cells during Gastrulation and the Organization of a Moristic Pattern in the Paraxial Mesoderm**
   Patrick P. L. Tam, Devorah Goldman, Anne Camus, and Gary C. Shoenwolf

2. **Retrospective Tracing of the Developmental Lineage of the Mouse Myotome**
   Sophie Eloy-Trinquet, Luc Mathis, and Jean-François Nicolas

3. **Segmentation of the Paraxial Mesoderm and Vertebrate Somitogenesis**
   Olivier Pourqulé

4. **Segmentation: A View from the Border**
   Claudio D. Stern and Daniel Vasiliauskas

5. **Genetic Regulation of Somite Formation**
   Alan Rawls, Jeanne Wilson-Rawls, and Eric N. Olsen

6. **Hox Genes and the Global Patterning of the Somitic Mesoderm**
   Ann Campbell Burke

7. **The Origin and Morphogenesis of Amphibian Somites**
   Ray Keller

8. **Somitogenesis in Zebrafish**
   Scott A. Halley and Christiana Nüsslain-Volhard

9. **Rostrocaudal Differences within the Somites Confer Segmental Pattern to Trunk Neural Crest Migration**
   Marianne Bronner-Fraser

Volume 48

1. **Evolution and Development of Distinct Cell Lineages Derived from Somites**
   Beate Brand-Saberi and Bodo Christ

2. **Duality of Molecular Signaling Involved in Vertebral Chondrogenesis**
   Anne-Hélène Monsoro-Burq and Nicole Le Douarin

3. **Sclerotome Induction and Differentiation**
   Jennifer L. Docker

4. **Genetics of Muscle Determination and Development**
   Hans-Henning Arnold and Thomas Braun

5. **Multiple Tissue Interactions and Signal Transduction Pathways Control Somite Myogenesis**
   Anne-Gaëlle Borycki and Charles P. Emerson, Jr.

6. **The Birth of Muscle Progenitor Cells in the Mouse: Spatiotemporal Considerations**
   Shahragim Tajbakhsh and Margaret Buckingham

7. **Mouse–Chick Chimera: An Experimental System for Study of Somite Development**
   Josiane Fontaine-Pérus

8. **Transcriptional Regulation during Somitogenesis**
   Dennis Summerbell and Peter W. J. Rigby

9. **Determination and Morphogenesis in Myogenic Progenitor Cells: An Experimental Embryological Approach**
   Charles P. Ordahl, Brian A. Williams, and Wilfred Denetclaw

Volume 49

1. **The Centrosome and Parthenogenesis**
   Thomas Küntziger and Michel Bornens

2. **γ-Tubulin**
   Berl R. Oakley

3 γ-Tubulin Complexes and Their Role in Microtubule Nucleation
Ruwanthi N. Gunawardane, Sofia B. Lizarraga, Christiane Wiese, Andrew Wilde, and Yixian Zheng

4 γ-Tubulin of Budding Yeast
Jackie Vogel and Michael Snyder

5 The Spindle Pole Body of *Saccharomyces cerevisiae*: Architecture and Assembly of the Core Components
Susan E. Francis and Trisha N. Davis

6 The Microtubule Organizing Centers of *Schizosaccharomyces pombe*
Iain M. Hagan and Janni Petersen

7 Comparative Structural, Molecular, and Functional Aspects of the *Dictyostelium discoideum* Centrosome
Ralph Gräf, Nicole Brusis, Christine Daunderer, Ursula Euteneuer, Andrea Hestermann, Manfred Schliwa, and Masahiro Ueda

8 Are There Nucleic Acids in the Centrosome?
Wallace F. Marshall and Joel L. Rosenbaum

9 Basal Bodies and Centrioles: Their Function and Structure
Andrea M. Preble, Thomas M. Giddings, Jr., and Susan K. Dutcher

10 Centriole Duplication and Maturation in Animal Cells
B. M. H. Lange, A. J. Faragher, P. March, and K. Gull

11 Centrosome Replication in Somatic Cells: The Significance of the G_1 Phase
Ron Balczon

12 The Coordination of Centrosome Reproduction with Nuclear Events during the Cell Cycle
Greenfield Sluder and Edward H. Hinchcliffe

13 Regulating Centrosomes by Protein Phosphorylation
Andrew M. Fry, Thibault Mayor, and Erich A. Nigg

14 The Role of the Centrosome in the Development of Malignant Tumors
Wilma L. Lingle and Jeffrey L. Salisbury

15 The Centrosome-Associated Aurora/Ipl-like Kinase Family
T. M. Goepfert and B. R. Brinkley

16 **Centrosome Reduction during Mammalian Spermiogenesis**
   G. Manandhar, C. Simerly, and G. Schatten

17 **The Centrosome of the Early *C. elegans* Embryo: Inheritance, Assembly, Replication, and Developmental Roles**
   Kevin F. O'Connell

18 **The Centrosome in *Drosophila* Oocyte Development**
   Timothy L. Megraw and Thomas C. Kaufman

19 **The Centrosome in Early *Drosophila* Embryogenesis**
   W. F. Rothwell and W. Sullivan

20 **Centrosome Maturation**
   Robert E. Palazzo, Jacalyn M. Vogel, Bradley J. Schnackenberg, Dawn R. Hull, and Xingyong Wu

Volume 50

1 **Patterning the Early Sea Urchin Embryo**
   Charles A. Ettensohn and Hyla C. Sweet

2 **Turning Mesoderm into Blood: The Formation of Hematopoietic Stem Cells during Embryogenesis**
   Alan J. Davidson and Leonard I. Zon

3 **Mechanisms of Plant Embryo Development**
   Shunong Bai, Lingjing Chen, Mary Alice Yund, and Zinmay Rence Sung

4 **Sperm-Mediated Gene Transfer**
   Anthony W. S. Chan, C. Marc Luetjens, and Gerald P. Schatten

5 **Gonocyte–Sertoli Cell Interactions during Development of the Neonatal Rodent Testis**
   Joanne M. Orth, William F. Jester, Ling-Hong Li, and Andrew L. Laslett

6 **Attributes and Dynamics of the Endoplasmic Reticulum in Mammalian Eggs**
   Douglas Kline

7 **Germ Plasm and Molecular Determinants of Germ Cell Fate**
   Douglas W. Houston and Mary Lou King

Volume 51

1. **Patterning and Lineage Specification in the Amphibian Embryo**
   Agnes P. Chan and Laurence D. Etkin

2. **Transcriptional Programs Regulating Vascular Smooth Muscle Cell Development and Differentiation**
   Michael S. Parmacek

3. **Myofibroblasts: Molecular Crossdressers**
   Gennyne A. Walker, Ivan A. Guerrero, and Leslie A. Leinwand

4. **Checkpoint and DNA-Repair Proteins Are Associated with the Cores of Mammalian Meiotic Chromosomes**
   Madalena Tarsounas and Peter B. Moens

5. **Cytoskeletal and Ca^{2+} Regulation of Hyphal Tip Growth and Initiation**
   Sara Torralba and I. Brent Heath

6. **Pattern Formation during *C. elegans* Vulval Induction**
   Minqin Wang and Paul W. Sternberg

7. **A Molecular Clock Involved in Somite Segmentation**
   Miguel Maroto and Olivier Pourquié

Volume 52

1. **Mechanism and Control of Meiotic Recombination Initiation**
   Scott Keeney

2. **Osmoregulation and Cell Volume Regulation in the Preimplantation Embryo**
   Jay M. Baltz

3. **Cell–Cell Interactions in Vascular Development**
   Diane C. Darland and Patricia A. D'Amore

4. **Genetic Regulation of Preimplantation Embryo Survival**
   Carol M. Warner and Carol A. Brenner

Volume 53

1. **Developmental Roles and Clinical Significance of Hedgehog Signaling**
   Andrew P. McMahon, Philip W. Ingham, and Clifford J. Tabin

2. **Genomic Imprinting: Could the Chromatin Structure Be the Driving Force?**
   Andras Paldi

3. **Ontogeny of Hematopoiesis: Examining the Emergence of Hematopoietic Cells in the Vertebrate Embryo**
   Jenna L. Galloway and Leonard I. Zon

4. **Patterning the Sea Urchin Embryo: Gene Regulatory Networks, Signaling Pathways, and Cellular Interactions**
   Lynne M. Angerer and Robert C. Angerer

Volume 54

1. **Membrane Type-Matrix Metalloproteinases (MT-MMP)**
   Stanley Zucker, Duanqing Pei, Jian Cao, and Carlos Lopez-Otin

2. **Surface Association of Secreted Matrix Metalloproteinases**
   Rafael Fridman

3. **Biochemical Properties and Functions of Membrane-Anchored Metalloprotease-Disintegrin Proteins (ADAMs)**
   J. David Becherer and Carl P. Blobel

4. **Shedding of Plasma Membrane Proteins**
   Joaquín Arribas and Anna Merlos-Suárez

5. **Expression of Meprins in Health and Disease**
   Lourdes P. Norman, Gail L. Matters, Jacqueline M. Crisman, and Judith S. Bond

6. **Type II Transmembrane Serine Proteases**
   Qingyu Wu

7. **DPPIV, Seprase, and Related Serine Peptidases in Multiple Cellular Functions**
   Wen-Tien Chen, Thomas Kelly, and Giulio Ghersi

Contents of Previous Volumes

8 **The Secretases of Alzheimer's Disease**
   Michael S. Wolfe

9 **Plasminogen Activation at the Cell Surface**
   Vincent Ellis

10 **Cell-Surface Cathepsin B: Understanding Its Functional Significance**
   Dora Cavallo-Medved and Bonnie F. Sloane

11 **Protease-Activated Receptors**
   Wadie F. Bahou

12 **Emmprin (CD147), a Cell Surface Regulator of Matrix Metalloproteinase Production and Function**
   Bryan P. Toole

13 **The Evolving Roles of Cell Surface Proteases in Health and Disease: Implications for Developmental, Adaptive, Inflammatory, and Neoplastic Processes**
   Joseph A. Madri

14 **Shed Membrane Vesicles and Clustering of Membrane-Bound Proteolytic Enzymes**
   M. Letizia Vittorelli

Volume 55

1 **The Dynamics of Chromosome Replication in Yeast**
   Isabelle A. Lucas and M. K. Raghuraman

2 **Micromechanical Studies of Mitotic Chromosomes**
   M. G. Poirier and John F. Marko

3 **Patterning of the Zebrafish Embryo by Nodal Signals**
   Jennifer O. Liang and Amy L. Rubinstein

4 **Folding Chromosomes in Bacteria: Examining the Role of Csp Proteins and Other Small Nucleic Acid-Binding Proteins**
   Nancy Trun and Danielle Johnston

Volume 56

1. **Selfishness in Moderation: Evolutionary Success of the Yeast Plasmid**
   Soundarapandian Velmurugan, Shwetal Mehta, and Makkuni Jayaram

2. **Nongenomic Actions of Androgen in Sertoli Cells**
   William H. Walker

3. **Regulation of Chromatin Structure and Gene Activity by Poly(ADP-Ribose) Polymerases**
   Alexei Tulin, Yurli Chinenov, and Allan Spradling

4. **Centrosomes and Kinetochores, Who needs 'Em? The Role of Noncentromeric Chromatin in Spindle Assembly**
   Priya Prakash Budde and Rebecca Heald

5. **Modeling Cardiogenesis: The Challenges and Promises of 3D Reconstruction**
   Jeffrey O. Penetcost, Claudio Silva, Maurice Pesticelli, Jr., and Kent L. Thornburg

6. **Plasmid and Chromosome Traffic Control: How ParA and ParB Drive Partition**
   Jennifer A. Surtees and Barbara E. Funnell

Volume 57

1. **Molecular Conservation and Novelties in Vertebrate Ear Development**
   B. Fritzsch and K. W. Beisel

2. **Use of Mouse Genetics for Studying Inner Ear Development**
   Elizabeth Quint and Karen P. Steel

3. **Formation of the Outer and Middle Ear, Molecular Mechanisms**
   Moisés Mallo

4. **Molecular Basis of Inner Ear Induction**
   Stephen T. Brown, Kareen Martin, and Andrew K. Groves

5. **Molecular Basis of Otic Commitment and Morphogenesis: A Role for Homeodomain-Containing Transcription Factors and Signaling Molecules**
   Eva Bober, Silke Rinkwitz, and Heike Herbrand

6 **Growth Factors and Early Development of Otic Neurons: Interactions between Intrinsic and Extrinsic Signals**
   Berta Alsina, Fernando Giraldez, and Isabel Varela-Nieto

7 **Neurotrophic Factors during Inner Ear Development**
   Ulla Pirvola and Jukka Ylikoski

8 **FGF Signaling in Ear Development and Innervation**
   Tracy J. Wright and Suzanne L. Mansour

9 **The Roles of Retinoic Acid during Inner Ear Development**
   Raymond Romand

10 **Hair Cell Development in Higher Vertebrates**
   Wei-Qiang Gao

11 **Cell Adhesion Molecules during Inner Ear and Hair Cell Development, Including Notch and Its Ligands**
   Matthew W. Kelley

12 **Genes Controlling the Development of the Zebrafish Inner Ear and Hair Cells**
   Bruce B. Riley

13 **Functional Development of Hair Cells**
   Ruth Anne Eatock and Karen M. Hurley

14 **The Cell Cycle and the Development and Regeneration of Hair Cells**
   Allen F. Ryan

Volume 58

1 **A Role for Endogenous Electric Fields in Wound Healing**
   Richard Nuccitelli

2 **The Role of Mitotic Checkpoint in Maintaining Genomic Stability**
   Song-Tao Liu, Jan M. van Deursen, and Tim J. Yen

3 **The Regulation of Oocyte Maturation**
   Ekaterina Voronina and Gary M. Wessel

4 **Stem Cells: A Promising Source of Pancreatic Islets for Transplantation in Type 1 Diabetes**
   Cale N. Street, Ray V. Rajotte, and Gregory S. Korbutt

5 Differentiation Potential of Adipose Derived Adult Stem (ADAS) Cells
  Jeffrey M. Gimble and Farshid Guilak

Volume 59

1 The Balbiani Body and Germ Cell Determinants: 150 Years Later
  Malgorzata Kloc, Szczepan Bilinski, and Laurence D. Etkin

2 Fetal–Maternal Interactions: Prenatal Psychobiological Precursors to Adaptive Infant Development
  Matthew F. S. X. Novak

3 Paradoxical Role of Methyl-CpG-Binding Protein 2 in Rett Syndrome
  Janine M. LaSalle

4 Genetic Approaches to Analyzing Mitochondrial Outer Membrane Permeability
  Brett H. Graham and William J. Craigen

5 Mitochondrial Dynamics in Mammals
  Hsiuchen Chen and David C. Chan

6 Histone Modification in Corepressor Functions
  Judith K. Davie and Sharon Y. R. Dent

7 Death by Abl: A Matter of Location
  Jiangyu Zhu and Jean Y. J. Wang

Volume 60

1 Therapeutic Cloning and Tissue Engineering
  Chester J. Koh and Anthony Atala

2 α-Synuclein: Normal Function and Role in Neurodegenerative Diseases
  Erin H. Norris, Benoit I. Giasson, and Virginia M.-Y. Lee

3 Structure and Function of Eukaryotic DNA Methyltransferases
  Taiping Chen and En Li

4 Mechanical Signals as Regulators of Stem Cell Fate
  Bradley T. Estes, Jeffrey M. Gimble, and Farshid Guilak

5 Origins of Mammalian Hematopoiesis: *In Vivo* Paradigms and *In Vitro* Models
  M. William Lensch and George Q. Daley

6 Regulation of Gene Activity and Repression: A Consideration of Unifying Themes
  Anne C. Ferguson-Smith, Shau-Ping Lin, and Neil Youngson

7 Molecular Basis for the Chloride Channel Activity of Cystic Fibrosis Transmembrane Conductance Regulator and the Consequences of Disease-Causing Mutations
  Jackie F. Kidd, Ilana Kogan, and Christine E. Bear

Volume 61

1 Hepatic Oval Cells: Helping Redefine a Paradigm in Stem Cell Biology
  P. N. Newsome, M. A. Hussain, and N. D. Theise

2 Meiotic DNA Replication
  Randy Strich

3 Pollen Tube Guidance: The Role of Adhesion and Chemotropic Molecules
  Sunran Kim, Juan Dong, and Elizabeth M. Lord

4 The Biology and Diagnostic Applications of Fetal DNA and RNA in Maternal Plasma
  Rossa W. K. Chiu and Y. M. Dennis Lo

5 Advances in Tissue Engineering
  Shulamit Levenberg and Robert Langer

6 Directions in Cell Migration Along the Rostral Migratory Stream: The Pathway for Migration in the Brain
  Shin-ichi Murase and Alan F. Horwitz

7 Retinoids in Lung Development and Regeneration
  Malcolm Maden

8 Structural Organization and Functions of the Nucleus in Development, Aging, and Disease
  Leslie Mounkes and Colin L. Stewart

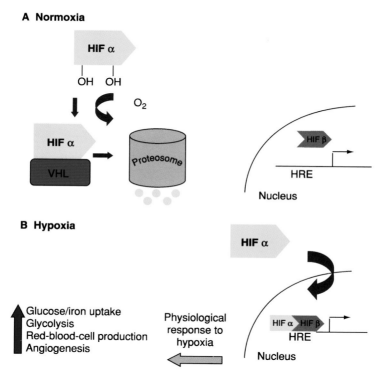

Chapter 2, Figure 1 Simplified model of HIF regulation. Under normoxic conditions, the ODD domain of HIF-α subunit is hydroxylated, which targets it for pVHL-mediated proteosomal degradation. Under hypoxia, HIF-α is stabilized and translocates into the nucleus. Here, HIF-α dimerizes with HIF-1β/ARNT to recruit cofactors and activate gene transcription.

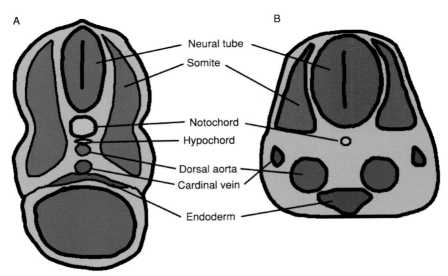

Chapter 3, Figure 1 Schematic cross section through the trunk of midgestation vertebrate embryos at limb level. (A) Representative amphibian/zebrafish embryo. (B) Representative avian/mammalian embryo.

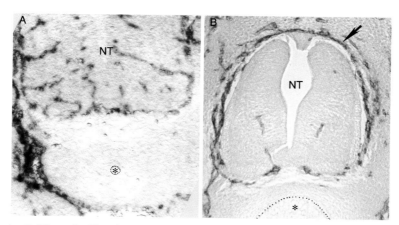

Chapter 3, Figure 2 Vascular patterning at the embryonic midline of amniotes. (A) The perinotochordal area is avascular as visualized by *in situ* hybridization for flt-1 mRNA in an E11.5 mouse embryo. The notochord is denoted by an asterisk and surrounded by dotted lines, and the vessels are stained purple. (B) The peri-neural vascular plexus (PNVP) (arrow) surrounds the neural tube (NT) as shown in this HH stage 24 quail embryo. The vessels are reacted with QH1 antibody and appear brown.

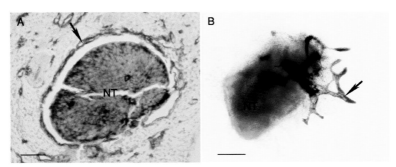

Chapter 3, Figure 3 The neural tube directs PNVP formation. (A) Transverse section through the trunk of a HH stage 24 quail embryo containing a grafted *ROSA* +/− mouse neural tube, 3 days postsurgery. Chimeric embryos were whole-mount stained for β-galactosidase (blue), then sectioned and reacted with QH1 antibody (brown). The arrow points to a host-derived vascular plexus surrounding the grafted neural tube. NC, notochord; NT, neural tube. (B) *Flt1* +/− mouse presomitic mesoderm grafts from E8.5 mouse embryos were placed beside stage HH 10–13 quail neural tubes in collagen gels and cultured for 72 h in basal medium. Mouse vascular cells were visualized by β-galactosidase staining (blue). Arrow indicates vascular plexus. Scale bars are 50 μm in A and 200 μm in B.

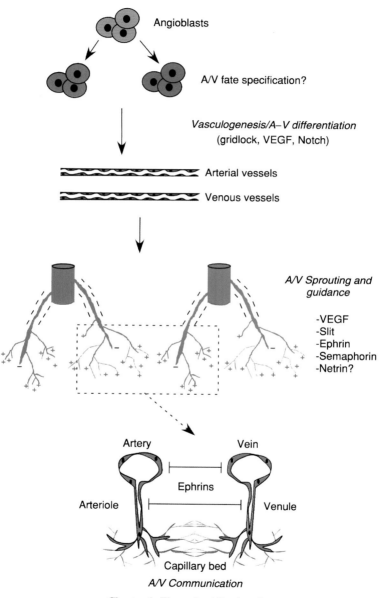

Chapter 4, Figure 1 *(Continued)*.

Chapter 4, Figure 1 Differentiation and patterning of the embryonic vascular circuit. Three steps are required for the formation of a mature vascular network. First, the central axial vessels coalesce from angioblasts and differentiate into arteries and veins. Although little is known about the initial segregation of angioblasts into arterial and venous subpopulations in mammals, key insights from zebrafish studies suggest that the formation of separate arterial and venous vessels relies on Notch signaling proteins and effectors. Evidence is accumulating for a role of VEGF in arterialization as well, perhaps as a result of Notch activation. Second, arteries and veins sprout from the central axial vessels during angiogenesis and are guided to common distal targets along parallel, but distinct, paths. It is this step in vascular network formation that is most analogous to neural guidance and appears to share many of the same regulatory molecules and mechanisms. Finally, molecular mechanisms, such as Eph:ephrin signaling, are required to maintain the separate domains of the larger arterial and venous vessels during angiogenesis. It is not known how these mechanisms are overcome or modified at the level of the capillary bed, the only border at which the arterial and venous networks intersect.

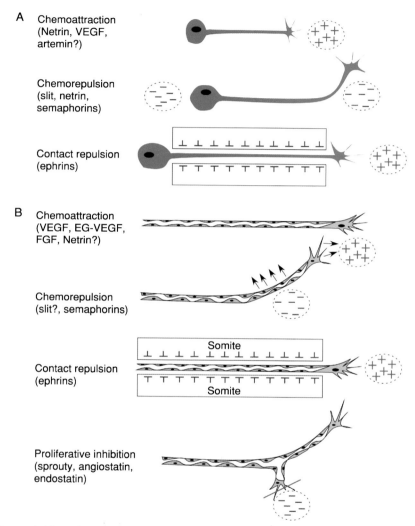

Chapter 4, Figure 2 Axonal guidance as a paradigm for vascular patterning. Evidence for several modes of directional axon growth or neuroblast migration have been described in the literature, including chemoattraction, chemorepulsion, and contact repulsion (inhibition) (A). Similar patterning mechanisms are also apparent in the vasculature (B), particularly with regard to chemoattraction and contact repulsion. Functional studies that definitively discriminate between chemorepulsion (progressive movement of a vessel away from a point source) and proliferative inhibition have not yet been reported.

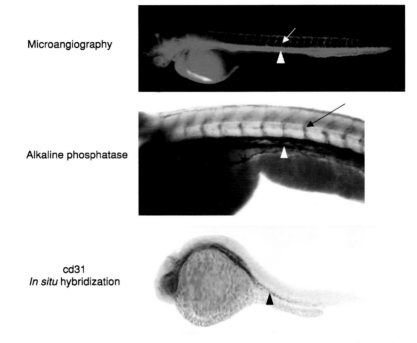

Chapter 5, Figure 1 Detection of the zebrafish zasculature. (Upper) Microangiography showing blood circulation. FITC-dextran (green color) was injected in the cardinal vein at 56 hpf. Green fluorescence visualizes functional blood flow in axial vessels (arrowhead) and ISV (arrow). (Middle) Alkaline phosphatase staining showing blood vessel structure. Zebrafish embryos (72 hpf) were stained with substrates of alkaline phosphatase (BCIP and NBT). The dark purple stained structures are axial vessels (arrowhead) and ISV (arrow). (Lower) *In situ* hybridization (ISH) with *cd31*, a specific EC marker. Whole-mount ISH with 24 hpf embryos was performed. Arrowhead indicates axial vessels. ISV are not formed yet at this 24 hpf stage.

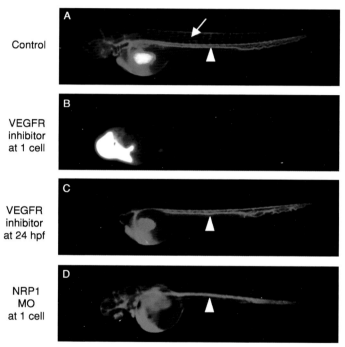

Chapter 5, Figure 2 Vasculogenesis and Angiogenesis analyzed temporally by FITC-dextran microangiography. (A) FITC-dextran visualizes functional blood flow in axial vessels (arrowhead) and ISV (arrow). (B) Administration of VEGFR kinase inhibitor at the one-cell stage blocks blood circulation via axial vessels and ISV. (C) Administration of VEGFR kinase inhibitor at 24 hpf blocks blood circulation via ISV but not axial vessels (arrowhead). (D) Nrp1 knockdown inhibits blood circulation via ISV but not axial vessels (arrowhead).

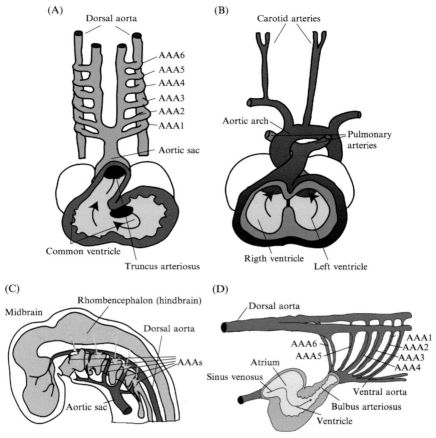

Chapter 7, Figure 1 Development of the heart and pharyngeal arch system (PAA) in mice and zebrafish. (A) The developing heart and the PAA system, before the remodeling of the arch artery system and before the formation of the aortico-pulmonary septum, with a common ventricle, a single outflow vessel called truncus arteriosus, and the aortic sac, which gives rise to six bilaterally symmetric vessels known as PAAs; (B) the mature heart configuration, in which the truncus arteriosus has become divided into an aortic and pulmonary trunk: the aortic trunk arises from the left ventricle, the pulmonary trunk from the right ventricle. The aortic arch system has extensively remodeled into the mature aortic arch and proximal pulmonary arteries; (C) lateral view of a murine embryo of 10.5 days old, showing the pharyngeal arches (1–6) and their corresponding arch arteries. Neural crest cells (yellow arrows) migrate from the neural folds in the rhombencephalon or hindbrain to the adjacent pharyngeal arches. The crest cells that fill the first arch arise from the caudal midbrain and rhombomeres 1 and 2 of the hindbrain; those that populate the second arch emerge primarily from rhombomere 4; and, finally, the more caudal arches are filled by crest cells generated by rhombomeres 6 and 7; (D) schematic overview of the heart and blood vascular systems of an early zebrafish embryo. At 48-h postfertilization, the zebrafish heart consists of a smooth-walled tube partitioned into all four segments with the definite structure of sinus venosus, atrium, ventricle, and bulbus arteriosus, which gives rise to six bilaterally symmetric vessels known as aortic arch arteries (AAA) that drain into the dorsal aorta.

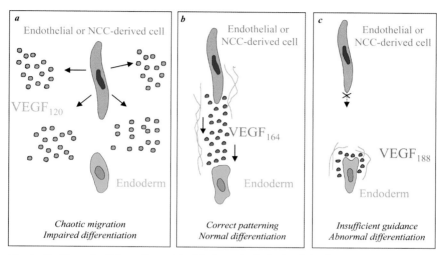

Chapter 7, Figure 2 Schematic model of the hypothetical role of the different VEGF isoforms. VEGF isoforms provides distinct guidance and differentiation signals for the remodeling of the pharyngeal arch arteries. In all panels, VEGF is produced by the pharyngeal arch endoderm, while the endothelial cells of the pharyngeal arch arteries or the neural crest cell (NCC)–derived periendothelial cells around the arteries are the VEGF-responsive target cells. Since VEGF is produced by the endoderm at the periphery of the arch and the pharyngeal arch arteries are located centrally in the arch, VEGF must diffuse over a finite distance to reach its target cells. VEGF could be necessary for providing spatial guidance cues as well as essential differentiation signals. (*a*) The VEGF120 isoform is too soluble and therefore diffuses randomly in a chaotic pattern and over too long distances, resulting in a lack of focused instructive spatial guidance and insufficient differentiation cues. (*b*) The VEGF164 isoform is soluble, yet binds sufficiently to the extracellular matrix so that it can reach its target cell but still provides spatial guidance through its association with the matrix and differentiation signals via binding to neuropilin-1. (*c*) The VEGF188 isoform is secreted but not soluble, as it is attached to the extracellular matrix. Since this isoform cannot reach its target cell, insufficient spatial guidance and differentiation signals are generated. Thus, either excessive or insufficient matrix binding results in imbalanced vascular remodeling. Similar mechanisms might be operational in other DiGeorge syndrome–affected organs, although the VEGF-responsive cell type may vary.

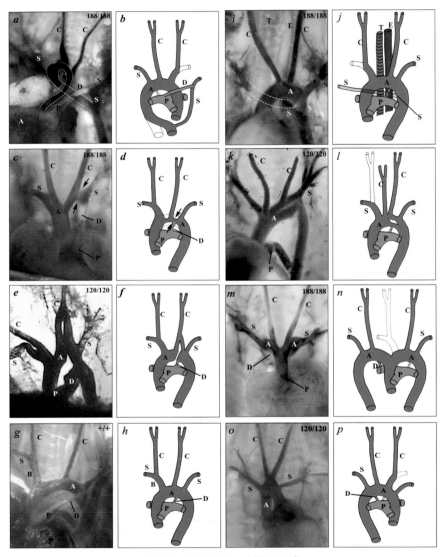

Chapter 7, Figure 3 *(Continued).*

Chapter 7, Figure 3 Aortic arch and conotruncal defects in VEGF120 and VEGF188 embryos and neonates. This shows intraventricular injection of ink or Evans blue to visualize the vascular malformations in VEGF120 and VEGF188 neonates. For each set of two panels, the left panel displays a micrograph of the thoracic vessels, while the right panel schematically illustrates the vascular malformation. (*a, b*) Normal aortic arch configuration in a wild-type neonate; (*c, d*) VEGF188 neonate with a coarctation of the aorta (indicated between arrows) and aberrant connection of the ductus arteriosus to left subclavian artery, resulting from abnormal partial regression of the left fourth arch artery. The ascending aorta is also enlarged. (*e, f*) Interruption of the aortic arch in E14.5 VEGF120 embryo, caused by abnormal regression of the left fourth arch artery. The blood flows from the heart into the carotid artery, and via an abnormally persisting carotid duct to the descending left aorta; (*g, h*) VEGF188 neonate (oblique view): the blood flows through the right-bending aortic arch and persistent right dorsal aorta into to the left-sided descending aorta (the left aortic arch is interrupted). The left subclavian artery and the ductus connect to the left dorsal aorta. A vascular ring around the trachea and esophagus (removed for clarity) is formed by the right-bending aorta, the right dorsal aorta, the left dorsal aorta, the ductus arteriosus, and the pulmonary trunk; (*i, j*) VEGF188 neonate: the right subclavian artery (S; traced with yellow dashed line in *a* and indicated in yellow in *b*) originates aberrantly from the descending aorta and crosses the midline behind the esophagus and trachea. This "artera lusoria" is caused by abnormal regression of the right fourth arch artery and persistence of the right distal dorsal aorta. For better visualization of the vascular defects, the ductus arteriosus and pulmonary trunk were removed; (*k, l*) E14.5 VEGF120 embryo, showing a "duplication" of the aortic arch segment between the left carotid and subclavian artery, resulting from a persistent left carotid duct and a hypoplasia of the left fourth arch artery. The ductus arteriosus is absent; (*m, n*) VEGF188 neonate: double-sided aortic arch with a right-sided ductus arteriosus, caused by abnormal persistence of the right dorsal aorta, abnormal regression of the left sixth arch artery, and persistence of the right sixth arch artery. (*o, p*) VEGF120 neonate with a right-sided aortic arch and descending aorta, a right-sided ductus arteriosus (D), and a left-sided brachiocephalic trunk. This condition is the "mirror image" of the normal aortic configuration. A (aorta); AS (aortic sac); AV (atrio-ventricular channel); B (brachio-cephalic trunk); C (carotid artery); CD (carotid duct); D (ductus arteriosus); DA (dorsal aorta); E (esophagus); LV (left ventricle); P (pulmonary trunk); PTA (persistent truncus arteriosus); S (subclavian artery); T (trachea).

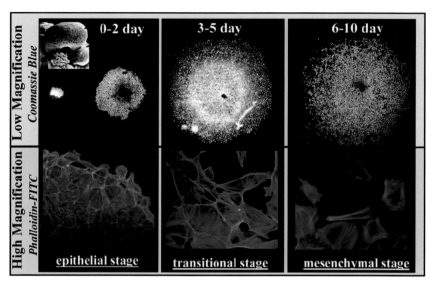

Chapter 8, Figure 3 Epithelial to mesenchymal transformation (EMT) of explanted proepicardium. Top row: Coomassie blue stain to show overall cell shapes and colony morphology. Bottom row: Phalloidin-FITC staining to show the organization of the actin cytoskeleton. Inset, upper left: HH17 quail PE (arrow) extending toward the looped heart tube. Note that proepicardial cells initially have a mesothelial phenotype with extensive cell–cell contacts (Top left), and cytoskeletal actin organized as prominent subcortical bundles (Bottom left). Between 3 and 5 days after explant, proepicardial cells undergo EMT (Top center) with loss of cell–cell contacts, and rearrangement of cytoskeletal actin into stress fibers (Bottom center). By 6–10 days after explant, EMT is completed and the vast majority of cells now display a mesenchymal phentoype.

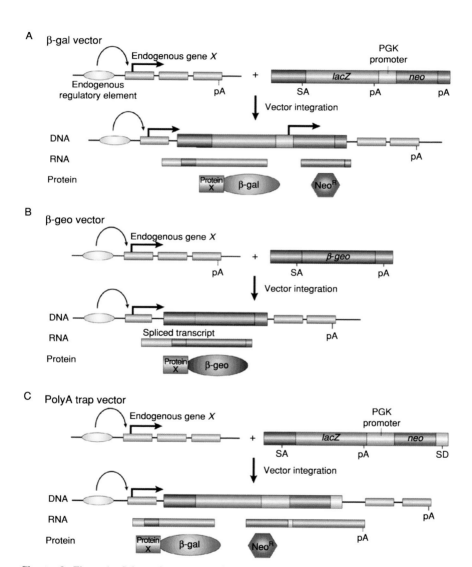

Chapter 9, Figure 1 Schematic structure of gene trap β-gal, β-geo, and polyA trap vectors.

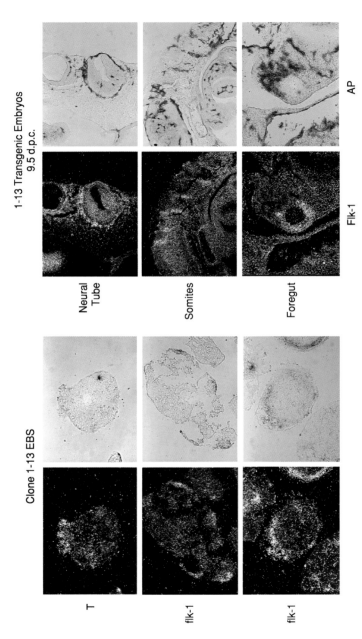

Chapter 9, Figure 2 Colocalization of *flk-1* and AP reporter gene expression in embryoid bodies and transgenic mice derived from ES cell clone 1–13. *In Situ* hybridization on sectioned, AP stained clone 1–13 EBs (Left panel) using *Brachyury T* and *flk-1* riboprobes; each section is shown as a dark-field (Left) and a bright-field image. Transverse sections through E9.5 transgenic embryos derived from clone 1–13 (Right panel); each section is shown as a dark-field (Left) and bright-field image.